About Island Press

Island Press is the only nonprofit organization in the United States whose principal purpose is the publication of books on environmental issues and natural resource management. We provide solutions-oriented information to professionals, public officials, business and community leaders, and concerned citizens who are shaping responses to environmental problems.

In 1999, Island Press celebrates its fifteenth anniversary as the leading provider of timely and practical books that take a multidisciplinary approach to critical environmental concerns. Our growing list of titles reflects our commitment to bringing the best of an expanding body of literature to the environmental community throughout North America and the world.

Support for Island Press is provided by The Jenifer Altman Foundation, The Bullitt Foundation, The Mary Flagler Cary Charitable Trust, The Nathan Cummings Foundation, The Geraldine R. Dodge Foundation, The Charles Engelhard Foundation, The Ford Foundation, The Vira I. Heinz Endowment, The W. Alton Jones Foundation, The John D. and Catherine T. MacArthur Foundation, The Andrew W. Mellon Foundation, The Charles Stewart Mott Foundation, The Curtis and Edith Munson Foundation, The National Fish and Wildlife Foundation, The National Science Foundation, The New-Land Foundation, The David and Lucile Packard Foundation, The Pew Charitable Trusts, The Surdna Foundation, The Winslow Foundation, and individual donors.

Bioregional Assessments

Bioregional Assessments
Science at the Crossroads of Management and Policy

edited by

K. Norman Johnson, Frederick J. Swanson,
Margaret Herring, and Sarah Greene

foreword by

Jerry F. Franklin

ISLAND PRESS
Washington, D.C. • Covelo, California

Library of Congress Cataloging-in-Publication Data
Bioregional assessments : science at the crossroads of management and
   policy / edited by K. Norman Johnson ... [et al.].
       p.   cm.
   Includes bibliographical references and index.
   ISBN 1–55963–657–2 (cloth). — ISBN 1–55963–658–0 (pbk.)
     1. Environmental management—United States—Case studies.
   2. Environmental policy—United States—Case studies.   I. Johnson,
   K. Norman.
   GE310.B56   1999   98–42547
   363.7'05—dc21   CIP

Printed on recycled, acid-free paper

Manufactured in the United States of America
10 9 8 7 6 5 4 3 2 1

Contents

Acknowledgments   ix

Foreword   xi

Introduction   *Margaret Herring*   1

Part One: Practice and Theory of Bioregional Assessments

1. Learning from the Past and Moving to the Future
    Jack Ward Thomas   11

2. Stepping Back: Assessing for Understanding in Complex
    Regional Systems   *Lance H. Gunderson*   27

Part Two: The Role of Science in Bioregional Assessments

3. History and Assessments: Punctuated Nonequilibrium
    John C. Gordon   43

4. Perspectives on Scientists and Science in Bioregional
    Assessments   *Frederick J. Swanson and Sarah Greene*   55

5. A Political Context Model for Bioregional Assessments
    *Hanna J. Cortner, Mary G. Wallace, and
    Margaret A. Moote*   71

Part Three: Case Studies and Reviews

6. Forest Ecosystem Management Assessment Team
    Assessment   85

Case Study	*K. Norman Johnson, Richard Holthausen,*
	Margaret A. Shannon, and James Sedell 87
Science Review	*Logan A. Norris* 117
Management Review	*Judy E. Nelson* 121
Policy Review	*Paula Burgess and Kristin Aldred Cheek* 127

7.   Great Lakes–St. Lawrence River Basin Assessments   133
        Case Study   *Henry A. Regier*   135
        Science Review   *Michael L. Jones*   153
        Management Review   *James Addis*   157
        Policy Review   *Michael Donahue*   162

8.   Everglades–South Florida Assessments   167
        Case Study   *John C. Ogden*   169
        Science Review   *Frank J. Mazzotti*   187
        Management Review   *Susan D. Jewell*   192
        Policy Review   *Maggy Hurchalla*   196

9.   Northern Forest Lands Assessments   201
        Case Study   *Perry R. Hagenstein*   203
        Science Review   *David E. Capen*   214
        Management Review   *Henry L. Whittemore*   218
        Policy Review   *Laura Falk McCarthy*   224

10. Southern California Natural Community Conservation
     Planning   229
        Case Study   *Dennis D. Murphy*   231
        Science Review   *Peter A. Stine*   248
        Management Review   *James E. Whalen*   255
        Policy Review   *Ron Rempel, Andrew H. McLeod, and
                                 Marc Luesebrink*   262

11. Interior Columbia Basin Ecosystem Management
     Project   269
        Case Study   *Thomas M. Quigley, Russell T. Graham,
                                 and Richard W. Haynes*   271
        Science Review   *James K. Agee*   288
        Management Review   *Martha G. Hahn*   293
        Policy Review   *John Howard*   299

12. Sierra Nevada Ecosystem Project   303
        Case Study   *Don C. Erman*   305
        Science Review   *Robert R. Curry*   321
        Management Review   *Jon D. Kennedy*   326
        Policy Review   *Dennis T. Machida*   330

Part Four: Synthesis

13. Understanding Bioregional Assessments
     K. Norman Johnson and Margaret Herring   341

        About the Contributors   377
        Index   385

Acknowledgments

Many of the ideas in this book were first discussed at the Crossroads Conference in Portland, Oregon, in November 1995. The historic buildings of a former home for the indigent poor provided a picturesque setting as 120 people gathered from across the country to compare their experiences with bioregional assessments. Jack Walstad, Mike Unsworth, and Scott Reed helped us find the wherewithal to bring these people together in a memorable conference. We appreciate the assistance given by Toni Gwin and the Forestry Conference Office at Oregon State University in making this conference possible. And we wish to acknowledge all the participants at the conference who offered their ideas to help define this work we call *bioregional assessments*. We are especially indebted to John Gordon, Tom Crow, Elizabeth Estill, Michael Mantell, and Don Erman, whose early conversations with us helped push us past our regional questions and introduced us to the larger world now represented in this book.

Each assessment in this review is presented through the filter of individual experience. We wish to thank all our contributors, who generously provided their candid appraisals of the case studies. In many cases, we were asking these people to critique their recent life's work. We received, in every instance, a clear-eyed evaluation of strengths and weaknesses. Their honesty and insights are the backbone of this book. The opinions they share here are their own and do not reflect official policy of the agencies represented.

We also wish to thank the many reviewers from around the country who read and commented on parts of this book. In particular, great thanks go to Denise Lach at the Center for Analysis of Environmental Change, and Tom Spies of the USDA Forest Service Pacific Northwest Research Station, whose careful reviews of the entire manuscript have made this a better book. And a tip of the hat to George Lienkaemper, whose beautiful GIS maps illustrate each case study.

Finally, we would like to acknowledge the support we have received for this project from College of Forestry, Oregon State University, the Center for Analysis of Environmental Change, and the Pacific Northwest Research Station and Washington Office of the USDA Forest Service.

Foreword

Jerry F. Franklin

Direct incorporation of science and scientists into natural resource policy development is a relatively recent phenomenon. It is a phenomenon that owes much of its development to the environmental legislation and litigative history of natural resource policy development during the second half of the twentieth century.

When I began my forestry career in the late 1950s, resource managers commonly viewed scientific research as occasionally useful and generally harmless, a pursuit most appropriate for intellectuals not suited to a more active career (and, perhaps, best kept off the streets). Managers were most likely to be comfortable with research focused on improved implementation of existing methods and policies ("domesticated science"). It was a rare manager who encouraged research that probed the basic assumptions underlying current policy ("wild science"), such as clear-cutting or ecological values of old-growth forest.

It is not surprising, therefore, that regular, meaningful collaboration between managers and scientists in development of natural resource policies has been the exception rather than the rule during most of the twentieth century. When science was incorporated into policy, it was filtered through the management organization; discussions between the scientist and the manager were "kept in the family." Direct involvement of scientists with development of policy or with decision makers was strongly discouraged—largely by mutual agreement. Scientists didn't want the pressures and potential influences of policy making in their work, and managers didn't believe scientists would understand the compromises and practicalities of the real world.

The isolation of science and scientists from natural resource policy began to change in the 1960s, however, with the publication and popular acceptance of the book *Silent Spring* by Rachel Carson and other symbols of the emergence of ecology as a cultural element. New environmental laws were passed, major controversies over resource management surfaced, and natural resource policy began to enter the courts.

As the legal battles commenced, an incredible thing happened—appellants began bringing science and scientists into the courtroom. Agencies such as the Forest Service found themselves facing testimony damaging to their projects and agendas from expert witnesses in their own employ. The agencies' own scientists and scientific data were being used to discredit current management practices and proposed new activities.

This was a most distasteful development. Consequently, in the 1970s and 1980s there were frequent comments regarding the need to "get the scientists under control" and to improve their appreciation of being "part of an agency team." However, the relative independence of research in agencies such as the Forest Service made control of the scientists and censorship of scientific publications difficult. Perhaps just as important was the transparency provided by a heightened public profile for science and more open flows of information stimulated by ever more sophisticated environmental analyses and counteranalyses and by the Freedom of Information Act. Natural resource policy was no longer the exclusive purview of resource professionals; indeed, policy was increasingly being made in a goldfish bowl.

By the mid-1980s, it was clear to most agency resource managers that they at least needed to be aware of the best and most current scientific information relevant to their projects. This information was going to surface, one way or another; they could try to incorporate it ahead of time, in their decision documents, or they could expect to encounter it in court. Still, an effective and acceptable method for systematically incorporating science and scientists in policy development and project analyses was not clear to agency managers. Scientists were, by their very nature, unpredictable and, therefore, risky participants in processes that had politically bounded, if not predetermined, outcomes. Agencies rarely could bring themselves to incorporate scientists directly in their decision-making processes, and decision screens—boundary conditions for acceptable management alternatives—sometimes existed unknown to either the scientific advisors or to the public (as described by Steven Yaffee in his book, *The Wisdom of the Spotted Owl*).

In the last decade, decisions were taken out of the hands of the resource professionals as the agencies lost their scientific credibility with the public and with all three branches of the federal government. Court injunctions stopped most timber harvesting activities in the Pacific Northwest and were threatened elsewhere. One judge directed the U.S. Fish and Wildlife Service to take a second, scientifically credible look at delisting the northern spotted owl, and another judge directed the Forest Service to develop a scientifically credible plan for management of the owl. Congress created its own scientific advisory group, the Scientific Panel on Late-Successional Forest Ecosystems (a.k.a. the Gang of Four). A newly elected President Clinton convened a forest summit and created his own scientific panel, the Forest Ecosystem Management Assessment Team (FEMAT).

Science and scientists were catapulted into central roles in natural resource policy analysis by the need—legal and social—for development of scientifically credible plans for management of wildland ecosystems over entire regions. This also resulted in significant direct interaction between decision makers—especially politicians—and natural resource scientists for the first time. While there have always been individual scientific advisors to high-level administrators and selected scientific testimony to Congress, the number of scientists and the extent of their direct involvement with decision makers greatly increased. As a whole, the scientific community began poorly prepared for such tasks by both training and inclination but quickly learned, on the job, how to develop and conduct meaningful analyses of various management alternatives.

We are now entering a new era, in which science and scientists—along with managers and stakeholders—will be intimately and continuously involved in natural resource policy development. Today, most resource managers genuinely want the most comprehensive and current scientific information available, whatever its implications for management. Scientists want to be a part of the process but not be co-opted by it. However, we are still very much at the stage of learning how the scientific, the technical, and the social can be integrated.

Bioregional assessments have provided our most important opportunities to learn how science can be systematically incorporated into resource policy development. This book is about our experiences in this new interaction, this new partnership of science with management and the public. It provides the student of policy development with the views of the scientist, the manager, and the policy maker, primarily drawing from seven important and contrasting case studies. These diverse studies and a synthesis chapter prepared by one of the world's foremost forest policy analysts make clear the ways in which we can improve how science is utilized in making decisions about natural resource policies and ecosystem management.

This is important because there is a great deal still to be learned about the best approaches to bioregional analyses with regard to both design of the process and usefulness of the products. For example, it is not clear that analyses developed with larger budgets and over longer periods are superior to those developed under very short time lines. Similarly, the relative merits of publicly open versus more closed processes are not clear; analyses conducted in a goldfish bowl may not necessarily produce results superior to those developed by scientific and technical personnel working in isolation. The diversity of approaches represented in this book will be very useful in helping to resolve such questions.

Introduction

Margaret Herring

Tension between development and conservation of natural resources exists in all parts of the country. Battles erupt over consumptive water use and ecological values in the Everglades, old-growth forest habitat and logging in the Pacific Northwest, and traditional use and potential development in the north woods of New England. These conflicts reflect a real limitation of resources and a growing list of the demands and values society puts on these resources.

Established planning systems failed to deal with these conflicts. Agency plans, local ordinances, and specific regulations seem to patch over spots while problems continue to spread well beyond the boundaries of recognized jurisdictions. Resource managers often receive the brunt of public anger as people are increasingly frustrated by decisions they feel are imposed on them that affect the well-being of their communities and their future. Opposing sides become polarized, and policy makers may see all choices as politically punishing. Lack of consensus causes delays, and the job of managing natural resources often passes to the courts.

An alternative to court-ordered, crisis-driven decision making is emerging across the country. This new tool has many names and many applications, as it develops not from explicit design but from necessity and innovation. We call this new tool *bioregional assessment*. It is the effort to build knowledge about a region prior to decision making and management action. Bioregional assessment is an essential part of ecosystem management. It is the first step taken in order to understand the condition and possible futures of a region. Just as a medical history and physical examination precede diagnosis and treatment, so a bioregional assessment precedes ecosystem management and conservation planning.

Bioregional assessments integrate a broad range of information about the social, economic, and ecological conditions within a region in order to provide a basis for making decisions and taking action. They are *bioregional*, which is to say they are ecosystem-based, delineated by natural processes and elements rather than by planning units and political jurisdiction. And they are *assessments*, including elements of study and evaluation rather than a full planning and implementation process. They examine natural and social systems at large landscape scales, and they consider changes to those systems that may be occurring over

1

long periods of time. Bioregional assessments turn traditional questions on their heads. Rather than asking, for example, what is the maximum sustainable yield of timber from the forest, they ask what are the trends in forest condition in relation to environmental, economic, and social systems within the region.

By providing a clearer understanding of current conditions and trends, bioregional assessments are meant to bring accountability to otherwise fractious issues of resource use and management. Instead of posing one side against the other, assessments attempt to bring together traditional opponents in a joint effort to assess conditions and possibilities within a shared bioregion. For example, a Republican governor and a Democratic secretary of interior have endorsed an assessment of the last undeveloped lands of coastal Southern California, encouraging the partnership of land developers and conservation groups in a region where rapid population growth conflicts with endangered species protection on private land. International agreements have commissioned a series of assessments in the Great Lakes region, where chronic environmental problems affect communities in two nations. In the West, where federal lands dominate the landscape, conflict over resource management prompted both Congress and the president to initiate assessments in the Pacific Northwest, the interior Columbia River basin, and the Sierra Nevada.

Reviewing Bioregional Assessments

This review began as conversations between the editors of this volume and leaders across the country about the use of science in guiding environmental policy. It includes an in-depth examination of seven case studies from around the country. The first case study is of the Forest Ecosystem Management Assessment Team, an assessment in which my fellow editors were deeply involved through its many iterations from the Gang of Four to the president's plan. What we learned in FEMAT, and what we witnessed in other assessments, made us question how science is used to make decisions about management and policy in ecosystems.

Our review takes that question across the country to examine the assessment experience in other bioregions. These assessments invite comparison, but defy reduction to a single methodology. Our purpose, therefore, is not to prescribe steps for others to follow, but rather to review the steps that we and others have taken in order to learn ways in which science can be better used to guide management and policy of ecosystems.

Although agencies have printed voluminous documentation of individual assessments, there has been little comparison among examples or integration of lessons from experience. Each assessment has proceeded with little reference to what has been learned elsewhere, yet mandates and money for new assessments continue to appear. What has been tried in each case? What has been learned? Are these assessments effective ways of guiding management and policy decisions? This book begins to address these questions. The lessons offered from these examples provide a starting place for new assessments and help set a course for all those involved in regional planning and ecosystem management.

The case studies reviewed here began as a way for decision makers to use scientific information as a basis to negotiate conflicting demands on natural resources. Many of these assessments have been very costly, but rarely has their effectiveness been reviewed. This book attempts such a review. We considered our experience with FEMAT and other assessments and observed regional differences in the way assessments were conducted and the problems they were meant to inform. We observed temporal differences, too, as each successive assessment struggled to distinguish itself from the perceived mistakes of the past.

At the time we began our inquiry, there were dozens of bioregional assessments to choose from, and many more have been commissioned since. In selecting the case studies for this review, our first questions were empirical. We compared the extent of their geography, the length of time they had taken to complete, the variety of participants, and the issues they were meant to inform. In order to learn more about the role of science in assessments, we chose examples that had been commissioned to provide information about specific resource problems in a specific region; that focused on a large landscape with multiple ownership or management; that represented different regions and a variety of resource concerns across the country; and that included a mix of knowledgeable, credible participants from whom we could draw meaningful reviews.

As a result, forty-five contributors provide a broad mix of experience from seven precedent-setting assessments. Of the seven case studies, three are concerned primarily with the management of private land, and four with public lands. Two are focused on aquatic systems, two on terrestrial systems, and three focused on both. Four assess resources within the boundaries or influence of large urban areas, and three assess primarily rural, resource-based economies. The assessments lasted from three months to three decades, and implementation of the recommendations derived from the work continues at various stages. Despite their variety, all the examples were driven by questions from policy makers in bioregions where conflicts over the management of natural resources were intense.

We centered our review around three key questions:

- What policy questions is science able to answer?
- What scientific information is most useful to decision makers?
- How can assessments be effective in guiding management decisions?

In part one, we introduce bioregional assessments in practice and in theory. Jack Ward Thomas, speaking from his experience as a scientist, manager, and policy maker, opens with a hard look at the practical challenges that bioregional assessments present to resource agencies and managers. For the editors of this book, and for several of its contributors, the idea of bioregional assessments began in 1989 when Congress asked Thomas to head the Interagency Scientific Committee in order to find a scientifically credible conservation plan for the northern spotted owl. Thomas understood the science of the forest and its inhabitants, as well as the divisive politics surrounding the owl. From the owl forests of the Pacific

Northwest to his tenure in Washington, D.C., Thomas shares the lessons he has learned regarding bioregional assessments.

In his chapter, Lance Gunderson invites us to step back, to look at the theory and practice of bioregional assessments and notice that underlying both is uncertainty. Bioregions, Gunderson asserts, are an appropriate scale for recognizing uncertainty in natural and social systems. Uncertainty is what assessments should use as a compass to guide new research, management action, and policies that build understanding in and about communities. And so, Gunderson recommends keeping a "ruthless hold on uncertainty . . . as part and parcel to all phases of the assessment."

Part two looks specifically at the role of science in bioregional assessments. John Gordon introduces readers to the cultures of science and policy and discusses several early examples of science-based assessments. He examines possible models to describe how science influences policy making. Frederick Swanson and Sarah Greene describe the framework that science provides to build understanding of a bioregion's conditions and possibilities. They discuss the emerging field of bioregional science and the challenges this work presents to traditional scientific inquiry.

In the final chapter of this section, Hanna Cortner and colleagues challenge the assumption that science can deliver unequivocal answers to political problems. Assessments, they assert, are tools for learning, one part of a politically democratic process. They should not be allowed to consume the time, money, and attention needed to influence social action.

Part three presents a wealth of experience and insight from the practice of assessments. Regional leaders provide their retrospective in seven case studies from across the country. Each case study is followed by reviews of the science, management, and policy implications in each bioregion.

The work of the Forest Ecosystem Management Assessment Team (FEMAT) that begins our case histories provided an important turning point in the history of the Pacific Northwest forests, marking an end to pioneering exploitation of seemingly boundless natural resources, and turned management of the federal forest on its head. Several of this book's contributors refer to their experience with FEMAT, or with its predecessors, the Interagency Scientific Committee and Gang of Four. As we look beyond the borders of the owl forests, we find similar experiences in other parts of the country.

The second case study, a history of science-based assessments in the Great Lakes–St. Lawrence River basin, is an analysis not of one discrete assessment as much as of an ongoing, evolving process of more than twenty-five years, a succession of assessments to deal with a succession of emerging natural resource problems. Henry Regier documents the changing process of science, decision making, and governance in the Great Lakes bioregion, part of a major transformation toward an "Emerging Era." Many assessments have begun to explore these new conventions, and the Great Lakes story introduces them in terms of a mature assessment process.

In contrast to the Great Lakes, where the institutional infrastructure seems to be evolving toward a capacity to plan for ecological sustainability, the Everglades has struggled with institutional gridlock. A century of technological applications has transformed the Everglades, to the point that no one can predict exactly how the ecosystem will respond to the proposed restoration. Up until recently, litigation, or its threat, has set the agenda for much of the research, planning, and management in South Florida. John Ogden describes an independent assessment by a consortium of scientists in the region, which has offered a scientific basis for the restoration planning process but has yet to be embraced by agencies still debating how to proceed with the restoration.

The assessment area for the Northern Forest Lands Study encompasses a sprawling forest, largely private timberland, in parts of four New England states, intuitively understood and valued by the public as the north woods. The two-part assessment responded to concern for the protection of this piece of New England heritage, which was believed to be poised for new housing developments and change. Forest economist Perry Hagenstein was one of the first to suspect what international investors already knew. In 1987, he warned, "The increasing spread between the value of this land for timber growing and for recreation and development puts pressure on current owners." He described large land holdings as profit centers and predicted a "sea of change in forest ownership in northern New England." In this chapter, Hagenstein describes an extensive social assessment that preceded a science-based analysis of the northern forest and its economy. Town hall meetings garnered public opinion and contributed to setting regional goals in a traditional process that reflects some of the oldest and newest thinking about civic participation and public understanding of science.

Natural Community Conservation Planning (NCCP) is a coordinated state and federal response to the Endangered Species Act (ESA) developed to resolve the escalating conflicts between land development and land conservation in California. Dennis Murphy describes its first application, in the coastal sage scrub of Southern California, where the endangered California gnatcatcher lives amid some of the nation's most expensive real estate. By establishing a portion of the most critical lands as a managed habitat reserve system under NCCP, other lands are released for development or resource extraction without further restrictions from the ESA. Such planning promises to provide comprehensive, habitat-based species protection as well as "one-stop, once-and-for-all" regulatory permitting for developers and provides a model for conservation planning elsewhere in the state and nation.

A bioregional assessment of truly epic scale, the interior Columbia River basin assessment, attempted to answer growing concerns about forest health, wildlife, and anadromous fish across nearly 145 million acres in parts of seven states by assessing the ecological, economic, and social outcomes associated with federal agencies' management and policy. Building from the FEMAT experience, this assessment has worked hard to include the participation of local communities and county governments. The assessment, a collaborative effort among federal agencies in the region, has produced an innovative, dynamic database but as yet no blueprint for coordinated revision of forest and district plans throughout the region.

The Sierra Nevada Ecosystem Project grew from concern for a much loved, and well-studied, region that was in peril. The Sierra has been closely studied for over a century, and with two national parks and several national forests within the region, it would seem to have adequate protection from some of the threats that other bioregions have faced. And yet questions about its sustainability, fanned by the controversies in the Cascades farther north, sparked a comprehensive, three-year study of the condition of the range.

In addition to the retrospectives offered in each case study, we have included provocative reviews from the perspectives of science, management, and policy that discuss the implications of these assessments in their bioregions and beyond. These reviews provide a more focused look at the issues that concern scientists, managers, and policy makers. The following three examples illustrate the caliber of these outstanding reviews.

For example, in his review of science in the Great Lakes assessments, Michael Jones articulates the concern of many scientists around the country. He describes a fundamental shift in how scientists bring their expertise to the policy table. "Scientists must convey to decision makers and stakeholders alike just how uncertain we are about these ecosystems we are trying to manage. We have to get over the hurdle that would have us believe that admitting uncertainty is to admit weakness, a hurdle that has deep roots in our adversarial system of decision making. . . . It is within the context of recognizing, indeed probing, uncertainty, that science can continue to play a pivotal role in environmental policy development."

In her management review of the FEMAT assessment, Judy Nelson points out the frustration that many managers feel with assessments that leave out practical questions of implementation. In the FEMAT process, as well as in other assessments, the effort was focused on resolving the legal and scientific debate, but it did not address the underlying value debate. The decision framework "became a simplistic trade-off between timber production and species protection," ignoring the larger questions of public values that have continued to hamper the decision-making process in the region.

In her review of policy in the Everglades, Maggy Hurchalla admonishes scientists to understand the political world in which assessments exist. "Scientists who remain detached from the political debate risk the detachment of their supporters. Scientific leadership, not just research, is necessary to achieve long-term goals in ecosystem restoration. Politicians are neither as stupid nor as clever as they seem. They grab for simple solutions and will need to be taught that the environment is not a decision of yes and no, but an ongoing acknowledgment of mosaics, diversity, patterns, and pulses."

In part four, Norman Johnson and I offer a synthesis of the collective experience within the case studies and an analysis of the assessment process. We discuss the effect these assessments have had on our understanding and management of bioregions and conclude with a list of twenty-five observations generalized from the experience demonstrated by the case studies.

Challenges for Bioregional Assessments

The bioregional assessments we chose to review began with questions posed from policy makers to a community of experts. There was not always a comfortable fit between the information requested by decision makers and the information provided by experts. Throughout these assessments you will find a mismatch between the desire for political certainty and the inherent uncertainty of natural systems. How can predictability in industry regulation, for example, be reconciled with unpredictability in nature and our own understanding? Policy makers are asked to provide predictable policies in an unpredictable world. Scientists reject the idea that knowledge is ever complete enough to ensure no surprises.

With or without a foundation of scientific understanding, management decisions will be made and are being made, even if by default. As all these case studies make clear, the choice of no action has specific consequences. Bioregional assessments offer a way of quantifying choices, so that consequences are better understood. Applying a scientific framework to the available information helps to organize what is known, and not known, about the condition of the bioregion; society has a stake in ensuring that we use the best available scientific information for forming policy.

There are no established criteria for success of bioregional assessments, so the temptation exists to study the problem until the money runs out, then hand off the report to someone else. We must be able to recognize where testable solutions lie and where there are additional researchable questions and what the difference is between the two. Decision makers can no longer afford to loop through interminable studies without resolution; nor can they wrap themselves in a single solution, closed off from new information. In these assessments there was rarely enough information to achieve scientific certainty, but often there was enough information to suggest options for policy action.

Sometimes scientists within assessments are perceived by some to have too much power developing policies. Conversely, sometimes they appear to have spent millions of dollars researching a question with no useful result. The boundary between expert advice and policy making has not always been clear, and often it is in this gray zone that bioregional assessments occur. Our hope is to shed light on this gray zone, to learn how policy makers pose questions, and to understand how science answers.

Much of what is discussed in this volume indicates a changing field for science, management, and policy. Bioregional assessments are a step toward managing land and resources in a new way, using an ecosystem approach to coordinate management across interconnected ecosystems and economies. They are a step toward conducting science in a new way, integrating information and techniques from many disciplines to help society solve some of its most difficult environmental problems. And they are a step toward new policies that recognize the inter-connection between healthy environments and healthy economies.

A review such as this is not without its limitations. Although it can be argued that bioregional assessments have been commissioned since the expedition of Lewis and Clark, the use of scientific information to make natural resource policy decisions still falls far short of being an exact science. Our understanding of ecosystem function is rapidly evolving. Our experiments at the crossroads are tentative at best. None of the case studies presented here provides a perfect road map to conflict-free resource management and policy; there are gaps between the aspirations and the accomplishments. Yet, we believe, the retrospectives shared by this remarkable collection of contributors provide both a milestone for measuring our progress and a marker along the path toward integrative science, ecosystem management, and collaborative decision making.

PART ONE

Practice and Theory of Bioregional Assessments

Learning from the Past and Moving to the Future

Jack Ward Thomas

I spent twenty-seven years in the Forest Service's (FS) research division, where I had the pleasure of being able to pontificate freely about natural resource management and not be responsible for the management itself. Then suddenly, upon appointment as Chief of the FS, I found myself being the recipient of pontification—voluminous and ceaseless—*and* being responsible for the outcome of management action. I learned, in short order, that pontification is considerably easier than responsibility. Now, as an academic, pontification on my part is back in style.

Our society, as owner of the public's land, is engaged in a great and growing debate over how and for what purposes those lands should be managed. The challenge to natural resource agencies is to make difficult resource decisions in light of diverse and often contradictory views held by those members of society with an interest in natural resource management. Land management agencies are asked to seek the simultaneous achievement of ecological and economic goals. There has been much pontification about the balancing of these goals, but the responsibility for their achievement lies with managers. Unfortunately, the old military saw that "Authority is commensurate with responsibility" has less and less truth in it as public land management becomes more and more politicized and increasing numbers of decisions are made at levels above that of agency heads. Whether or not such is a desirable turn of events is another issue.

Failures of Past Planning Systems

Natural resource management is governed by a plethora of laws and regulations, variously written to address either economic or ecological goals. Compliance with these laws is neither easy nor straightforward—a gauntlet that managers run in making decisions. Managers "must muddle through a haphazardly developed morass of public land laws and the functional, target-oriented institutional culture that they fostered" (Schlager and Freimund 1994).

The National Environmental Policy Act (NEPA), at first glance, is one of the strongest laws that connect economic and ecological goals. However, Schlager and Freimund (1994) point out several procedural difficulties. First, they note that because an environmental impact statement must examine the environmental impacts of an array of alternatives, and since a thorough review of ecosystem possibilities could produce an exhaustive list of alternatives, the opportunity to challenge assumptions in the alternatives increases exponentially. This is likely to produce interminable delay, decision-making gridlock, and increased management by federal judges. Second, although NEPA requires agency consultation, it does not provide a mechanism for resolving disagreements among agencies. Such consultation is apt to be increasingly exacerbated by differences in agency missions. Multiple-use land management agencies, because of their missions, routinely opt for slightly greater risk to biodiversity over a shorter period. Regulatory agencies, with a simple mandate to protect biodiversity, will likely opt for minimal risk over a longer period. There is, ordinarily, an enormous difference in the resulting "decision space" for land managers. Further, the agencies have different constituencies with which they are concerned and their employees have very different mind sets. To paraphrase the title of a popular book, multiple-use managers are from Venus and regulators are from Mars.

Both NEPA and the National Forest Management Act (NFMA) require early public involvement. This too is a laudable goal that has turned sour and ended up with more, not less, land management decisions being made by federal judges. The current situation has resulted from two sets of requirements, one for public involvement and the other for more technically complex and voluminous assessments and decision documents in response to appeals and litigation. This has produced a new and thriving private-sector industry composed of analysts, lawyers, lobbyists, publicists, fund-raisers, and political organizers who are involved with public land management decision making and management activity. They represent opposite extremes of the issues, poised as either the "environmental" or "industry" positions. There are no similar representatives for positions in between.

This new "conflict industry" thrives on polarization. Agencies caught in the middle are savaged from both sides, and in some cases, attacks are made on the integrity of individuals. To keep the money rolling in, it may be necessary to present more and better "threats" to the welfare of their benefactors. Therefore, this well-intentioned effort in law to "come, let us reason together" has produced what, at worst, might be deemed semi-anarchy and, at best, referred to as much "sound and fury signifying nothing."

An alternative strategy for involving the public is being tried by local communities and agencies to establish collaborative groups to work together on public land management problems. Predictably, the controversy industry has attacked these efforts. After all, with no fight there is no need for gladiators—and no receipts. Regulatory agencies and land managers are working collaboratively to overcome some of these problems, but there are no clear, unambiguous goals set through legislation that integrate the needs of society with the protection of ecosystems. Managers are left to muddle through, balancing goals to obtain the best possible outcomes for social, economic, and ecological conditions. This takes unprecedented interagency cooperation and new ways of doing business. The insertion of periodic political agendas of elected officials complicates these processes to a marked degree.

A New Way of Doing Business

My experience with, and evolving view of, bioregional assessments began in 1989 in the Pacific Northwest. I was drafted by the agency heads of the FS, Fish and Wildlife Service (FWS), National Park Service (NPS), and Bureau of Land Management (BLM) to head a team (the Interagency Scientific Committee, or ISC) charged to produce a scientifically credible plan to assure continued viability of the northern spotted owl well distributed across federal lands within its range.

It had become clear that the owl was significantly associated with old-growth forest conditions and that its habitats were being rapidly removed and fragmented through timber cutting. As a result, owl numbers were thought to be declining proportionally—or even more rapidly. The FS and BLM were not in compliance with either the Endangered Species Act or the regulations issued pursuant to the NFMA. The "diversity clause" of those regulations is actually more specific and more demanding than the ESA. That clause requires that viable populations of all native and desirable non-native species be maintained well distributed within the planning areas. Understanding and acknowledging the consequence of this regulation came slowly to the debate surrounding management of the national forests.

The FS made repeated attempts to put together a plan to meet the requirements of this regulation—and failed legal tests each time. These failures resulted from unwillingness, or simple inability, to accept the political and economic consequences of any significant reduction in timber yields from national forests. These failures to satisfactorily comply with the "viability regulation" and the increasing probability of a listing of the subspecies as threatened by the FWS led to the appointment of the ISC. The team was given carte blanche in regard to resources and technical approaches to produce a scientifically credible plan. This would prove to be but one step in a five-year drama replete with bizarre twists and turns—both political and technical. Yet this process was to bring the FS to the present point of embracing ecosystem management. And, from there, to the necessity for bioregional assessments. To understand the need for and evolving requirements of bioregional assessments, it is well to understand the journey of evolving issues from spotted owls to old-growth to ecosystems.

From Spotted Owls to Ecosystems

Following the release of the ISC report (Thomas et al. 1990), there was a sequence of events (Marcot and Thomas 1997) that mirrored the confusion resulting from the unwillingness, or political inability, to undertake the management actions necessary to come to grips with the spotted owl issue. This reluctance can be attributed, at least in part, to dire predictions by some economists of social and economic chaos for the Pacific Northwest if the ISC strategy were adopted (Bueter 1990). Images of destitute timber communities and harsh criticism from industry groups (Northwest Timber Association 1990) planted seeds of doubt (or hope) in the minds of elected and appointed officials.

The political playing field quickly shifted to analyses of the economic and social impact of the application of the ISC strategy (Bueter 1990; Gorte 1990; Greber, Johnson, and Lettman 1990; Hamilton et al. 1990; Lee 1990; Olson 1990; Satchell 1990; Maki and Olson 1991; Rubin, Helfand, and Loomis 1990; U.S. Department of Interior 1991; Montgomery et al. c. 1992; Sample and Le Master 1992). Results varied widely, from predictions of over fifty thousand jobs lost to essentially no jobs lost. This muddied the political waters, and the outcome was seen as grim. It must be noted that the effects predicted by most of these analyses proved to have been, on balance, hugely exaggerated as the economies of western Oregon and Washington "boomed" during the period 1991–1997.

These analyses were not broad based nor inclusive, focusing rather on short-term potentially adverse affects. What seemed to be ignored was that the ESA was the trump in the debate. There was no evading that fact. Political figures simply could not bring themselves to face the full intent and consequences of the law. The FS announced that national forests would be managed in a manner "not inconsistent with" the ISC strategy. BLM Director Cy Jamison announced that BLM would follow the "Jamison Plan," a much less protective approach (Marcot and Thomas 1997).

At this juncture the federal courts lost patience. Agency timber-harvesting operations were shut down pending the adoption of a plan that would assure viability of the owl through prescribed NEPA processes (Marcot and Thomas 1997). This was quickly followed by committees of the House asking a team of four scientists to develop alternatives for management approaches that might be established in law. The team included John Gordon of Yale University, Jerry Franklin of the University of Washington, K. Norman Johnson of Oregon State University, and myself. Perjoratively dubbed the "Gang of Four" by a timber industry spokesman, the team produced the required assessments and alternatives and, for the first time, considered the "old-growth ecosystem" and anadromous fish species. Congress would not bring itself to act (Marcot and Thomas 1997).

In 1992, in a follow-up case, Judge William Dwyer again enjoined timber sales until the government analyzed the potential effects of the agency's plan on other species that might be associated with old-growth ecosystems. In response, the Forest Service established the Scientific Assessment Team (SAT) to perform that task. BLM Director Jamison, with approval by Secretary of Interior Lujan, tried another approach to escape the dilemma and asked the Endangered Species Committee (the "God Squad") to exempt forty-four timber sales from the injunction. The strategy backfired when the God Squad freed up several sales but ordered BLM to comply with the draft recovery plan—which was essentially the ISC strategy (Thomas et al. 1990).

A New Kind of Assessment

The SAT moved the FS and BLM into a new arena of assessment. Some 667 species were cataloged as closely associated with old-growth habitat, including fungi, lichens, bryophytes and other nonvascular plants, invertebrates, fish, amphibians, reptiles, birds, and mammals. Each management alternative was evaluated as to its probability of providing for the variety of viability of these species (Thomas et al. 1993). The work of the SAT was a pivotal event that changed resource management to include a full consideration of the complete species spectrum and associated habitats. This was a step closer to "ecosystem management"—and things would never be the same again (Marcot and Thomas 1997).

Managers had no role to play in the activities of the ISC, SAT, or the Gang of Four. This was due, in part at least, to the perception that managers had inappropriately influenced earlier efforts of agency technical experts to address the issues. In reality, the more significant concern was to isolate management issues and produce results within very tight time lines.

As the presidential election campaign of 1992 intensified, action gave way to rhetoric. President George Bush deplored the situation and promised change (undefined) after the election. Candidate Bill Clinton promised to convene a "forest summit" to address the issue. Clinton won, convened the promised summit, and announced at that time the formation of FEMAT with instructions to provide him alternatives for solutions to the impasse within sixty days.

The instructions issued to FEMAT (FEMAT 1993, i–iv) were a turning point in natural resource policy for the management of public lands. Key points included the following:

- "Identify management alternatives that attain the greatest economic and social contribution . . . and meet the applicable laws and regulations . . ."
- "Take an ecosystem approach to forest management and . . . particularly address maintenance and restoration of biological diversity . . ."
- "The impact of protection and recovery of threatened and endangered species on nonfederal lands . . . should be minimized."
- "To achieve similar treatment on all federal lands . . . you should apply the viability standard to the Bureau of Land Management lands."
- "In addressing biological diversity you should not limit your consideration to any one species . . . develop alternatives that meet the following objectives . . . maintenance and/or restoration of habitat known (or reasonably expected) to be associated with old-growth forest conditions . . ."

This required an assessment that would allow decision makers to consider costs and benefits in the following areas: timber sales, short and long term; production of other commodities; effects on public area and values; effect on environmental and ecological values; jobs; economic and social effects on local committees; economic and social policies to aid in transition; economic and social benefits from ecological services; and regional, national, and international effects related to timber supply, prices, and other key economic and social variables.

FEMAT, obviously, was obligated to perform a significant level of regionwide assessment to undergird the development and analysis of management alternatives to be presented for consideration by the president. Instructions from the administration, including the pursuit of an "ecosystem management" approach, began to define ecosystem management and embodied several key attributes. First, the area under consideration had to be meaningful in terms of ecological, economic, social, political, and geographic context (this implies a much larger area than was considered in land-use planning up to this point). Second, the time frames had to include an adequate span to accommodate concepts of sustainability. And third, people and their needs and desires had to be a mandatory component of such a management assessment.

The efforts evolved from consideration of the habitat requirements of one subspecies (the northern spotted owl) in the ISC effort to several species considered likely to be "threatened" in the Gang of Four effort, to about seven hundred species thought to be associated with old-growth in the SAT exercise, to over one thousand species (including invertebrates and plants) in FEMAT's report. The level of consideration given to the spotted owl would likely be technically and economically infeasible if universally applied to all species. Hence, the move toward ecosystem considerations was based on habitat classifications and occurrence over time.

FEMAT achieved its assigned mission within the prescribed time frame. An alternative for management was selected that was considered to make the tightest fit to obedience to environmental laws and maximum allowable timber harvest. As might be expected, the process and decision were challenged in federal court—by both sides. The court upheld the adequacy of the FEMAT process and the alternative chosen. One key point of challenge was to the legality of "ecosystem management." Judge William Dwyer ruled that, "Given the current condition of the forests, there is no way the agencies could comply with environmental laws without planning on an ecosystem basis" (U.S. District Court 1994).

President Clinton, recognizing that a similar drama was destined to play out in Oregon and Washington east of the Cascades, ordered an assessment of the east-side conditions. Understanding the checkered history of the Pacific Northwest, the director of BLM, Michael Dombeck, and I persuaded the administration to extend the assessment to a more meaningful ecological, economic, and social region. Because many runs of anadromous fish were headed for listing as threatened or endangered, the logical area was determined to be the interior Columbia River basin.

Other assessments were simultaneously in preparation for the Sierra Nevada and the Southern Appalachians. New standards and approaches for federal land-use planning were set with these assessments. It does not seem likely that the trends toward ecosystem management, with its inherent demand for overarching assessments, will, or can, be reversed under current law and case law precedents.

An Opportunity to Consider Lessons

These new assessments have necessitated a review of the lessons learned over the previous fifteen years in dealing with such matters. First among these is that an ecosystem management approach is more appropriate than a species-by-species approach triggered by the application of ESA, NFMA, and NEPA. Second, an assessment should include an evaluation of social and economic effects along with ecological considerations. Third, it is inappropriate to consider only one option, and scientists should not make decisions. Fourth, instead, scientists should provide decision makers with potential management alternatives and appropriate information on which to base a decision. Fifth, all the agencies involved with the application of the myriad pertinent laws should be involved from the beginning. Sixth, it is essential to include an adequate array of technical specialists—including economists and social scientists. Seventh, it is important to remember that ecosystem management is as much about people as anything else. Eighth, managers should be involved in the process. And, ninth, there should be opportunity for elected officials to be informed along with the public—i.e., the process should be "transparent."

Plans and Promises

One lesson may come as a shock to the souls of the innocent and the idealistic. Bioregional assessments must lead to a plan that can work, and that plan almost certainly will not satisfy those on either extreme of a contentious argument. There will always be detractors. Part of the challenge of conducting an assessment is to allow scientific disagreement to strengthen the analysis, not to derail it. Many people outside of the science community misinterpret scientific criticism as debunking results. And there are always those who will capitalize on dis-agreement in order to attempt to alter the outcome of the assessment one way or another in order to better advance their own ends. There may be attacks on personal credibility. It is here, once again, that scientific credibility and pro-cedural correctness provide the bastion of defense on the inevitable day of judgment.

Another part of the challenge is provision of a comprehensive assessment of the defined bioregion that can be used to make local decisions. The assessment must lead to a plan, and that plan must be useful to those with decisions to make. But one plan will not fit every location. Our intention with FEMAT was not that the default prescriptions take over the management world, but, rather, that they provide guidelines within which decisions could be modified locally with minimum debate.

Unfortunately, default prescriptions have prevailed, as it has proven politically difficult to produce locally adapted prescriptions. Any shift from the default posture constraining managers brought attack from the "green" constituencies— internal or external to the agencies. Managers simply found it too difficult to prepare and make such changes from the political standpoint. Flexibility is an effective management tool only to the extent that managers can, or will, exercise that flexibility.

FEMAT finished its assignment in ninety days by presenting an assessment of ecological, economic, and social factors and ten options with their associated attributes and risks. These options were incorporated into an environmental impact statement (EIS) by a separate team. President Clinton chose "Option 9" as prescribed by the EIS team—which had been considerably modified from the original Option 9 prepared by the science team. Numerous "bells and whistles" were added to Option 9 in the process, I believe, primarily to meet concerns of agency biologists and environmentalists and to "assure" compliance by the land management agencies. I, and a number of other scientists involved with the development of the original Option 9, were concerned that with the bells and whistles appended to the final version, timber projections could not be achieved due to technical, personnel, budget, and time constraints. Experience over the past several years has shown that concern to be valid.

There was yet another lesson to be learned. Plans should not be larded with promises to increase their political palatability—especially if they cannot be executed without dramatically increasing the costs of management. This is a poor mechanism to address the lack of trust between agencies and among technical experts. In retrospect, the revised Option 9 should have been reviewed by the original science team and a panel of managers to assure conformity with the science, the use of all pertinent information, the statement of confidence intervals, and the projections of results.

Technically Credible and Procedurally Correct

Throughout these assessments, it is always necessary to be technically credible and procedurally correct. The credibility of the science and the correctness of the procedures used to determine the defensibility of the assessments are essential, even while the policy questions create a moving though expanding target. But the necessary skills to produce useful assessments extend beyond the realm of traditional science, with its standards of hypotheses testing and reporting results. These assessments are an emerging field of science wherein existing information from myriad sources is integrated to produce a new understanding and basis for action. The assessment of information over a wide array of sources requires technical experts with an ability to integrate across disciplines, to synthesize ideas, and to convert synthesized knowledge into a form that is useful to managers and policy makers.

This emerging field of science does not replace the need for specific, highly technical research. Rather, it takes the information created from basic research and methodical collection and integrates it with knowledge gained from many other disciplines in order to create new understanding of complex, inter-related functions of ecosystems. From there, the consequences of certain decisions on the sustainability of those interrelated functions can be analyzed. In the FEMAT case, the different technical specialty groups conducted their assess-ments in parallel efforts, and it was learned how difficult it can be to snap together findings at the end to form a coherent and integrated whole. In future assessments I would expect to see more cross-disciplinary exchange of informa-tion throughout the process, conducted by scientists trained or experienced in synthesis and integration. The evolving capabilities of geographic information systems (GIS) and multiple-scale data management may help streamline this integration.

Managers must constantly upgrade their understanding of the assessment science or be increasingly dependent on technical experts. Conversely, it would be well for assessment teams to include members with management experience in order to make assessments useful to managers and other decision makers.

Assessing Risk

Assessments such as FEMAT and the interior Columbia Basin represent a syn-thesis of current scientific knowledge including a description of uncertainties and assumptions. For federal land managers, these assessments are not decision doc-uments. They *do not* provide direct answers to specific problems. Rather, these assessments provide the foundation for proposed changes to existing land man-agement policy, with the necessary information for policy discussions and deci-sions.

Considering risk is an important part of an assessment because of the sto-chastic nature of ecosystem processes. Risk associated with predicting outcomes can be partially mitigated by revealing the underlying relationships of cause and effects. Science plays a large role in disclosing these relationships. But scientists should not be asked to define how much risk is acceptable. This is a question for decision makers. How much risk society chooses to bear is a managerial, rather than a technical or scientific, decision. Technical assessment, then, assesses risk and points out perceived consequences of various decisions. Then, most decid-edly, the decision maker must weigh the social, political, and ethical factors involved in the decision.

More Lessons

It is unwise to delay dealing with the issue of the preservation of biodiversity— no matter how much "politics" comes to bear. Delays exacerbate the ecological situation by "eating up" management options resulting from ongoing alteration, removal, or fragmentation of habitat. As options are reduced, adjustments to the economic status quo to meet requirements of law and technical validity become increasingly severe.

Agencies must act in a coordinated and collaborative fashion from the beginning in the assessment and development of alternatives for management. That cooperation must carry through into management. "Splits" between agencies produce lack of management cohesion and provide easy targets for adverse publicity and attacks on legal grounds.

Any action based on less than a full undergirding assessment of pertinent legal, ecological, social, and economic factors will be subject to continuous—and likely successful—legal and political assault.

"New knowledge" such as that which emerges from assessment or from ongoing research and monitoring cannot be ignored without legal consequences.

The role of science is not to describe trade-offs and enter the realm of making trade-offs that are, of necessity, based on value judgments.

It is unwise to assume that what agency administrators and their technical and legal advisors consider full compliance with applicable law will not draw legal challenge from some quarter. It is, likewise, unwise to assume that judges will see things the agency's way—or be logical or consistent in their judgments. That is why we have appeals courts.

It is unwise to "pull the tail" of a federal judge. Judges lack a certain sense of humor when it comes to compliance with the law and the respect they believe due to the separation of powers. Once irritated, federal judges can and will issue injunctions and court orders that will shut down resource extraction from federal lands until full compliance with the law and case law has been achieved. The law means what the judge says it means.

Those judges who will decide the cases, as well as the advocates that will bring the cases, are totally familiar with the entire history of any issue as background for discussion and decision. Judges can deal severely with those who insult their intelligence by repeating mistakes (as already determined by other judges). Judges are aware of and sensitive to precedents—whether they make sense or not.

Politicians who promise their constituents that they can get around the consequences of the law, case law, and experiences elsewhere by manipulations of budget or budget language to guide agency actions are likely to do those constituents significant harm in the longer term. Such manipulations through the means of the "quick fix" will only delay compliance. Experience shows that delays will increase the impact on those who rely on resource extraction. This will force reliance on recovery plans, produced one at a time, for threatened or endangered species as the drivers for land-use plans or modifications in those plans. Experience, again, demonstrates the consequences of that approach.

The process should be as "open"or "transparent"as possible. There is, after all, nothing but technical work in progress. Of course, in observing such a give-and-take process, it is well to remember the old adage that "Like sausage, you will enjoy it a lot more if you don't watch it being made." Such a process will take longer and cost more. But the alleviation of associated paranoia is adequate justification. This open approach should be thought of as a significant contribution to democratic processes.

State and local officials should be kept fully informed and fully involved. At best, this will enhance opportunities to achieve consensus. At worst, there is the opportunity to fully understand what is going on and to remove the mystery from the process.

The actual "hired hands"doing the day-to-day work of the assessment should all be federal employees to avoid any potential conflicts with the Federal Advisory Committee Act. Routine consultation with other state or local government entities is acceptable and should be encouraged.

Peer review of "scientific"assessments should be carried out, and all files concerning such reviews should be made available.

"Science"assessments and reviews should be conducted separately from other parts of the process such as EISs, decision documents, and records of decisions.

Limits to Science and Management

While the application of science and the involvement of scientists have a role to play in developing and describing the consequences of management options and estimating risks, they cannot deliver stability. Listen carefully to those who pick at the science involved in a related decision to natural resources. Frequently, they are not so much upset with the science as they are with the changes that are in the wind.

To long for "stability," to want things to stay the same, is a common human characteristic. During my school days, I was taught the mantra of a fully regulated forest that would provide forest products in such a sustained and predictable fashion that "community stability" would be a reality. This, in turn, became a beacon for federal land managers—and for foresters in general.

My dream faded slowly and then finally collapsed in the focused glare of reality. We now recognize (or should) that we are one drought, one insect or disease outbreak, one dramatic fire season, one new law, one pivotal court case, one power-shifting election, one budget, one press campaign, one propaganda blitz, one shift in demographics, one change in public opinion, one shift in market demand, one shift in price, one loss of a management tool (say, DDT or clear-cutting), or one piece of "new information" away from stability at all times. And these variable influences do not emerge one at a time. They ripen in bunches—much like bananas.

In fact, the guts of this exacerbating controversy over natural resource management are not primarily technical in nature. These are human issues rooted in differences in morality, ethics, belief, and political position—most of all, they revolve around whose ox is getting gored. These human factors sometimes become embodied in laws passed at different times by different Congresses for different purposes. It is not surprising that these laws mesh poorly.

But when these expressions of national will embodied in national law are applied in regional and local arenas where there is deep concern with the social, political, and economic consequences of the compliance to those laws, there is apt to be reluctance to comply—and even resistance. Initially, there is room and reason to argue exactly what the law requires and what constitutes compliance. But as the case law piles up higher and higher, there is less and less rationale for arguing these matters in the arena of the judiciary.

Such is a fine line to walk as the pressure to satisfy the requirements of law is counterbalanced by the pressure to satisfy local social, economic, and political demands. The maximum probability for being judged in compliance with myriad laws, with ESA as the trump, is to do nothing in terms of active management. The maximum chance to satisfy local, social, economic, and political demands is to exercise as much active management as possible and to remain in compliance with the law. These countervailing pressures produce a razor-thin decision space.

There is little that agency administrators can do to alter that situation except to learn from the past and modify approaches to ensure compliance with the law(s) in a manner that produces the least social disruption and the most overall efficiency. Such conflicts cannot be solved through application of science—ecological, social, political, or economic. Clearly, the costs of compliance with the laws have escalated to the point where it is becoming more and more unlikely that "harvesting" of natural resources from public lands can be achieved in an "above cost" fashion—particularly when the "books are cooked" more and more to increase costs, or the appearance of costs. While there are always some efficiencies that can be captured, this is an insidious problem of the extant "crazy quilt" of law and regulations. To the extent that this is a problem, it is one that can only be solved by Congress and the administration. The two mantras of "below cost" and "subsidy" are potent propaganda tools. One sometimes wonders about the provision of recreation, including hunting and fishing, in the light of "below cost" and "subsidy" arguments.

Temptations and Consequences

Insanity has been described as doing the same thing over and over and expecting a different result. If so, sanity involves learning from the past and appropriately modifying behavior while moving to the future. The aim of regional assessments, to a large degree, is to provide knowledge for decision making without repeating the politically induced foibles that occurred in earlier struggles over federal land management in the Pacific Northwest forests, the interior Columbia Basin, the Sierra Nevada, and elsewhere.

Forty-two years in the natural resources business—and particularly those when I was involved in the issues in the Northwest and then as chief of the FS— have led me to formulate facetiously a number of "Thomas's Laws of Natural Resources Management." One of my more recently formulated postulates is "Quick fixes to real or perceived problems resulting from compliance with natural resources laws lead inevitably to insanity."

There are, of course, a number of corollaries to this "immutable law." They are as follows:

Corollary 1. The interactions of numerous laws (formulated in different congressional committees and passed by different Congresses over a century) and regulations (issued by different agencies with different missions) with little or no consideration of the interactions of those laws and regulations, when simultaneously applied, produce politically undesirable results. These results evolve from overlaps in agency jurisdiction and the certainty of clarification (i.e., confusion) from frequently conflicting and always evolving case law. Those problems are confounded by the political unacceptability of the management outcomes, which vary markedly by geographic locale and constituency. This, inevitably, produces a situation that is unacceptable to some member(s) of Congress.

Corollary 2. This member, if powerful enough, will propose a "quick fix"to the perceived problem. Knowing that the proposed quick fix would never receive approval if introduced and debated as a bill, the member inserts the quick fix as a rider to some piece of legislation—usually a budget bill that must be passed, or where it will not be noticed.

Corollary 3. The quick fix, upon application, will inevitably cause more political, legal, and operational problems than it solves.

Corollary 4. The problems caused by the quick fix will, more often than not, produce the perceived need for a quick fix of the quick fix.

Corollary 5. The ramifications of the interactions of a series of quick fixes are apt to be multiplicative as opposed to additive.

Corollary 6. At every iteration of the quick fix syndrome, it is possible to search for, address, and bring forth a solution to the underlying problem.

Corollary 7. Addressing underlying problems, such as conflicting and overlapping laws, requires understanding, power—of both influence and persuasion—and hard work over a prolonged period with small chance of success. These, of course, are excellent reasons to avoid this approach.

Corollary 8. Therefore, another try at a quick fix will evolve as "the answer."

Corollary 9. Doing the same thing over and over and expecting a different result is insanity.

Conclusion

The mandate for management on federal lands is changing. It would be a lot easier to manage resources the old way and produce a "regulated forest." Managers would cut so many board feet of lumber, build so many miles of road, plant so many acres, and cut again at maturity. Of course, "ologists" would advise on how this would be done to assure attention to other multiple-uses. A generation ago, these were the tasks resource professionals were trained to do.

But resource management is more complex now, made so by a rapidly increasing human population, decreasing natural resources per capita, rapidly increasing knowledge, and many more laws, regulations, and court decisions.

Managers find themselves attempting to manage society's conflict over the value of natural resources (for which they are poorly equipped) as much as they manage the resources themselves (for which they are well equipped).

Increasingly, managers are expected to be cognizant not only of the latest research and use of rapidly increasing technology but also of evolving social values and management of escalating conflict. Regional assessments are the latest attempt to organize some approach to this complexity and compare the effect of different actions in a more meaningful framework. Assessments provide a means to examine and explain the complex, interrelated functions of ecosystems and the consequences of certain management choices on the sustainability of those functions.

These changes in natural resource management are evolutionary in nature—not revolutionary. They are part of an evolution of environmental values that has been moving along in much the same trajectory throughout my forty-two-year career. Events have caused our progress to rise above or dip below the trajectory of this change, but no revolutions have shaken us irretrievably off course. The direction still points toward sustainability, as our understanding expands of just how complex is the task that lies ahead to achieve that vision.

LITERATURE CITED

Bueter, J. H. 1990. *Social and economic impacts of the spotted owl conservation strategy.* Technical Bulletin 9003. Washington, DC: American Forest Resource Alliance. 36 pp.

FEMAT (Forest Ecosystem Management Assessment Team). 1993. *Forest ecosystem management: An ecological, economic, and social assessment.* Portland, OR: U.S. Department of Agriculture, U.S. Department of Interior, and others.

Gorte, R. W. 1990. *Economic impacts of protecting the northern spotted owl.* CRS report for Congress. 90-74 ENR. Washington, DC: Library of Congress, Congressional Research Service. 48 pp.

Greber, B. J., K. N. Johnson, and G. Lettman. 1990. Conservation plans for the northern spotted owl and other forest management proposals in Oregon: The economics of changing timber availability. *Papers in Forest Policy* 1. Corvallis, OR: Oregon State University, College of Forestry, Forest Research Laboratory. 50 pp.

Hamilton, T. et al. 1990. *Economic effects of implementing a conservation strategy for the northern spotted owl.* Washington, DC: U.S. Department of Agriculture, Forest Service; U.S. Department of the Interior, Bureau of Land Management. 65 pp.

Lee, R. G. 1990. *Social and cultural implications of implementing "a conservation strategy for the northern spotted owl."* Seattle, WA: University of Washington, College of Forest Resources. 42 pp.

Maki, W. R., and Olson, D. C. 1990. *Economic and social impacts of preserving current forests in the Pacific Northwest.* Technical Bulletin 91-07. Washington, DC: American Forest Resource Alliance. 19 pp.

Marcot, Bruce G., and Jack Ward Thomas. 1997. *Of spotted owls, old growth, and new policies: A history since the Interagency Scientific Committee Report.* General Technical Report PNW-GTR-408. Portland, OR: USDA Forest Service, Pacific Northwest Research Station. 34 pp.

Montgomery, C. A., G. M. Brown, Jr., and D. M. Adams. c. 1992. *The marginal cost of species preservation: The northern spotted owl.* Missoula, MT: School of Forestry, University of Montana.

Northwest Timber Association. 1990. *Summary and critique of the spotted owl conservation strategy proposed by the Interagency Spotted Owl Scientific Committee final report issued May 1990: "An industry review of the Thomas report."* [Place of publication unknown]: Northwest Timber Association. 21 pp.

Olson, D. C. 1990. *Economic impacts of the ISC northern spotted owl conservation strategy for Washington, Oregon, and northern California.* Portland, OR: Mason, Bruce, and Girard. 37 pp.

Rubin, J., G. Helfand, and J. Loomis. 1991. A benefit-cost analysis of the northern spotted owl. *Journal of Forestry* (December):25–30.

Sample, V. A., and D. C. Le Master. 1992. Economic effects of northern spotted owl protection: An examination of four studies. *Journal of Forestry* 90(8):31–35.

Satchell, M. 1990. The endangered logger: Big business and little bird threaten a northwest way of life. *U.S. News and World Report* 108(25):27–29.

Schlager, D. B., and W. A. Freimund. 1994. *Institutional and legal barriers to ecosystem management.* Submitted to the Eastside Ecosystem Management Project. 87 pp.

Thomas, J. W., E. D. Forsman, J. B. Lint, et al. 1990. *A conservation strategy for the northern spotted owl: A report of the Interagency Scientific Committee to address the conservation of the northern spotted owl.* Portland, OR: U.S. Department of Agriculture, Forest Service; U.S. Department of the Interior, Bureau of Land Management, Fish and Wildlife Service, National Park Service. 427 pp.

Thomas, J. W., M. G. Raphael, R. G. Anthony, et al. 1993. *Viability assessments and management considerations for species associated with late-successional and old-growth forests of the Pacific Northwest.* Portland, OR: U.S. Department of Agriculture, Forest Service. 530 pp.

U.S. Department of Interior, Fish and Wildlife Service. 1991. *Economic analysis of designation of critical habitat for the northern spotted owl.* Draft rep. Washington, DC: USDI. 49 p (plus appendix).

U.S. District Court. 1994. *SAS v. Lyons,* no. 92 479 WD, December 21.

CHAPTER 2

Stepping Back: Assessing for Understanding in Complex Regional Systems

Lance H. Gunderson

In assessing the issues of bioregional systems, uncertainty is everywhere—from a complexity of interactions inherent in the scale of natural and social systems to ever changing sets of interactions as these systems evolve. The rule, rather than the exception, is that regional-scale resource systems behave in surprising ways. Observations of the ecological system don't always agree with human expectations, thereby generating crises of policy. Certainly, it would be easier and more tractable if regional systems behaved in normative, predictable ways, allowing us to detect and avoid such crises. But the experience in attempting to manage these systems has shown that failures are common and crises appear unavoidable.

So how does one assess the unpredictable in order to manage the unmanageable? This chapter makes some suggestions by stepping back to look at the question from two directions. The first looks at some useful theory, and the second involves some hard-won lessons from experience in regional-scale systems. This is not a new idea; indeed, the Greeks argued that only through integrating theory and praxis can we begin to understand. The primary theme in this chapter is that understanding and learning form the foundations for both assessment and management in a world full of uncertainty.

The first direction that this chapter will look in will be to review nascent theory on the dynamics of complex systems, theory that captures some of the patterns displayed in the adaptive dance between people and nature. These theoretical concepts were developed and weakly tested by applying them to a series of case histories of regional-scale ecosystems (Gunderson, Holling, and Light 1995). The next direction of this chapter will be to recount lessons from practical experience, building on themes of assessing for surprises, and assessing and managing for the property of resilience in these coupled systems of people and nature. This chapter ends with some conclusions learned from "stepping back." But first, some theoretical propositions on the dynamics of complex, adaptive systems.

Dynamics of Complex Systems

Ecosystems and the people who utilize, manage, and depend on those ecosystems form inherently complex systems. They are complex, due in part to the myriad of components, i.e., ecologic, economic, social, and institutional and connections among those parts. But they are also complex because of their behavior over time. That is, the behavior of these systems alternates between periods of ordered, predictable change and periods of disordered, difficult-to-predict change. The alternation between predictability and unpredictability has been at the center of many ecological debates, particularly those surrounding ecosystem succession.

Much of the temporal dynamics displayed by ecosystems has been captured in a conceptual, heuristic model of Holling (1986, 1992) that includes four phases (figure 2.1). Ecosystem succession has been described in two phases: *exploitation,* in which rapid colonization of recently disturbed areas occurs, and *conservation,* in which slow accumulation and storage of energy and material are emphasized (Odum 1969). In ecologic literature, the species in the exploitative phase have been characterized as r-strategists (with attributes of rapid growth in an arena of scramble competition) and in the conservation phase as K-strategists (with attributes of slower growth rates and survival in an arena of exclusive competition). To an economist or organization theorist, those functions could be seen as

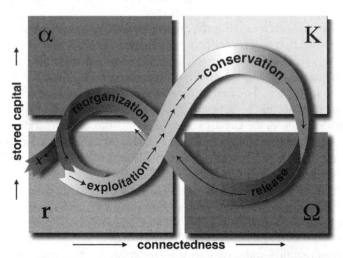

Figure 2.1. *Four phases of complex, adaptive systems (Holling 1986, 1992) and the flow of events among them. The arrows represent time; the short arrows represent slow change, while the long arrows represent fast change. The cycle represents changes in connectivity in the system and in stored capital. The system may exit from the cycle at point "x."*

equivalent to an entrepreneurial market for the exploitation phase and a bureau-cratic hierarchy for the conservation phase. These two phases comprise much understanding and predictability. That is, with knowledge of initial conditions, the trajectories of succession can be broadly defined and predicted. However, to capture changes in many terrestrial systems, two additional phases are observed. One is a *release*, or "creative destruction," phase, in which tightly bound accumu-lations of biomass and nutrients that have become increasingly susceptible to dis-turbance (*overconnected*, in systems terms) suddenly are released by agents such as forest fires, insect pests, or intense pulses of grazing. The fourth phase is *reor-ganization*, in which processes come into play to reorder the system for the next phase of exploitation. This last function is essentially equivalent to processes of innovation and restructuring in an industry or in a society—the kinds of eco-nomic processes and policies that come to practical attention at times of economic recession or social transformation. This is designated as the *alpha* phase.

Time, stability, and predictability all vary as the system moves through these four phases. The progression in the ecosystem cycle moves from the exploitation phase slowly into the conservation phase, very rapidly through release, reorgani-zation, and back into exploitation. During the slow sequence from exploitation to conservation, connectedness and stability increase and ecological "capital" such as nutrients and biomass slowly accumulate. In an economic or social system, the accumulating capital could be infrastructure capital, levels in the organization, or standard operating techniques that are incrementally refined and improved. Resilience and recovery are determined by the fast release-and-reorganization sequence, whereas stability and productivity are determined by the slow exploita-tion-and-conservation sequence. Instabilities trigger the release phase, which then proceeds to the reorganization phase, where weak connections allow loosely controlled chaotic behavior and the unpredictable consequences that can result. Stability begins to be reestablished in the exploitative phase. The system then, in short, exhibits a pattern of chaos erupting from order, and order emerging from chaos, not unlike dynamics described by Prigogine (1980).

Just as ecosystems exhibit four phases over time, so do human organizations. Public policy, such as that practiced in the United States at various scales and by various organizations, exhibits similar phases and transitions. Most policy (defined broadly as a set of rules, operations, etc.) becomes increasingly mature and conservative over time and in the process of implementation. At some point, it is recognized that existing policy has undeniably failed. Often the failure is the result of the latent or intrinsic dynamic of the ecosystem within which the policy is applied. For example, the shortcomings of water policies in the Everglades that focused on flood control were revealed during a period of drought in the early 1970s (Light, Gunderson, and Holling 1995). That pattern is chronicled in a vari-ety of cases elsewhere (Gunderson, Holling, and Light 1995) and in the cases of this volume. The crisis of the policy or release phase is rapidly dealt with by a series of approaches that involve reorganization. That reorganization phase generally involves creating a temporary institutional structure, such as the Gang of Four in the FEMAT story (chapter 6) or the 1989 restoration symposium in the Everglades (chapter 8). That temporary institution assesses the system (human and ecological) and then presents a series of options to a formal decision-making body (such as the federal legislature or a state water management agency) that will decide on the new or modified policy to be implemented during the next cycle.

In nature, there is a nested set of such cycles, each occurring over its own range of scales. For example, in many forest systems, fresh needles cycle yearly; the crown of foliage cycles with a decadal period; and trees, gaps, and stands cycle at close to a century or longer periods. The result is an ecosystem hierarchy, in which each level has its own distinct spatial and temporal attributes. A critical feature of such hierarchies is the asymmetric interactions between levels (Allen and Starr 1982; O'Neill et al. 1986). In particular, the larger, slower levels constrain the behavior of faster levels. In that sense, therefore, slower levels control faster ones. If that was the only asymmetry, however, then hierarchies would be static structures, and it would be impossible for organisms to exert control over slower environmental variables. However, these hierarchies are not static. They are transitory structures maintained by interaction across scales.

The birth, growth, death, and renewal cycle, shown in figure 2.1, transforms hierarchies from fixed static structures to dynamic adaptive entities whose levels are sensitive to small disturbances at the transition from growth to collapse (the omega phase) and the transition from reorganization to rapid growth (the alpha phase). It is at the two phase transitions between gradual and rapid change that the large and slow entities become sensitive to change from the small and fast ones. When the system is reaching the limits to its conservative growth, it becomes increasingly brittle and its accumulated capital is ready to fuel rapid structural changes. The system is very stable, but that stability is local and narrow. A small disturbance can push it out of that stable domain into catastrophe. Collapse can be initiated by either internal conditions or external events, but typically it is internally induced brittleness (linked to overconnected and accumulated capital) that sets the conditions for collapse. Another opportunity for small-scale processes to cause system change is during the transition from reorganization to exploitation. During this reorganization phase, the system is in a state opposite to the conservation phase previously described. There is little local regulation and stability, so that the system can easily be moved from one state to another. Resources for growth are present, but they are disconnected from the processes that facilitate and control growth. In such a weakly connected state, a small-scale change can nucleate a structure amid the sea of chaos. This structure can then use the available resources to grow explosively and to establish the exploitative path along which the system develops and which it then locks into.

In review, the accumulating body of evidence from studies of ecosystems indicates key features of discontinuities in processes and structures and reveals the appearance of multiple stable states. The four-phase cycle of adaptive renewal captures many of these dynamics for ecological systems. Linking the adaptive cycles across scales develops a heuristic model that has been called panarchy (figure 2.2). Panarchy is the word used to describe dynamic symmetries across hierarchical scales rather than an asymmetrical and static relationship across scales.

Two key features of panarchy are emerging that help understand the cross-scale nature of resilience from small-scale processes that contribute to disturbance vulnerability to broader-scale processes that contribute to recovery. As systems mature, key variables change and generate an increasing vulnerability of that system to disturbances through the processes that generate revolts. "Revolt" refers to processes that aggregate from smaller, shorter scales to broader, longer scales. One example is fuel accretion in fire-prone forests. Fuel accumulates at local scales, and it is the spatial connectivity of these local processes that allows for forest fires to occur over broader scales. Revolt elements determine how vulnerable a system is to changing stability domains following a disturbance (i.e., how big a revolt will occur—a minor or major transformation). Following periods of creative destruction, it is the "remembrance" of the system that fosters renewal and reorganization. The property of remembrance may be due to residual elements following a disturbance, such as serotinous seeds or nutrients after a fire. Renewal also relies on exogenous inputs, such as seed rain from unburned

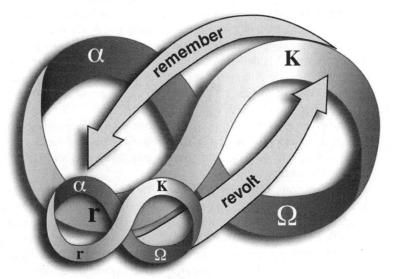

Figure 2.2. *Diagram of cross-scale dynamics, or panarchy, showing two coupled adaptive cycles (four phases from figure 2.1), each representative of distinct structural elements, and each with different domains in space and time. For example, the small cycle might represent a dynamics of a pine needle and the large cycle a tree crown. At key points, the cycles are strongly linked, either during revolt phase, describing contagious processes such as fires or pest outbreak, or remember phase, where the broader scale provides resources for the smaller-scale reorganization.*

areas, or human capital for replanting. Hence, resilience or adaptive capacity is linked across scales by these processes of revolt and remembrance.

Cross-scale linkages integrate the ecological and human facets of these complex systems. The human systems that monitor, assess, and manage ecological ones must do so at a variety of scales. Many surprises perceived by the human part of the system are related to ecosystem dynamics occurring at broad, long time frames. The next sections expand and describe the coupling between people and nature, starting with the notions about ubiquitous uncertainty.

Confronting the Unknown

Ecosystems and the people whose fates are intertwined with them form inherently complex systems that display periods of stability and instability, as well as unexpected behavior due to a variety of factors. The periods of predictable and unpredictable behavior suggest that humans are likely to be always searching for understanding in these complex resource systems. The struggle for understanding the uncertainty of resource issues has a certain parallelism with basic laws of thermodynamics. A crudely paraphrased version of these laws that pertain to the inherent unknowability of resource systems is: (1) we can't win—we don't know enough to predict with any confidence what is going to happen in these systems; (2) we can't break even—learning by trial and error is unlikely to reveal causes and effects; and (3) we can't get out of the game—therefore, we develop some sort of system for management, based on traditions, religions, or science.

In these coevolving systems of humans and natures, surprises are the rule not the exception. A surprise is defined as a qualitative disagreement with expectations. Brooks (1986), in describing the interaction between technology and society, defines three types of surprises: (1) unexpected discrete events, (2) discontinuities in long-term trends, and (3) emergence of new information. These categories can be broadened and placed in the context of the previously mentioned theories of change in ecological systems by redescribing Brooks's types as local surprise, cross-scale surprise, and true novelty. Examples and elaboration are described in the next paragraphs.

Local surprises can often be addressed by recognizing broader-scale processes. Unexpected events can be part of broader-scale fluctuations or variation of which there is little or no local knowledge. An ecological example of this is the cycle of flood and drought over the southeastern United States that is part of the global atmospheric and oceanic coupling known as El Niño/Southern Oscillation. An example in economic systems was the world oil crisis of 1979 in the United States, created by a variety of interacting global factors, but which involved political turmoil and disruption of oil production in Iran. In these cases, it is the ignorance of broader- and longer-term processes that contributes to the local surprise.

The next class of surprise deals with abrupt, nonlinear, or discontinuous behavior of a system that, after analysis, can be attributed to an interaction between key variables that operate at distinctly different scale ranges. That is, the surprise is due to a faster variable interacting with a changing, but slower variable much like the panarchy previously described. In ecological systems, examples include fire in ecosystems, where ignition frequency is a "fast" variable and stand age is a "slow" variable. In human systems, the surprise of the AIDS virus may be attributed to accelerated changes in variables such as human connectivity and land-use transformation (Ewald 1993).

The final type of surprise is genuine novelty—something truly unique and new, not previously experienced by humans, or at least outside the experience of a culture in a new situation (Gunn 1994). These types of surprise can generate evolutionary change, change whose consequences are inherently unpredictable. Most examples of new technologies (such as the personal computer in the late 1970s) fall into this category. In resource systems, invasions by exotic species are an example of this type of surprise. The invasion of alien trees such as *Melaleuca quinquenervia* in Florida (Myers 1983) alters key ecosystem processes such as nutrient cycling, water relations, and fire patterns.

Surprises can be either positive or negative. With surprise being the rule rather than the exception in ecologic systems, resilience in human systems is the key property with which to cope with the unexpected nature of ecological systems. The development of adaptive capacity in human institutions is suggested as one way to deal with environmental uncertainties (such as global climate change) rather than a focus on certitude of future predictions (Rayner and Malone 1997). It follows therefore that robust responses from assessments are more important than correct responses from assessments.

Robust Responses

Given that humans will continue to cope with systems that are partly knowable and partly unknowable, the responses that people make to uncertainty become part of "stepping back." The relationship between uncertainty and resilience is key; how people choose to deal with uncertainty either increases or decreases resilience of a system. In a weak typology, there are three kinds of responses that humans (or many other animals, for that matter) have in dealing with the unknown. These response can be described as: (1) do nothing by ignoring or assuming away the uncertainty, (2) replace the uncertainty with a faith that something else (religion, tradition, institutions, technocratic elite) will resolve the uncertainty, or (3) confront the uncertainty in a systematic manner. The ignorance response won't be discussed here, but rather I use the latter two classes as an entrée into a deeper discussion of resilience and uncertainty. To begin, I start with examples of how displacement of uncertainty may help to erode system resilience.

During most of this century, the goal of technologically based resource management has been to control the external sources of variability in order to seek a singular goal, such as maximization of yield (trees, fish) or controlling levels of pollution. This approach, also called "command and control," focuses on the control of a target variable, which is successful at first, but then slowly changes other parts of the system. That is, by isolating and controlling the variables of concern (i.e., assuming that the uncertainty of nature can be replaced with the certainty of control), resilience is eroded. The manifestation of that erosion is the pattern of policy crisis and reformation, as mentioned above and elsewhere (Gunderson, Holling, and Light 1995). One such example involves water management in the Everglades of Florida.

Since the early 1900s, concomitant with the development of water resources in the Everglades, there has been an erosion of resilience in the management institutions. During this period, new technologies (levees, canals, pumps) and new institutions were designed and implemented to control unwanted hydrologic variation. This development ultimately led to a loss of institutional resilience (sensu Holling 1973) as indicated by the inability of extant institutions to deal with ecological crises (Light, Gunderson, and Holling 1995). A major flood in 1947 resulted in the current configuration of federally funded structural elements (Central and Southern Florida Project, designed by the U.S. Army Corps of Engineers) and managed by a state Flood Control Board. That board was eliminated in 1971, when a drought crisis (low rainfall intersecting with a burgeoning population) led to a new institution—the South Florida Water Management District—whose implicit goal was to reestablish flexibility of options for water supply. Unforeseen algae blooms in Lake Okeechobee and vegetation changes in the Everglades (associated with nutrients moving through a nutrient-poor landscape) during the 1980s led to lawsuits among government agencies and, eventually, to new means of intergovernmental interaction such as the Everglades Coalition. Currently, the management institutions have little flexibility to try alternative management policies and seem more intent on protecting turf than adapting to deal with an uncertain future.

There is growing evidence that acknowledgment and confrontation of uncertainty add resilience to managed systems. People involved in the practice of resource management are all linked by the need for understanding. The experience and practice during this century have been to turn to scientists, the heart and soul of technology, and to technologic solutions as the fountains of understanding. But there has been a growing sense that traditional scientific approaches are not working, and indeed can make the problem worse (Ludwig, Hilborn, and Walters 1993). Two reasons rigid scientific and technological approaches fail are that because they tend to focus on the wrong types of uncertainty and on narrow types of scientific practice. Many formal techniques of assessment and policy analysis presume a system near equilibrium, with a constancy of relationships, in which uncertainties arise not from errors in tools or models, but from lack of appropriate information to go into the models. Picking up on the Everglades example from the prior paragraph, a team of resource scientists used such a process to build an integrated understanding of options based on a broader assessment of uncertainties regarding chronic ecologic problems (Davis and Ogden 1994). That understanding reestablished management resilience, in the form of policy options to deal with the next ecological crisis rather than to resolve the last crisis. That resilience is manifest as a set of composite, regional restoration goals that incorporate environmental and social goals (chapter 8, Ogden).

Adaptive management is an integrated, multidisciplinary method for natural resource management (Holling 1978; Walters 1986). It is adaptive because it acknowledges that the natural resources being managed will always change; therefore, humans must respond by adjusting and conforming. There is and always will be uncertainty and unpredictability in managed ecosystems, both as humans experience new situations and as these systems change through management. Surprises are inevitable. Active learning is the way in which this uncertainty is winnowed. Adaptive management acknowledges that policies must satisfy social objectives but also must be continually modified and flexible for adaptation to surprises. Adaptive management therefore views policy as hypotheses— that is, most policies are really questions masquerading as answers (Light, personal communication). If policies are questions, then management actions become treatments in the experimental sense. The process of adaptive management includes highlighting uncertainties, developing and evaluating hypotheses around a set of desired system outcomes, and structuring actions to evaluate or "test" these ideas. Although learning occurs regardless of the management approach, adaptive management is structured to make that learning more efficient. Trial-and-error is a default model for learning while managing; people are going to learn and adapt by the simple process of experience. Just as the scientific method promotes efficient learning through articulating hypotheses and putting those hypotheses at risk through a test, adaptive management proposes a similar structure.

In the few cases where adaptive management has been applied, it has failed to varying degrees. There are numerous reasons, but they distill to the issue of flexibility in either the ecological or human system. That is, if the risk of failure during experimentation is not acceptable, then adaptive management is not possible. Witness the example of the Columbia River Basin (Lee 1993), where invocation of the Endangered Species Act halted any actively adaptive management experimentation (Volkman and McConnaha 1993). In other systems, such as the Everglades, a rigidity or lack of flexibility in management institutions and extant political power relationships has precluded adaptive experiments when there exists flexibility in the ecological system for such experimentation (Walters and Gunderson 1994).

In spite of limitations in management applications, adaptive environmental assessment has been quite successful when applied in bioregional-scale systems. The heart of this success has been a ruthless hold on uncertainty, that is, not assuming it away, but keeping it as part and parcel to all phases of the assessment. One way of communicating the scientific uncertainty of resource issues is to articulate alternative arguments as competing hypotheses. By doing this, research, models, and data can be mobilized from the assessment to generate useful policy probes. Although management actions have a history of being compromised in the social or political arena, many adaptive assessments have been effective in transforming the understanding of the ecological system and policy options in a number of systems. In New Brunswick, the modeling and assessment team transformed understanding of the forest dynamics (Clark, Jones, and Holling 1979) and set the foundation for a new system of property and lease arrangements following a crisis in the late 1980s (Baskerville 1995). In the Everglades, a similar result was noticed; the adaptive assessment group activities set the conceptual framework for much of the ongoing ecosystem restoration (Davis and Ogden 1994; chapter 8, Ogden). The assessments in these cases have resulted in dramatic shifts in the perception and evaluation of policy options.

Assessing and Managing for Resilience

Referring to the four-phase and cross-scale model mentioned above, resilience is the key property that helps a system through phases of crisis and reformation. In this section, notions about how resilience can be maintained or restored in resource systems are discussed. Resilience is defined by the adaptive capacity of a system; i.e., how much of a disturbance a system can absorb before it configures in a qualitatively different stability domain (Holling 1973; Walker et al. 1969; Walker 1995). People create resilience in resource systems in two ways: (1) they develop strategies that increase the resilience of a managed resource system (keeping it from flipping into an undesirable stable state); and (2) they promote flexibility and learning to improve the resilience in human organizations. Since many assessments involve both ecological and institutional dimensions, resilience is a key ingredient in transforming and creating more robust systems. In order to add resilience to managed systems, a number of strategies are employed: increase the buffering capacity of the system; manage for processes at multiple scales; and nurture sources of renewal.

Buffering tends to remove the effects of unwanted variation in the system in order to facilitate a return to a desired equilibrium. In many agricultural systems, resistance to change is dealt with by a combination of barriers to outside forces (tariffs, fences, etc.) and internal adjustments such as water-control or cost-control mechanisms (Conway 1993). Water resource systems can be designed for resilience by increasing buffering capacity or robustness through a redundancy of structures (and flexibility of operations) rather than fewer, larger structures and rigid operational schemes (Fiering 1982; Light 1983).

Resilience in resource systems may be increased by managing processes at multiple scales. In the Everglades water management system during the mid 1970s, water deliveries to Everglades National Park were based on a seasonally variable but annually constant volume of water. This system was changed in the mid-1980s to recognize interannual variation in the calculation of water delivery (Light and Dineen 1994; MacVicar 1985). Folke and Berkes (1995) argue that local communities or institutions may coevolve by trial and error at time scales in tune with the key processes that structure ecosystems, within which the groups are embedded. Many of the crises chronicled in this review and others (Gunderson, Holling, and Light 1995) were created from an inherent focus on a single scale for management and new understanding of the multiple scales by which the ecosystem was functioning.

Another way in which people manage for resilience in resource systems is by concentrating on sources of renewal. Many forms of catastrophic insurance provide this function, by creating a fiscal reservoir that can be tapped should structures need to be replaced. Another mechanism that explicitly plans for renewal in resource systems is a scheme of market-based property-rights systems developed for Australia. Young and McCay (1995) argue that adding flexibility and renewable structure to property-rights regimes will increase resilience. They indicate that market-based property-rights schemes (licenses, leases, quotas, or permits) should have built-in sunset (termination) to the scheme, within stability of arrangements (entitlements, obligations) in the interim years.

Institutions (defined broadly as the set of rules and structures that allow people to organize for collective action) can add resilience to a system. A few key ingredients appear necessary to facilitate the movement of systems out of crisis through a reformation. These include functions of learning, as well as the ability to tap into deeper understanding and trust (Gunderson, Holling, and Light 1995). Lee (1993) calls this process "social learning," by combining adaptive management frameworks within an institutional framework of collective choice. Other authors (Folke and Berkes 1995) argue that social capital, comprised of the institutions, traditional knowledge, and common property systems, provides resilience. Putnam (1994) bemoans the demise of social capital, being those norms and networks that facilitate cooperation through trust in pluralistic democracies such as the United States, with the wonderful metaphor of bowling alone in America. It is such linkages and connectivity across time and among people that help navigate transitions through periods of uncertainty to restore resilience.

Stepping Back and Looking Forward

So far, this chapter has used the Holling (1986, 1992) conceptual model of adaptive cycles to describe patterns of management histories from regional ecosystems. That model captures much of the nonlinear coevolution between people and ecosystems. Ecosystems and management institutions appear to undergo periods of exploitative growth, brittle conservatism, creative destruction, and renewal. Policy, defined as a principle or plan of action, follows a similar time course: implementation, failure, generation of alternative proposals, and reconfiguration. Various groups dominate the stage of these dynamics; bureaucracies implement policies, extremists declare policy failure, shadow or epistemic networks develop alternative choices, and formal decision makers decide new policies (Gunderson, Holling, and Light 1995). This model sets the framework for some tentative propositions about the nature of bioregional assessments. These propositions are described in the following paragraphs and focus on confronting uncertainty, building resilience, and creating new institutional structures.

Assessing for Uncertainty

Evaluating uncertainty should always be a key component of bioregional assessments. The assessment should focus on sources of uncertainty, either in attempting to understand how ecological systems operate (such as assumptions, myths, or competing explanations or hypotheses of resource issues) or in institutional or social arrangements. Adaptive environmental assessments have made a contribution to this approach, with over a quarter of a century of experience in using computer models to help communicate among a diverse set of actors, highlight uncertainties, screen alternative hypotheses, and propose management actions to help achieve understanding as much as other social objectives.

Managing for Resilience

In many systems, management actions and economic activities erode resilience of ecosystems and reduce flexibility of institutions. The signals of that erosion are often detected too late for considered action. An example is the ability to place a species on an endangered species list but the inability to institute incentives that maintain species diversity. When resilience has been eroded, recovery plans and actions may be the very things that push the species to extinction. Therefore, it is critical in assessments to understand the resilience and flexibility of both ecological and social systems, in order to make robust recommendations for policy.

Reframing Our Institutions

In this volume and other reviews of regional case studies, changes in policy and practice rarely come from existing bureaucracies and extant top-down hierarchies. The dynamic was facilitated by other players, including activists intent on declaring existing policy as a failure, and a transient group (a "shadow network") that developed an integrated understanding and created new policy opportunities for the future. The implication is that new types of bridging devices should be designed to combine people from inside agencies (loyal heretics) with those outside to facilitate flexible and adaptive management. We can no longer count on the "technocratic elite" to solve environmental issues. Cases of successful resource policy renewal involve participation of the people affected by the old and new policies. Regional policies and political action become possible when enough of the citizenry develops a knowledgeable sense of the region as their place. A huge uncertainty remains, though, on how to develop foundations for an informed citizenry to judge among competing arguments, objectives, and policies at a regional scale. There may be an opportunity to tap existing experiences in citizen science (e.g., various citizen monitoring and policy-design projects) to propose or launch a new generation of experiments that recognize the linkages between home, region, and planet. This proposition is not just a retreat to community-based management, which assumes a placid ecosystem, but is a call for new institutions that actively learn and respond to their environment.

Acknowledgments

C. S. Holling, Steve Light, Garry Peterson, and Rusty Pritchard contributed to and enriched many of the ideas presented in this chapter; I am grateful to all of them.

Literature Cited

Allen, T. F. H., and T. B. Starr. 1982. *Hierarchy: Perspectives for ecological complexity.* Chicago: University of Chicago Press.

Baskerville, G. 1995. The forestry problem: Adaptive lurches of renewal. In *Barriers and bridges to the renewal of ecosystems and institutions,* edited by L. H. Gunderson, C. S. Holling, and S. S. Light. New York: Columbia University Press.

Brooks, H. 1986. The typology of surprises in technology, institutions and development. In *Sustainable development of the biosphere,* edited by W. C. Clark and R. E. Munn. Cambridge: Cambridge University Press.

Clark, W. C., D. D. Jones, and C. S. Holling. 1979. Lessons for ecological policy design: A case study of ecosystem management. *Ecological Modelling* 7:1–53.

Conway, G. 1993. Sustainable agriculture: The trade-offs with productivity, stability and equitability. In *Economics and ecology, new frontiers and sustainable development,* edited by E. B. Barbier. London: Chapman and Hall.

Davis, S. M., and J. C. Ogden. 1994. *Everglades, the ecosystem and its restoration.* Delray Beach, FL: St. Lucie Press.

Ewald, P. W. 1993. The evolution of virulence. *Scientific American* 273: 86–93.

Fiering, M. B. 1982. Alternative indices of resilience. *Water Resources Research* 18:33–39.

Folke, C., and F. Berkes. 1995. Mechanisms that link property rights to ecological systems. In *Property rights and the environment,* edited by S. H. A. M. Munasinghe. Washington, DC: Beijer International Institute and World Bank.

Gunderson, L. H., C. S. Holling, and S. S. Light. 1995. *Barriers and bridges to renewal of ecosystems and institutions.* New York: Columbia University Press.

Gunn, J. D. 1994. Global climate and regional biocultural diversity. In *Historical ecology: Cultural knowledge and changing landscapes,* edited by C. L. Crumley. Santa Fe, NM: School of American Research Press.

Holling, C. S. 1973. Resilience and stability of ecological systems. *Annual Review of Ecology and Systematics* 4:1–23.

———. 1978. *Adaptive environmental assessment and management.* London: John Wiley and Sons.

———. 1986. Resilience of ecosystems: Local surprise and global change. In *Sustainable development of the biosphere,* edited by W. C. Clark and R. E. Munn. Cambridge: Cambridge University Press.

———.1992. Cross-scale morphology, geometry and dynamics of ecosystems. *Ecological Monographs* 62(4):447–502.

Lee, K. 1993. *Compass and gyroscope.* Washington, DC: Island Press.

Light, S. S. 1983. *Resilience and surprise in water management institutions.* Ph.D. thesis, University of Michigan, Ann Arbor.

Light, S. S., and W. Dineen. 1994. Water control in the Everglades: An historical perspective. In *Everglades: The ecosystem and its restoration,* edited by S. M. Davis and J. C. Ogden. Delray Beach, FL: St. Lucie Press.

Light, S. S., L. H. Gunderson, and C. S. Holling. 1995. The Everglades: Evolution of management in a turbulent environment. In *Barriers and bridges to the renewal of ecosystems and institutions,* edited by L. H. Gunderson, C. S. Holling, and S. S. Light. New York: Columbia University Press.

Ludwig, D., R. Hilborn, and C. Walters. 1993. Uncertainty, resource exploitation, and conservation: Lessons from history. *Science* 260:17, 36.

MacVicar, T. K. 1985. *A wet season field test of experimental water deliveries to northeast shark river slough.* Technical Publication 85-3. West Palm Beach, FL: South Florida Water Management District.

Myers, R. L. 1983. Site susceptibility to invasion by the exotic tree Melaleuca quinquenervia in South Florida. *Journal of Applied Ecology* 20:645–658,

Odum, E. P. 1969. *Ecology.* Philadelphia: Saunders.

O'Neill, R. V., D. L. DeAngelis, J. B. Waide, and T. F. H. Allen. 1986. *A hierarchical concept of ecosystems.* Princeton: Princeton University Press.

Prigogine, I. 1980. *From being to becoming: Time and complexity in the physical sciences.* New York: W.H. Freeman.

Putnam, R. D. 1994. Bowling alone: America's declining social capital. *Journal of Democracy* 6:65–78.

Rayner, S., and E. L. Malone. 1997. Zen and the art of climate maintenance. *Nature* 390: 332–334.

Volkman, J., and W. E. McConnaha. 1993. Through a glass darkly: Columbia River salmon, the Endangered Species Act, and adaptive mangement. *Environmental Law* 23:1249–1272.

Walker, B. 1995. Conserving biological diversity through ecosystem resilience. *Conservation Biology* 9(4):747–752.

Walker, B. H., D. Ludwig, C. S. Holling, and R. M. Peterman. 1969. Stability of semi-arid Savanna grazing systems. *Ecology* 69:473–498.

Walters, C. J., and L. Gunderson. 1994. A screening of water policy alternatives for ecological restoration in the Everglades. In *Everglades: The ecosystem and its restoration,* edited by S. M. Davis and J. C. Ogden. Delray Beach, FL: St. Lucie Press.

Walters, C. J., 1986. *Adaptive management of renewable resources.* New York: Macmillan.

Young, M., and B. J. McCay. 1995. Building equity, stewardship and resilience into market-based property rights systems. In *Property rights and the environment,* edited by S. H. A. M. Munasinghe. Washington, D.C.: Beijer International Institute and World Bank.

PART TWO

The Role of Science in
Bioregional Assessments

History and Assessments: Punctuated Nonequilibrium

John C. Gordon

We need a new model for linking science, management, and policy, a structure that synthesizes science and management, and by which that synthesis can inform public policy. Science must change to be more directed to the needs of managers, and managers must accept that science implies experimentation and that with experimentation comes uncertainty. Policy makers, in particular, need to realize the costs and limits of science as a policy tool.

According to Professor John Tukey of Princeton University,

> A major problem . . . lies in the 'non-comparability' of the comparative outcomes of different policy decisions. The eventual policy decisions will implicitly compare the noncomparable and will, in crucial instances, have to be made, on behalf of all the people of the United States, by the Congress and the administration (or the courts). The realistic problem . . . is to go as far as reasonably possible in providing the quantitative facts about consequences while avoiding as much as possible the conflicts whose resolution will require a national decision-making mechanism.

Natural resource management is changing rapidly and radically; a paradigm shift is occurring with many of the characteristics described by Thomas Kuhn in his book, *The Structure of Scientific Revolutions* (Kuhn 1962). Kuhn describes a paradigm shift in science as recognizable by ever more frantic attempts to make the old paradigm solve new problems and explain anomalies. A period of confusion and contending schools of thought occurs as a new paradigm arises as a result of the "revolution." I believe we are passing through such a period now (NRC 1990) and have been since the early 1980s.

Natural resource management is undergoing a rapid and deep-seated change. Our past paradigm, concocted of land allocation, sustained yield, and multiple use, no longer provides us with the mechanisms we need to manage adequately (Gordon 1992). Previously, if we worried about wilderness, we would allocate more land; if we worried about resource supply, we would plant faster-growing hybrids; if we worried about competing uses, we would point out picnic tables as well as stumps. But, eventually, problems occurred that could not be solved by these methods, and anomalies mounted that could not be explained by the reigning logic. Knowledge outstripped the ability of our paradigm to explain the world and forecast cause and effect. There turned out to be not enough land to allocate for every human use and every species need, and sustained-yield management was not able to sustain a healthy environment. We tried every means to keep the old view in place yet failed utterly.

Change is now not only necessary but unavoidable. Attempts to revive the old paradigm will fail and be marginalized. Ecosystem management and adaptive management are emerging as possible new approaches, particularly on public lands. At their simplest, these are approaches to management that recognize local differences in systems and goals, and they admit that managing natural resources is complex and therefore a continuous learning process. These approaches to management are essentially responses to scarcity. Ecosystem management responds to a scarcity of land to allocate to specific, exclusive uses and recognizes that the production potential of ecosystems is finite, while adaptive management is a response to a scarcity of information and knowledge. These responses to scarcity point out the need to make comparisons among Tukey's "noncomparables."

This new natural resource management paradigm signals changes in our basic worldview, changes in the techniques we employ, changes in what constitutes effective leadership (Berry and Gordon 1993), and changes in how we see and make policy. One frequently overlooked aspect of the changes is the increase in the sheer number of policies in the form of laws, regulations, and guidelines that must be taken into account in management decisions. The rise of interest in "industrial ecology," at least in part due to multiplying regulatory challenges, may signal a similar change in how we do and view business and manufacturing operations (Socolow 1992).

History

Enlisting science in the service of resource management and policy making has a relatively long history. In 1878, John Wesley Powell published his *Report on the Lands of the Arid Region,* an assessment of what was known and unknown about the arid western landscape. His scientific survey, authorized and funded by Congress, was driven by comprehensive questions about a bioregion defined by aridity and provides an early and formidable example of science informing policy. His report recommended to Congress a science-based policy for settling the West, according to which, scientists would assess which lands should be classified for irrigation, for grazing, and for timber. Powell intended the assessment to guide settlers' choices toward places where they would have the best chance of making a living.

Congress ignored his report.

In 1890, Powell began a series of hearings before the House Select Committee on Irrigation of Arid Lands, hoping to suggest a new political structure based on scientific understanding of these arid lands. Historian Donald Worster (1994) recounts the hearings:

"Powell wanted the congressmen to know the country as he knew it, both from high up on the canyon rim and down there on the river, and to understand its patterns of water, climate, and geology that had been interacting over so many eons. He wanted them to realize how little of the country could ever grow crops. . . . He wanted the men in the hearing room to appreciate the conflicts that were already brewing in those western valleys, as upstream developers diverted the water that downstream farmers depended on . . . . Above all, he wanted them to see that all the natural resources of the West were connected into a single integrated whole, so that what was done to the mountain forests affected the lowland streams, and the lands without water were intricately related to those with water."

Again he was ignored.

The history of development of the West from that point on is well-known. Powell's recommendations, which to us now seem visionary, were ahead of their time, and his concern for a landscape approach to development was not a concern others shared at the time. Powell's assessment of the condition and future of the arid western lands found no place with those who believed that rain would follow the plow. The America of the 1890s was not ready to put constraints on resources that seemed boundless.

Additional examples can be drawn from two much more recent assessments, in which I have participated: the Society of American Foresters' (SAF) Task Force on Scheduling the Harvest of Old-Growth Timber, and the Scientific Panel on Late-Successional Forest Ecosystems ("Gang of Four"). These are mature enough to assess, and they show something of the evolution of assessments focused on forest subjects.

Old-Growth Forests, the SAF Task Force, and the Scientific Panel

It was clear as early as the 1970s that the fate of remaining old-growth forests in the Pacific Northwest was to be a subject of considerable political debate. The Society of American Foresters thus established a task force and produced both a report and a "position" on old-growth forest issues. The task force included three of the four principal participants in the much later Scientific Panel on Late-Successional Forest Ecosystems. The conclusions of the task force were that old-growth needed to be more precisely described and better inventoried; that there were sound reasons, including but not limited to the protection of the northern spotted owl and other potentially endangered species, for establishing a system of old growth reserves; and that harvest scheduling of the kind then done on public forests needed to be changed to accommodate the realities of forests in transition from old-growth to second growth. It seems fair to say, in hindsight, that these were sound recommendations. It is also fair to say that the report and position had little impact, for several reasons:

First, because policy makers were not involved in the creation of the task force, or in posing the specific questions to be pursued, we produced an answer to a question that no one (outside SAF) had specifically asked. Second, some of the participants who had produced the report, and signed off on the final version, publicly repudiated the report soon after its publication, severely damaging its credibility. And third, there was no planned or sustained effort to make the content of the report known to those who might use it. It was published, but it was not distributed effectively.

One overriding lesson from the task force assessment is the importance of implementation. It is clear in hindsight that implementation begins with initial communication between the policy makers and the science actors, which leads to a careful framing of the questions to be answered. This initial communication was omitted. It is equally clear now that commitment by those who produce the report to a planned, funded program of transmission and explanation of the product, and to supporting it with additions and revisions as knowledge and questions change over time, is necessary if the assessment is to be used by those who request the information.

The Scientific Panel on Late-Successional Forest Ecosystems

The assessment of the Scientific Panel on Late-Successional Forest Ecosystems had the benefit of several previous efforts, most notably the Thomas Report from the Interagency Scientific Committee. It also had what in hindsight, at least, seemed to be several advantages. First, the questions to be answered were discussed thoroughly by policy makers and researchers and written down clearly in a letter. Second, no ponderous organizations or protocols separated the scientists involved from the policy makers. Scientists worked directly with Congress and agencies. And third, agreement was reached at the outset that the final product would be information arrayed as a series of choices and their probable consequences, not as a single recommended solution.

The Scientific Panel, however, had severe limitations. Old-growth had not been effectively mapped on all federal lands; time was extremely short, and it was thus impossible to fully access all sources of existing information. Most important, although there had been a great acceleration of research on old-growth and related species a few years before, we still had a woefully inadequate information base, particularly on old-growth-related species other than spotted owls and on the effects of silvicultural practices on wildlife populations.

Specifically, we learned that if scientific knowledge is plentiful and well agreed upon before policy questions arise, it is probably easier to use in policy making. However, in this instance, the questions and the knowledge emerged together. We learned that science pursued by interest groups in the heat of policy battles does not seem to be an important ingredient in resolving conflicts. Once the panel began its work, advocacy groups from opposite camps used partially developed scientific "discourse" to support their agendas. And, finally, despite honest and energetic attempts, broadly effective communication does not yet seem to have been achieved between scientists and the policy community.

Thus, in the old-growth debate, there have been several attempts at science-based assessment, but the science base was not well developed when the general policy questions matured. The much larger and more expensive Forest Ecosystem Management Assessment became the next iteration and set a new standard for assessments, at least in terms of scientific scope.

Science and Policy Issues in Natural Resources

Science has never been formally and directly linked to natural resource policy making in this country, although such a linkage has been much discussed and even legislated (for example, see the National Forest Management Act). John Wesley Powell's scientific prognostications about the arid West in the nineteenth century seem now to have been very much on the mark. But they had little or no effect on such key public policy matters as homestead size and shape (Herrick 1991—92). Thus, science has had a haphazard effect on public policy (figure 3.1). Public policy's feedback to science has "tended to be unidimensional, denominated largely in terms of resource transfers for research, and not by a continuous search for new and better ideas" (NYAS 1995). A more sensible model might provide the institutions and relationships to continuously feedback to policy institutions a synthesis of the observed outcomes of the interaction of science and management. Assessments are attempts to use science-derived information and tech-

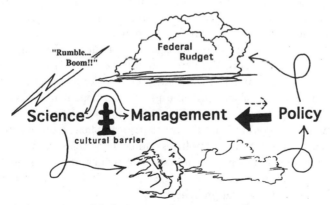

Figure 3.1. *A model of science at the crossroads of management and policy. Communication between science and management takes a long detour around a cultural barrier; communication between management and policy is largely a one-way street. Communication from science to policy is in the form of hot air. Scientific information collides with existing policy and convects into thunderstorms of crisis. These storms spark budget adjustments to fund, or not to fund, science. In this model, the "rumble-boom" of environmental crisis always precedes the flash of funding.*

niques to answer, or help answer, questions formulated outside science by policy makers, thus to provide the synthesis feedback suggested previously.

All substantive questions about natural resources and their management are now "issues" at some level. Most managers think that science has something to contribute to their resolution. Thus, all managers want "their own scientist" (John Martin, USDA Forest Service, district ranger, personal communication) in the hope that science can improve the effectiveness and credibility of decision making and implementation. Many (but not all) of the largest, most contentious issues are rooted directly in scientific advances and in advocacy based on science. Thus, arguments about the preservation of biological diversity, the preservation of old-growth or ancient forests, and the fate of anadromous fish all involve relatively new scientific findings and writings. Other issues, such as the role of federal regulatory activity on private property, and what constitutes a "taking" of private property, are not directly rooted in scientific findings but draw on science for the development of principles and standards and often look to science for the resolution of specific disputes.

Even though science as a culture has tried to divorce itself from questions of value, it is sometimes used as a surrogate for debates over values. Those who, for example, oppose clear-cutting may do so principally on aesthetic grounds but adduce science-based arguments about loss of diversity, soil erosion, and water quality to support their cause. Managers therefore have a tendency to believe that if they address the science questions adequately, the value questions will be covered and the management action that follows will be acceptable to most, or enough, of those people who consider themselves affected by the decision.

Similarly, policy makers tend to believe that policies (laws, regulations, and large-scale standing answers to recurring questions) will be stronger and better accepted if grounded in the best available information, most of which is seen to be derived from science. Thus, both managers and policy makers may tend to believe in the efficacy of science as an aid to doing their jobs. However, the relationship between science, management, and policy has often lacked a structure that allows the integration of scientific findings into management decisions, particularly in a timely fashion (NYAS 1995). Often, policy is made with "old" science, while interest groups confront the manager of public resources with "new" science. Thus, caught between rigid policies and evolving science, managers may find themselves restricted to a narrow, and often to them and their clients, unsatisfactory set of options. Perhaps worse, in the long run, is the resulting reining-in of managers' desire to try new approaches.

Another paradox surfaces when managers and policy makers all want their own scientists to accredit decisions, yet advances in scientific understanding spark new, contentious issues and controversy. This suggests a relationship between public uncertainty and scientific uncertainty (Bernabo 1995): you can make policy with public uncertainty if the science seems certain, or you can make policy with scientific uncertainty if the public seems certain; but when both science and the public are uncertain, policy making is extremely difficult. Unfortunately, in this moment of shifting paradigms, both public and scientific uncertainty are common.

Bioregional and Science-Based Assessments

The notion that assessments are legitimate policy tools has grown slowly, along with science and its uneasy (but almost wholly dependent) relationship with government. "The idea that science should be the basis for resolution of public resource policy issues and conflicts over the application of technology to the management of natural resources is relatively new (last 150 years). For most of that time the science base has been inadequate to the tasks presented it, so its role has not been large. . . . However, issues are now generally acknowledged to be complex enough so that a scientific basis for decisions would be desirable, but the notions of how to go about this are not well developed" (Gordon 1992).

Brewer (1984) presented a set of tentative rules (in the form of questions to be asked in review or, better, at the outset) for carrying out assessments:

1. Was the policy question or issue carefully posed, specified, and well understood by both . . . researcher and policy maker?
2. Did the policy question "translate" reliably into researchable terms, and, if not, what compromises were made and whose preferences were favored (and why)?
3. Were sufficient resources allocated to carry out the work, where "resources" are defined in terms of time, money, and talent?
4. Was the research keyed to the realities of the policy maker's world, in the sense of reflecting the resources over which the user exercises control and of incorporating the political constraints within which the user operates?
5. Were results [are results to be] results communicated in an intelligible fashion? And to whom, via which media?
6. Was the knowledge created in the work misused, abused, or not used, and why in each case?

These questions seem to have the effect of protecting scientists, but they only tangentially indicate the central issue in the provision of assessments to policy processes: timeliness. Science and scientists have most often responded suspiciously, and therefore slowly, if at all, to questions from outside science.

Faulty Models

In pursuing a science or science-management synthesis to aid policy making, faulty models are frequently used. Two of the most common used (although seldom stated) faulty approaches are:

- Science will decide what is right, and then that will be legislated, and
- For any policy position there exists a set of scientists willing to support it, and because all scientists are equal, science-into-policy is simply a matter of finding the right scientists.

Typically, scientists hold the former view, and policy makers the latter. Good reasons for the differing views exist. Often, policy issues arise because of scientific advances. For example, the controversies over old-growth timber allocation have been triggered in part by research showing that many old-growth forests will persist longer and function differently than previously thought. Concern about acid rain came to national attention following the development of technology to detect and assess its consequences. In both cases, the scientific advances that triggered policy attention were the subject of heated debate within the scientific community, as are most new scientific findings. Scientific debate, however, tends to reinforce the nonscientist's view (indicated previously). The scientific community often does little to clarify the nature of its internal debates, which usually ask the question "What is happening?" rather than "What can be done about it?" Its internal agenda thus tends to isolate scientists from policy processes. Also, science does not smoothly move to a consensus view; it advances jerkily and is usually incapable of producing results, let alone consensus, on demand.

Two Cultures

> One of the most bizarre features of any advanced industrial society in our time is that . . . , some of the most important choices about a nation's physical health are made, or not made, by a handful of men, in secret, who normally are not able to comprehend the arguments in depth.   —C. P. Snow, 1961

Scientists and policy makers often are products of quite different educational and organizational cultures and respond to very different reward systems, and this "two-cultures" dichotomy can itself be a barrier to communication. Functionally, science proceeds by testing hypotheses and thus by proving things wrong. Policy making is by definition choosing that which is right and rendering it into law or regulation. Thus, at their most fundamental levels, the processes of science do not fit well with the processes of policy making. This is rarely recognized, and it is too often glossed over when it is recognized, by asserting that both science and politics "seek truth." Even if this is so in some general sense, the methods of seeking are so dissimilar that even political scientists have trouble finding parallels.

What can change the traditional ways of communicating at the crossroads? Most of our natural resource laws were written when we were an isolated, sparsely populated country. Even the relatively recent dictum of multiple use suggests that we all can have whatever we want. But the system of managing natural resources as a big piñata has run its course. We are now a fully populated, globally connected country, and the piñata is battered, its contents scarce. Ecosystem management is a response to that scarcity and an acknowledgment of the complexity of the sources of natural goods and services. It suggests that none of us will get all we want. Bioregional assessments arose as a strategy to initiate ecosystem management and to allocate support before the rumble-boom of crisis.

Institutional Model for Research Access and Assessment Organizations

Currently, there is no smoothly functioning machinery through which politicians can readily access scientific research in a timely way (Carnegie Commission 1991). Sophisticated research access requires detailed knowledge of the policy and technical questions, in priority order, that are most critical to achieving objectives and similarly detailed knowledge of research sources. Policy-making bodies, especially Congress, need better research access than they now have.

Brewer (1984) indicates the primacy of framing questions, as does Russell (1992), so a sophisticated research access and assessment effort must provide a way to continuously forecast issues and frame questions. Questions, derived from conversation with decision makers and policy makers, are framed formally. The research access and assessment organization works with Congress or another policy body to clarify objectives and to define the boundary conditions. The assessment team works with the policy maker to prioritize questions and searches the scientific community for specific information and individuals to help answer them. The assessment team will decide on the legitimate range of scientific information and options relevant to the questions and will ensure that that range is included in the scientific portion of the group supplying answers.

Particular attention and skill must be focused on the durability of problems; those likely to remain unresolved for a long time allow a more leisurely pursuit of answers. More commonly, time pressure will be acute, and a solution that is a long time coming will be no solution. Thus, the ability to forecast future problems and to act quickly to produce competent assessments will be central to the success of a research access organization. One of the costs of doing business in this way will be the creation of answers to questions that are forecast to, but do not, become urgent policy issues. Perhaps, just as fire departments sometimes regard false alarms as legitimate training, these could be regarded as warranted practice in doing directed scientific syntheses.

A research access organization must be staffed with people who are capable of gaining the attention and respect of policy makers, managers, and scientists. Most will have considerable science credentials, because of the insular nature of science language and institutions.

Several approximate models of this kind of organization serve Congress, among them the National Academy of Sciences, the Congressional Research Service, and the Office of Technology Assessment. Although each has a considerable record of positive achievement, none consistently delivers timely assessments specifically addressing natural resource policy issues. Procedures and processes through which scientists can provide effective advice on specific pieces of environmental legislation often have been considered to be inadequate by both scientists and legislators. The record of the Congress in developing science-based policies on natural resource and environmental issues is, therefore, particularly mixed, perhaps in part because assessments continue without conclusion beyond the point when decisions need to be made. With more tightly bound time lines and budgets, and with carefully posed questions to pursue, assessments may become a more effective and reliable source of information for natural resource policy making.

Successful science assessment can be defined as marshaling results and people to give science-based answers to carefully defined policy questions, including making comparisons between noncomparable outcomes insofar as possible, and carefully describing the limits of the answers. Based on past experience, the requirements for a higher probability of success seem to be clear. Thus:

1. Questions will be carefully formulated with the active participation of managers and policy makers, as well as scientists, before assessment is begun;

2. Ideally, the assessment will use existing, peer-reviewed science; when new research is required during the assessment, it will be carefully time-bounded and structured for maximum credibility.

3. All those involved, particularly the scientists, will realize that a range of policy options and consequences must be considered and effectively communicated, along with formal and informal estimates of error.

4. Effective (clear and timely) communication will be as important as good science. In the political context of bioregional assessments, late or misunderstood science is bad science.

5. "Magic" options are rarely uncovered, no matter how carefully the assessment is done; "technofix" options for resolving policy questions are rare, and scientific agreement is almost never universal.

The time and trouble needed to do an assessment make sense only if a policy forum exists to accept it. This is often ignored by scientists and scientific societies, who, probably with impeccable motives, may prepare "white papers" on "issues" without working with policy makers on questions or managers on forecasts of results. While these are sometimes useful, they are often ignored. Acceptance is much more likely if the sequence of events begins with the formulation of questions, followed by an assessment, formulation of options (forecast outcomes for different answers to the questions), and making a choice. Politically formed options posed at the outset restrict the assessment unnecessarily and tend to produce advocacy groups within the assessment team. The process will rarely work if the assessment precedes the questions, formulation of options, and choice, for the simple reason that the assessment may not address the right questions. It is precisely this sequence, however, that the advocates of "science first" perhaps unwittingly advocate.

A Collaborative Role for Universities

Several attributes make universities good places to do science-based assessments to aid the formulation of policy. First, they are the only organizations likely to contain anything like the range of expertise needed to address modern environmental and natural resources questions. Second, they have a reputation as neutral ground dedicated to scholarly inquiry, with a record (although not a perfect one) of protecting investigators from political reprisals and "kill the messenger" behavior. Third, they are relatively stable institutions: they persist, at the same address, over long times. Universities, and particularly the environmental and natural resources colleges and departments within them, should make the doing of science-based assessments one of their major businesses, out of both duty to society and self-interest.

Society now has no ready place to turn for assessments. The help that the United States Congress receives does not provide for state legislatures or other public bodies and is, as stated, not fully adequate for Congress. Thus, universities could supply an increasingly acute societal need if they directed some of their energy to assessments. In terms of self-interest, beyond the general notion that universities do have a duty to the larger society around them, the payoff lies in integration. Because assessments are conducted to answer questions from outside science, they almost never can be answered by the practitioners of a single, or even a few, academic disciplines. The collaboration required almost always crosses the boundary between the natural and social sciences, because policy questions about the natural environment almost always have social, as well as biological and physical, dimensions. In this way, policy questions become a mechanism to lure faculty from widely different areas into discussion and synthesis. Universities, and even colleges within them, suffer from disciplinary isolation, and conducting assessments could help overcome this.

To successfully take on this role, several actions will be necessary. First, universities or colleges contemplating this action should review their existing experience. Most have conducted assessments, or have faculty who have participated in them under some other aegis. Second, they will need to create a "question formulation" group that, as indicated above, moves freely in scientific, managerial, and political company and is empowered to help policy people frame, launch, complete, and communicate assessments. Third, there will have to be new incentives for participation in assessments by faculty and students. Students can be useful in the assessment process in a variety of way and can at the same time begin to learn integrations skills that are increasingly important. Faculty must be induced to participate through pay, promotion credit, and the intellectual challenge of working across both disciplinary boundaries and the boundaries of science.

There are many potential problems: development of adequate review to insure intellectual soundness; exposure to an unprecedented level of political pressure, and time taken from more traditional activities, to name a few. But if a university, or a piece of one, were to be successful, it would be a powerful positive influence on the construction of better science-based natural resources policy, a thing we urgently need.

Conclusion

Ecosystem management of natural resources will require a quantum increase in the use of interactive science and adaptive management. This increase will be possible only in a "science friendly" policy environment that will arise only if the scientific and policy communities establish a culture, protocols, and institutional arrangements that allow them to work closely and continuously together without weakening the basic purpose of either. Science-based assessments are a step in this evolutionary change.

LITERATURE CITED

Bernabo, Christopher. 1995. Remarks presented at the workshop "At the Crossroads of Science, Management, and Policy," November 6–8, 1995, Portland, Oregon.

Berry, J., and J. Gordon, eds. 1993. *Environmental leadership: Developing effective skills and styles.* Washington, DC: Island Press.

Brewer, G. 1984. *Creating and using policy research.* Unpublished manuscript, Yale School of Forestry and Environmental Studies.

Carnegie Commission. 1991. *Science, technology and Congress: Expert advice and the decision-making process.* Report of the Carnegie Commission on Science, Technology and Government, New York, February.

Gordon, J. 1992. The role of science in resolving key natural resource issues. In *Seeking common ground: A forum on Pacific Northwest natural resources,* edited by T. Nygren. Proceedings, USDA Forest Service, Region 6, Portland, Oregon. February 24–25.

———. 1994. From vision to policy: A role for foresters. *Journal of Forestry* 92(7):16–19.

Herrick, C. N. 1991–92. Science and climate policy: A history lesson. *Issues in Science and Technology* (Winter):56–59.

Kuhn, Thomas S. 1962. *The structure of scientific revolutions.* Chicago: University of Chicago Press.

NRC (National Research Council). 1990. *Forestry research: A mandate for change.* Washington, DC: National Academy Press.

NYAS (New York Academy of Sciences). 1995. *Science and endangered species preservation: rethinking the environmental policy process, a sustainable forests exchange.* Special report from the New York Academy of Sciences Science Policy Program. New York: New York Academy of Sciences, 40 pp.

Russell, M. 1992. Lessons from NAPAP. *Ecological Applications* 2(2):107–111.

Snow, C. P. 1959. *The two cultures and the scientific revolution.* New York: Cambridge University Press.

———1961. *Science and government.* Cambridge: Harvard University Press.

Society of American Foresters. 1984. *Scheduling the harvest of old-growth.* SAF Resources Policy Series. Bethesda, MD: Society of American Foresters.

Socolow, R. 1992. Six perspectives from industrial ecology. In *Industrial ecology and global change,* edited by R. Socolow, C. Andrews, F. Berkhout, and V. Thomas. New York: Cambridge University Press.

Worster, D. 1994. *An unsettled country: Changing landscapes of the American West.* Albuquerque, NM: University of New Mexico Press.

CHAPTER 4

Perspectives on Scientists and Science in Bioregional Assessments

Frederick J. Swanson and Sarah Greene

The emergence of bioregional assessments as important steps in dealing with difficult natural resource issues comes at an interesting time for scientists and science. The participation of biophysical and social scientists in helping shape policy and management of natural resources is not new. Venues for science input to policy and management have included traditional technology transfer, congressional testimony, National Academy of Sciences committee reports, and a variety of other advisory mechanisms. What is new in recent years is the pressure Congress and science leaders are putting on the U.S. scientific establishment to focus more heavily on problems of immediate social relevance. Bioregional assessments have been an important medium for scientists responding to this charge.

Scientific study of region-scale biological phenomena has also developed substantially over past decades, with roots in bioregional sciences such as biogeography, regional economics of natural resources, and water resources of large basins. More recent bioregional science studies have addressed wide-ranging topics such as the history of land use (Foster 1993), patterns of nitrogen cycling in response to atmospheric pollution (Aber 1993), and ecological effects of urbanization (McDonnell, Pickett, and Pouyat 1993). Bioregional assessments, however, have a distinctive emphasis on interactions between development and conservation, including biological and water resources. Bioregional assessments, therefore, commonly require both greater breadth of understanding of interactions of humans and ecosystems and more focus on immediate social needs for information than considered in these examples of bioregional science. Computer simulation capabilities that might help to meet those needs have not been developed adequately (Groffman and Likens 1994). Bioregional assessments may boost development of bioregional science by forcing formation of interdisciplinary teams, assembling relevant databases, and highlighting the important needs of society for sound science at this scale. Bioregional science has grown in part by efforts to fill a critical gap in the difficult problem of scaling our understanding from local to global on the question of how human activities interact with the atmosphere and with terrestrial and aquatic ecosystems.

Participation in bioregional assessments creates both opportunities and challenges for scientists. Gunderson, Holling, and Light (1995) propose that natural resource problems commonly operate in the context of crises that link social conflict with thorny natural resource problems. This conflict, common in situations leading to bioregional assessments, typically removes the scientists from their traditional milieu of bounded experimentation and hypothesis testing. The traditional roles of scientists generally lack the highly interdisciplinary, broad geographic scope and political profile demanded of scientists in bioregional assessments. The questions posed by policy makers are less likely to be answered with the techniques of traditional science.

Individual scientists play critical roles in bioregional assessments but are often poorly prepared for those roles, which are set in policy and political contexts. Such contexts may thrust scientists into conflict with the public, policy makers, and land managers, all critical partners in the bioregional assessment process. As Cortner et al. (chapter 5) argue, bioregional assessments should "begin and end with the policy problems, not the scientific problems." The tension between political and scientific perspectives is likely to be unfamiliar and uncomfortable for scientists, but this is part of science in service to society.

This chapter addresses the roles of scientists in bioregional assessments by first considering their instincts, developed through training and traditional science experience. The challenges encountered by scientists are viewed in light of conflicts with the traditions and standard methods of science. We next examine the roles scientists may encounter during the stages of initiation and conduct of a bioregional assessment and the subsequent setting and implementation of policy. At each of these stages, important tools are being developed to help deal with information transfer and uncertainty. The chapter ends by looking at the challenges faced by scientists and the implications that bioregional assessments have for science and scientists, including contributions to the emerging field of bioregional science. We draw these perspectives from assessments addressed in this book and some others that have occurred in the forestry and watershed arenas.

Traditional Work of Scientists

Science is conducted traditionally using various approaches to hypothesis testing, including manipulative experiments, field observations, simulation modeling, and historical reconstruction. Scientists are trained to think critically and cautiously, and they seek repeatable tests of their theories and interpretations. Much of science is an ongoing debate to constantly check levels of agreement and credibility of both the practitioners and products of science. The important self-policing process in science relies strongly on formal peer review of grant proposals and manuscripts submitted for publication. Other credibility checks include reproducing experiments and data analysis. The scientific and public literature, grant proposal review processes, scientific meetings, and even litigation provide forums, albeit limited, for debate.

Uncertainty in science is treated by attempting to define its causes, levels, and consequences using methods such as statistics and sensitivity analysis. In most cases, uncertainty increases in more complex and larger-scale systems. Ecological and social sciences—fields central to bioregional assessments—traditionally have emphasized work at fine spatial scales and short time scales (seasons to years). Scientists are commonly cautious to limit their interpretations to the range of data. Bioregional assessments challenge these traditions by asking scientists for credible, objective information where the grand scale and complexity of issues preclude application of standard approaches to dealing with uncertainty.

Bioregional assessments have important implications for bioregional science, which has developed in recent years as new technologies and issues have fostered advances over earlier, largely descriptive regional science, such as in the field of biogeography. Geographic information systems, remote sensing, and computer simulation modeling, for example, make possible sophisticated analysis of air pollution patterns and changes in carbon sequestration resulting from land use. In some cases, such bioregional science reflects an effort to bridge analyses across local to global scales for particular issues, such as carbon dynamics and trace gas exchanges. The recent spate of bioregional assessments brings new, increasingly complex issues and data sets to the field of bioregional science. In most cases, the scientists participating in bioregional assessments were not conducting region-scale science before joining assessment teams, so the cadre of bioregional scientists may grow as well.

Roles of Scientists in Bioregional Assessment

Bioregional assessments are part of a larger process of developing management policy. This process includes initiation, the assessment itself, policy formulation using assessment results, and implementation of new policy through management action. Each assessment presents scientists with distinctive circumstances, resulting in substantial variation in the roles of scientists both between assessments and at each of these stages within an assessment.

Initiation of Assessments

Two conditions must be met for initiation of a bioregional assessment: (1) a problem must exist in natural resource management that has not been solved by past attempts or has not been addressed because of its broad scale and transjurisdictional nature, and (2) a need must exist for policy making, generally because major social conflict is involved. Scientists may affect both of these essential conditions leading to bioregional assessments. Scientists may identify natural resource problems and their potential causes (e.g., threatened loss of native species, potential impacts of invading exotic species, declining water quality), and they may contribute to the need for new policy by "ringing the fire bell in the night," warning society of natural resource problems they believe need to be addressed with a shift in policy and management.

Identifying and Recognizing the "Problem"

Although most assessments reviewed in this book were initiated by policy makers, the starting conditions that affected participating scientists differed greatly among assessments. In the case of FEMAT, for example, the highest policy maker in the land, the president, stepped in to break a policy and management stalemate, but the ensuing assessment was built on a sequence of prior science assessments (chapter 6, Johnson et al.). The Sierra Nevada Ecosystem Project (SNEP) and the Interior Columbia Basin Ecosystem Management Project (ICBEMP) assessments, on the other hand, were intended to head off a similar stalemate, so the science teams had more flexibility and opportunity to anticipate critical issues rather than just reacting to issues raised by past conflict. However, in each case, scientists were involved in alerting managers, regulators, and the public of pending ecosystem problems. It is important to note that scientists raise many concerns about ecosystem degradation that do not result in responses from policy makers—especially at the broad scale of bioregional assessments.

Developing the Charge and Identifying the Participants

Scientists have a potentially critical role in defining the charge for the assessment to assure that it can be met with the tools at hand. But before those tools can be put to work, it may be necessary for scientists to negotiate with policy makers to ensure that questions posed can be answered by scientists. An important step at this stage is crafting an explicit charter for the assessment. A negotiation among policy makers, scientists, and possibly other participants may be the genesis of a charter and the basis for matching expectations. Charters have taken many forms—from a single sentence (in a rider to a congressional appropriations bill that established SNEP) to long, substantive, even philosophical memoranda (ICBEMP; Quigley, Haynes, and Graham 1996). Although sometimes forgotten, a charter can be a useful guide as the heat of an assessment process leads participants onto potentially divergent paths.

Though identification of assessment team members is frequently a political decision, scientists may be involved in picking the team, since they know the skills needed and who can provide them. Requisites for team participation include appropriate disciplinary mix, professional credibility, and ability to function in teams. Participation in previous assessments may lead to recruitment for subsequent ones.

Team selection involves a critical tension between having a shared commitment to completing an assessment and having diverse specialties, institutional perspectives, and personal values within the team. The compromises that necessarily ensue often receive the criticism, following assessments, that one mind-set was represented in the process while others were excluded. An assessment team with members of narrowly similar background and points of view may limit the scope of analysis, framing of questions and assessment approach, and identification of alternatives (chapter 6, Norris). This is particularly true when the policy questions lead an assessment team to address issues well outside the bounds of scientific certainty. This leads to a domain where personal values and institutional perspectives have the potential to become more prominent than science per se in reaching conclusions. These circumstances set the stage for later public criticism.

These points emphasize the critical importance of strong, thoughtful science leadership in bioregional assessments. Effective leadership requires credibility, effective social skills, and good instincts for balancing the tensions of competing interests in broad- vs. fine-scale approaches and the varying personal values and worldviews of team members. The leader of an assessment may broker scientists' interests with those of policy makers, managers, interest groups, and the public at various stages from formation of the team and its charter to the handoff to policy makers and implementers. Few people have been groomed for such roles.

Conducting the Assessment

An initial step in conducting an assessment involves identifying the basic approach for addressing the policy issues. Conservation biology, with its species focus; ecosystem dynamics, with its focus on whole systems; and the perspective that technology has the answer with its anthropocentric focus offer examples of contrasting approaches to natural resource issues. An issue may itself define the dominant viewpoint, such as using a conservation biology approach to protect an endangered species. But in some situations alternative approaches are possible, so scientists may compete to have their approach adopted by the assessment and ultimately in natural resource policy and management. More complex policy problems may require a blending of approaches, such as conservation biology to deal with selected species and a systems approach for water issues, all within a single assessment, such as in the case of the Great Lakes (chapter 7, Regier).

Technical Aspects and Constraints of Conducting an Assessment

The essential work of scientists in assessments is to produce an objective, value-neutral synthesis of scientifically credible information to determine current conditions, trends, and their causes, and perhaps consequences of possible future courses of action. But this work may trigger argument. Even a task as apparently straightforward as objectively describing current conditions may be controversial, as in the case of assessing the population of an obscure species, or obvious, such as when a polluted river catches fire. Interpretation of patterns and causes of temporal trends opens yet additional sources of conflict and uncertainty. The broad scale of assessments may require that data be combined across time, space, and institutions, thus raising concerns about data quality and comparability. Real or perceived lack of data is a common problem. Use of technologies, such as geographic information systems, that at first appear to save time, may, in the end, prove costly and time consuming and their effectiveness be limited by data-quality issues.

An important technical problem for scientists is finding a useful balance between assembly of high-resolution data, a common approach in traditional science work, and adoption of a broad-perspective synthesis based on qualitative interpretations. Compilation of fine-scale data can be excessively time consuming; the data may be fraught with inconsistencies; and analysis of high-resolution data may not be very useful in dealing with broad-scale issues. Excessive attention to dealing with voluminous data can distract from synthesis and integration, which require broad perspectives. These scaling issues, addressed in part by hierarchy theory (Allen and Starr 1982), are common in many areas of science today.

The types of problems and their solutions, if there are any, depend not only on the issues but also on the level of funding and the duration of the assessment. Short, low-budgeted assessments must rely on straightforward synthesis of existing information. Longer-term, better-funded assessments may generate new information and models. For assessments as large as the ICBEMP, which lasted several years and covered 145,000,000 acres (58,000,000 ha), the data accumulation task may become daunting and impede synthesis. Limits of time and funds compounded by the broad scale of analysis make it very difficult to conduct more traditional science by framing and testing hypotheses, but this may occur in exceptional cases (Murphy and Noon 1992).

Analysis of multiple assessments has led scientists to develop general theoretical frameworks for understanding behavior of interacting, complex biophysical and social systems, including circumstances leading to conflict addressed through bioregional assessments. Holling (1978); Gunderson, Holling, and Light (1995); and Gunderson (chapter 2) propose a model of how exploitation of natural resources can lead to social crisis and ecosystem degradation. They argue that traditional science is not up to the task of dealing with natural resource crises at the regional scale, that a different type of integrative interdisciplinary science is needed, and that natural resource institutions must shift their focus from exploitation technology to learning and adaptive management in order to practice sustainable resource management. The evolution of issues and institutions in the Great Lakes bioregion over more than two decades offers an example. The general theoretical framework derived in part from analysis of case studies (chapter 2, Gunderson) provides a useful frame of reference for participants in any future assessments.

Dealing with Uncertainty, Credibility, and Integration

The credibility of assessments rests in part on their treatment of uncertainty common to the broad geographic extent, temporal scope, and complexity of bioregional natural resource systems. The issues of uncertainty and credibility are made more acute because scientists are commonly asked to provide objective assessment at scales beyond their experience and beyond the limits of traditional science. Scientists have developed statistical techniques to quantify uncertainty, and in traditional science they try to reduce uncertainty by conducting experiments to test and extend the limits of knowledge (chapter 2, Gunderson). However, at the scale of bioregions, traditional experimentation is impossible in terms of achieving replication, similarity of experimental units, and randomized application of treatment. Furthermore, if ecosystems and social systems are perceived to be beyond their abilities to respond resiliently to disturbance, it may be socially impossible to conduct experiments that may fail, causing loss of an endangered species or triggering some other unacceptable outcome.

Scientists have developed a number of techniques for dealing with uncertainty of the types critical to the credibility of bioregional assessments. In a broad sense, development of systems models and application of sensitivity analyses provide a basis for identifying major and minor sources of uncertainty. Risk analysis also provides some basis for examining scientific and decision-making uncertainties and their consequences (Lackey 1996). Species-specific analysis techniques, such as population viability assessments, consider scientific uncertainty in population trends using both parameter estimates and measures of their uncertainty in models used to examine population trends under alternative scenarios (Boyce 1992). Recent developments in coupling spatially explicit population and habitat models hold promise for assessing species response to land management alternatives (Turner et al. 1995). A technique somewhat analogous to species-level analyses that has been employed at the ecosystem level is the use of historic range of conditions as a basis for identifying future landscape conditions that may sustain native habitat conditions (Morgan et al. 1994; Swanson et al. 1994; Landres, Morgan, and Swanson, in prep.). However, each of these techniques is subject to current debate within the scientific community over its general appropriateness for application to complex, large-scale social and ecological systems (e.g., Lackey 1996).

The scientific credibility of an assessment's findings about the state and productive capacity of an ecosystem is critical in legal and other social contexts. Standard tests of credibility in the traditions of science include the science credentials of the scientists themselves, as reflected in their personal records of productivity and recognition by peers, and also peer review of assessments. Issues of credibility of science interpretations and scientists may not be readily appreciated by nonscientists, including policy makers and managers who are scientists' new partners in these assessments. Furthermore, a long record of scientific accomplishment does not assure suitability of the scientist for the broad, political scope of bioregional assessments. Credibility of scientists may be strongly tested in the heady atmosphere of decision making with potential for wide-reaching impact.

Peer review can be critical in checking the credibility of bioregional assessments since they rely heavily on expert judgment. However, peer review in this context is much more challenging than in the traditional cases of manuscript and grant proposal reviews. The subject matter for peer review of bioregional assessments is diverse—the concepts behind the assessment, the quality and appropriateness of data used, the range and consequences of alternatives, the description and degree of uncertainty, and the assessment documents themselves. The rigor and thoroughness of such review is frequently limited by time and funding, which are commonly exacerbated for large assessments involving huge documents and databases. It may be desirable to have peer review take several forms. For example, combining reviews from knowledgeable people known to the assessment team with blind reviews commissioned through scientific societies provides opportunity for open discussion with reviewers and for candid, anonymous comment addressing both credibility of the evidence and of the interpretations. The pool of candidate reviewers is likely limited because few people are experienced in region-scale thinking, and those that are may have been involved in the assessment already.

Scientists commonly seek detail, while regional policies must be built on broad, integrated perspectives. The process of integration across diverse disciplinary points of view, especially social and biophysical, may be the least developed but most important aspect of many bioregional assessments. In some cases, no integration is called for because the assessment is narrowly framed (e.g., around a single species) or involves several unintegrated themes (e.g., caves and karst terrane, salmon, and selected terrestrial species in the Tongass National Forest land management plan). Attempts at integration have followed quite different courses. A form of geographic integration, for example, was used to form Option 9 in FEMAT by attempting to maximize overlap of terrestrial (late-successional) and aquatic (key watershed) reserves (chapter 6, Johnson et al.). To meet a greater challenge to integrate social and biophysical considerations, the ICBEMP used measures of ecological integrity and social and economic resiliency in million acre analysis areas (chapter 11, Quigley et al.). The notions of risk and opportunity used in the ICBEMP can be applied to both the biophysical and social realms in a semiquantitative manner, thus providing a basis for attempting a broadly integrated assessment.

Transition from Assessment to Policy

The setting of policy generally occurs after an assessment is complete and is, therefore, beyond the focus of this book. However, transition from assessment to policy is critical and has important implications for how the assessment is conducted and presented. For federal lands, the transition itself may involve an environmental impact statement team, which, operating within NEPA requirements, prepares and evaluates a range of alternatives for decision makers to examine. Or it could involve a series of land-use planning maps as in Natural Community Conservation Plan (chapter 10), setting up state councils as in Northern Forest Lands (chapter 9), or restoration planning as in the Everglades (chapter 8, Ogden). In unusual cases, the initial formulation and evaluation of alternatives may be conducted by the assessment team (FEMAT) (chapter 6, Johnson et al.). Scientist involvement is often critical in policy development because scientists are highly informed about the scope of technical matters and so are well equipped to help formulate and describe the consequences of alternative future management scenarios.

The distinction between roles of scientists and the decision makers who set policy may be subtle but critical. "Science information alone does not 'make' a decision: Decision makers make decisions" (Everest et al. 1997; preface). A sharp distinction between the roles of scientists and policy makers leads to the question: How do scientists do their work without making policy? The scientists' role is to present information on the social systems and ecosystems in question, scientific uncertainty, and risks. The decision makers' role is to determine the level of socially acceptable risk. A tool for objective communication between scientists and policy makers is the policy analysis framework used to present attributes of future management alternatives as points on a continuum of cost, risk, and benefit. Scientists may not make policy, but they can influence policy alternatives through design of alternatives and criteria used to evaluate them.

Scientists may also be asked to evaluate if the science available was used appropriately in setting policy options and selecting the preferred alternative. A science consistency check, recently pioneered in the Tongass National Forest land management plan (Everest et al. 1997), is a technique for evaluating if the information was transferred from scientists to policy formulators and understood by them. The science consistency check can be used to achieve consistency through iterative application that may involve successive improvements in how the scientists state their findings and how the framers of management policy interpret the implications of those findings. Such an iterative communication process becomes an "adaptive decision-making process" (Everest et al. 1997). Criteria to achieve a "consistent" call are: (1) Is all relevant information used? (2) Is it interpreted correctly? (3) Is the level of risk displayed? Further, a finding of lack of consistency can be a point of appeal or legal challenge; a thoughtful, thorough check may sidestep that outcome. The science consistency check may itself be subjected to scientific peer review, as was the case in the Tongass National Forest planning effort. The science consistency check is a new process; implications of its use are not yet clear because the record of test cases is so limited.

Implementing the Assessment—From Policy to Management

The potential roles of scientists in helping managers implement policy based on bioregional assessments fall on a spectrum of involvement from zero to immersion. In the case of zero involvement, the scientists may simply withdraw from implementation. However, they may be called back to deal with implementation problems as they arise. A somewhat higher level of involvement has scientists available to help managers understand principles underlying the policy, to interpret implications of the assessment and policy direction in terms of on-the-ground management, or to deal with new analytical tools or methods. Yet higher levels of scientist involvement occur where policy directs an adaptive management approach to the evolution of management (Holling 1978; Walters 1986; Lee 1993; chapter 6, Johnson et al.). Scientists working at this level of involvement may conduct studies to test assumptions underlying proposed practices and to test their effectiveness. This can be extended to the case where scientists and managers explore alternative management systems beyond the scope of current policy and management practice. Scientists may participate intensively in implementing policy derived from bioregional assessments by providing continuous technical input to managers and by conducting a research component of adaptive management. Monitoring and evaluating effectiveness of new management practices and overall social and ecosystem conditions are critically important components to developing sound natural resource management, and they require substantial involvement of scientists, yet rewards are few and funding scarce. Development of a close working relationship with land managers may lead a scientist to a substantial, lasting career change. Scientists working with large management organizations may have difficulty choosing among the overwhelming diversity of roles they can play in management units ranging from local, on-the-ground to national-office levels of organizations.

The sense of responsibility of scientists to participate in implementation of policy arising from bioregional assessments can be either personal or institutional. Their motivations may be to honor the ecosystem, honor the assessment, honor the institution, or simply follow orders.

Challenges for Scientists and the Science

Advocacy

Scientists involved in bioregional assessments may be accused of advocating particular positions, in possible conflict with the tenet of maintaining scientific objectivity. In a provocative article, "Science Advocacy Is Inevitable: Deal with It," Shannon, Meidinger, and Clark (1996) identify various forms of advocacy that scientists may find difficult, if not impossible, to avoid. Silviculturists, for example, have committed to a professional life of growing trees for human use, while conservation biologists specializing in rare species are likely to approach management issues with an emphasis on species protection. This may lead to differences in personal values between disciplines (Brown and Harris 1998). As Shannon et al. (1996) argue, scientific disciplines and their practitioners embody purposes and values that can be interpreted as advocated positions in some circumstances.

In the past few years, many science leaders in the United States have encouraged scientists to advocate the use of science in dealing with societal problems (e.g., Lubchenco 1998). Yet even this advocacy may have a political aspect when, as in the case of Northern Forest Lands, there was significant pressure to exclude scientists from an assessment (Trombulak 1994).

The major focus of assessments commonly forces scientists to address issues beyond tested information and enter into the realm of professional judgment. Because of the scope and social conflict of bioregional assessments, these judgments may go beyond professional expertise to include intuition, belief, and personal and discipline values. Policy makers should know the context of the information they use to make decisions, so they can distinguish scientifically credible information from personal opinion. Policy makers may wish to know both but should know the distinction when they make their decisions. However, bioregional assessments commonly rely on a blend of science and expert, personal opinion; the distinction may not be clear.

Discipline differences may carry over to attitudes about managers and management activities. Brown and Harris (1998) observe a tendency for some disciplines to consider management actions as potentially beneficial, while other disciplines view management as potentially harmful. In their study of Forest Service employees, Brown and Harris find that silviculturists hold a more utilitarian land ethic than ecologists, who favor more naturalistic management. The matter of trust of managers to implement policies may also differ with discipline. The values and opinions ingrained in scientists may be manifest in various aspects of assessment results and in the natural resource policy developed from assessments. Lack of trust in managers' intent or skill in carrying out new policy may lead to restrictive prescriptions.

Professional Risks

Participation in bioregional assessments may have important short- and long-term impacts on scientists' careers. Brief, high-profile assessments may have little negative impact, but protracted involvement may interrupt a scientist's regular research and publication routine. Extended (multiyear) involvement may actually cause a scientist to abandon his or her original position and have to seek a new position. Scientists who return to pre-assessment work may do so without providing any follow-through during implementation of the management plan derived from the assessment. A consequence may be weak implementation that may blemish the assessment and participating scientists. Scientists are likely to work outside their home institutions and in a team with members from other institutions. This has the benefit of working with people with shared passions, but working outside the area of traditional reward systems may be a disadvantage, so some professional sacrifice may result.

The political atmosphere of assessments can create various risks for scientists. The taste of power and influence in the political realm or other less obvious interests in influencing decisions may lead a scientist to step outside the scientist role to advocate a particular policy and management outcome. Scientists must also be wary of being used as weapons in political warfare. The sole defense may be to speak up loudly at appropriate times to defend the credibility and proper interpretations of science results when they are misused.

Major personal moral and ethical issues may arise for scientists participating in assessments. Some assessments may have enormous implications for people's lives. At what point do scientists say, "We just don't know"? How should scientists deal with positive and negative attention of interest groups? What relationship should scientists have with land managers whose actions, at the behest of society, are seen by some as the cause of environmental problems when cooperation with them is the key to adaptive management and crafting an effective policy on the ground? These are difficult questions without simple answers. Our personal view is that scientists should not be distracted by minor, immediate conflicts, but instead they should develop long-term working relations with land managers, policy makers, and the public. This requires balancing scientific independence with good cooperative spirit.

Implications for Natural Resource Science and Scientists

Many participating scientists find bioregional assessments very important and stimulating parts of their careers, especially in terms of serving society, benefiting ecosystems, and broadening their perspectives. Bioregional assessments commonly create needs and opportunities for scientists and research organizations to participate in implementation and testing of the new management policies. However, reward systems for individual scientists and the priorities of research organizations strongly discourage these follow-up activities in many cases.

Scientists who started their careers examining the lives of a few individual species or the workings of small watersheds are pushed and pulled to address much larger issues and areas in bioregional assessments. Participation in bioregional assessments commonly changes the course of scientists' careers. Their scientific horizons are greatly broadened, as is recognition of the impact their work and knowledge can have on ecosystems and on society. Many assessment veterans continue their science careers with increased scope of interdisciplinary consideration, geographic and temporal scales of work, links with land managers, and appreciation of science applied in policy and management.

Change in research institutions has been much more limited. In a few cases, such as the Great Lakes (chapter 7, Regier), research institutions have been formed or changed to better institutionalize scientists' participation in bioregional assessments and implementation of resulting policies and management plans. However, in many cases, research and management institutions have made only limited changes to accommodate the dramatic changes in management policy arising from crises and catalyzed by bioregional assessments. Two areas of common shortcoming are commitment to monitoring and research to test assumptions used in assessments and new policy. This institutional inertia creates problems for scientists through lack of rewards and resources for continued regional work and for the institutions that are thereby limited in their service to society.

Cumulatively, the numerous bioregional assessments, each with its individual traits, are accelerating development of the emerging science applied at the scale of bioregions (Aber 1993; Foster 1993; McDonnell, Pickett, and Pouyat 1993; Groffman and Likens 1994). Science at regional scales concerns interactions among phenomena operating at finer and coarser scales and deals with a host of topics, such as effects of air and water pollution, land-use change, and human demography. Bioregional assessments contribute to bioregional science by highlighting regional issues of great social importance, providing a medium for integration across diverse disciplines, introducing a new group of scientists to regional studies, and assembling valuable databases for further analysis.

Bioregional assessments are advancing both science and public thinking about natural resources and ecosystems at the regional scale. Institutional adjustments in both the science and public sectors may transform assessments from discrete, infrequent events to ongoing processes that can anticipate and even head off crises. The long-term outcome may be social systems, assisted appropriately by science, that guide sustainable management of natural resources at the regional scale.

Conclusion

Participation in bioregional assessments offers both great challenge and opportunity for scientists. The opportunity commonly leads scientists to work at broader temporal and spatial scales of inquiry with greater links between social and biophysical worlds. Bioregional assessments are an important medium for translating scientific information and instincts to address societal needs. Challenges to scientists include working outside the traditional methods of science and beyond accustomed temporal and spatial scales. The work setting is interdisciplinary and interinstitutional, with high levels of public and political scrutiny. Traditional science experimentation must give way to methods including synthesis from case studies, simulation modeling, and monitoring. All this creates a variety of professional risks for scientists.

Future bioregional assessments will be more effective if they begin with reflection on the successes and shortcomings of past efforts. Key lessons from the past include the importance of strong, effective leadership and well-crafted balances in the tensions between the instincts of scientists and those of policy makers and between emphasis on detail and emphasis on broad perspectives and integration of social and biophysical considerations. The history of bioregional assessments and their consequences has led to development of a series of techniques and tools designed to add clarity and rigor at critical stages in the conduct of assessments. These include a charter negotiated at the point of initiation; risk analysis and species viability assessment techniques during an assessment; policy analysis and science consistency checks at the point of handing assessment results to decision makers; and various forms of peer review used at several stages of the process.

The continuing emergence of bioregional assessments is enhancing development of bioregional science linking the social and biophysical realms of science.

ACKNOWLEDGMENTS

Our views expressed here have been greatly informed by many discussions with Norm Johnson, Margaret Herring, Jim Sedell, Fred Everest, Gordon Grant, Tom Spies, Tom Mills, and many others. Limitations, omissions, and commissions of error are our responsibility alone.

LITERATURE CITED

Aber, J. D. 1993. Modification of nitrogen cycling at the regional scale: The subtle effects of atmospheric deposition. In *Humans as components of ecosystems: The ecology of subtle human effects and populated areas,* edited by Mark J. McDonnell and Stewart T. A. Pickett. New York: Springer-Verlag.

Allen, T. F. H., and T. B. Starr. 1982. *Hierarchy: Perspectives for ecology and complexity.* Chicago: University of Chicago Press.

Boyce, M. S. 1992. Population viability analysis. *Annual Review of Ecology and Systematics* 23:481–506.

Brown, G., and C. Harris. 1998. Professional foresters and the land ethic, revisited. *Journal of Forestry* 96(1):4–12.

Everest, F. J., D. Swanston, C. G. Shaw III, W. P. Smith, K. R. Julin, and S. D. Allen. 1997. *Evaluation of the use of scientific information in developing the 1997 forest plan for the Tongass National Forest.* General Technical Report. GTR-PNW-415. Portland, OR: USDA Forest Service, Pacific Northwest Research Station.

Foster, D. R. 1993. Land-use history and forest transformations in central New England. In *Humans as components of ecosystems: The ecology of subtle human effects and populated areas,* edited by Mark J. McDonnell and Stewart T. A. Pickett. New York: Springer-Verlag.

Groffman, P. M., and G. E. Likens. 1994. *Integrated regional models: Interactions between humans and their environment.* New York: Chapman and Hall.

Gunderson, L. H., C. S. Holling, and S. S. Light. 1995. *Barriers and bridges to renewal of ecosystems and institutions.* New York: Columbia University Press.

Holling, C. S. 1978. *Adaptive environmental assessment and management.* London: John Wiley and Sons.

Lackey, R. T. 1996. Is ecological risk assessment useful for resolving complex ecological problems? In *Pacific salmon and their ecosystems: Status and future options.* New York: Chapman and Hall.

Landres, P., P. Morgan, and F. J. Swanson. Evaluating the usefulness of natural variability concepts in managing ecological systems. *Ecological Applications,* in prep.

Lee, K. 1993. *Compass and gyroscope.* Washington, DC: Island Press.

Lubchenco, J. 1998. Entering the century of the environment: A new social contract for science. *Science* 279:491–497.

McDonnell, M. J., S. T. A. Pickett, and R. V. Pouyat. 1993. The application of the ecological gradient paradigm to the study of urban effects. In *Humans as components of ecosystems: The ecology of subtle human effects and populated areas.* New York: Springer-Verlag.

Morgan P., G. H. Aplet, J. B. Haufler, H. C. Humphries, M. M. Moore, and W. D. Wilson. 1994. Historical range of variability: A useful tool for evaluating ecosystem change. *Journal of Sustainable Forestry* 2:97–111.

Murphy, D. D., and B. R. Noon. 1992. Integrating scientific methods with habitat conservation planning: Reserve design for northern spotted owls. *Ecological Applications* 2:3–17.

Quigley, T. M., R. W. Haynes, and R. T. Graham, tech. eds. 1996. *Integrated scientific assessment for ecosystem management in the Interior Columbia Basin and portions of the Klamath and Great Basins.* General Technical Report. PNW-GTR-382. Portland, OR: USDA Forest Service, Pacific Northwest Research Station.

Shannon, M. A., E. E. Meidinger, and R. N. Clark. 1996. *Science advocacy is inevitable: Deal with it.* Society of American Foresters annual meeting, November 11, 1996, Albuquerque, New Mexico.

Swanson, F. J., J. A. Jones, D. O. Wallin, and J. H. Cissel. 1994. Natural variability—Implications for ecosystem management. In *Ecosystem management: Principles and applications, Vol. 2: Eastside forest ecosystem health assessment.* General Technical Report

PNW-GTR-318. Portland, OR: USDA Forest Service Pacific Northwest Research Station.

Trombulak, S. C. 1994. The northern forest: Conservation biology, public policy, and a failure of regional planning. *Endangered Species UPDATE* 11(12):7–16.

Turner, M. G., G. J. Arthaud, R. T. Engstrom, S. J. Hejl, J. Liu, S. Loeb, and K. McKelvey. 1995. Usefulness of spatially explicit population models in land management. *Ecological Applications* 5:12–16.

Walters, C. J. 1986. *Adaptive management of renewable resources.* New York: Macmillan.

CHAPTER 5

A Political Context Model for Bioregional Assessments

Hanna J. Cortner, Mary G. Wallace, and Margaret A. Moote

Bioregional assessments undertaken specifically to impact public policy issues must be placed in a political context; otherwise, they are unlikely to be of much utility. Public policy issues arise and are acted upon in a political context in which authoritative action is sought on issues affecting significant societal values. Science is part of the political context, and interest in the linkages between science, public policy, and politics has been long-standing (Brewer 1983; Caldwell 1987; Fischer 1990; Jasanoff 1990; Lee 1993; Mukerji 1989; Price 1965; Schooler 1971).

We acknowledge that science is, and should be, concerned with research undertaken solely to increase the body of scientific knowledge (science for science's sake), to explore new and uncharted areas of inquiry, or to further introspection or retrospection. We also acknowledge that research in which scientists define their own goals and objectives and bound research according to their own perceptions of public policy problems can provide critical information for policy decisions. This discussion, however, focuses particularly on assessments undertaken specifically with the intent that science inform public policy making. In this instance, the kinds of assessments that contribute most to environmental decision making are those that engage all participants in a process of mutual learning to address policy-relevant issues and questions in a timely manner.

Political Context Criteria

To place bioregional assessments in their political context, we offer six policy guidelines: (1) begin and end with the policy problems, not the scientific problems; (2) take a broad perspective; (3) keep in mind the human elements; (4) carefully evaluate analytical approaches and techniques; (5) define the assessment team's role; and (6) treat assessments as tools for learning. These criteria (drawn from Wallace, Cortner, and Burke 1995) are congruent with a political model of bioregional assessment that locates the assessment in an ongoing and publicly open process of evaluation and adaptive learning through which collective decisions are made about how best to deal democratically with resource problems.

Begin and End with the Policy Problems

The goals and objectives of a bioregional assessment need to be bound by the policy issues and the sociopolitical conflicts policy makers and stakeholders want addressed. The essential task is not how to adapt scientific research to policy and management, but how best to shape the assessment's agenda to provide information and data useful to address questions that the public and policy makers deem important. Otherwise, assessments can fall into the trap of being driven more by scientific curiosity—sensitive neither to time, to costs, to policy need. Past assessments, for example, have been criticized for failure to keep focused on what is important from the perspective of policy making. Take the example of the National Acid Precipitation Assessment Program (NAPAP), an effort that lasted over ten years and cost a half billion dollars. While considerable scientific information was collected, several critiques concluded that the assessment was defined more by scientific curiosity than by policy relevance.

> Instead of asking What do we really need to know to make the wisdom-type calls Congress will be called on to answer over the next 10 years, NAPAP managers asked, What are the intriguing and seminal scientific questions we can answer in 10 years? What's more, they seemed to operate on the naive assumption that Congress could wait for their answers. (Roberts 1991, 1303)

The last NAPAP director agreed that it made scientific contributions "at the expense of greater policy relevance" (Roberts 1991, 1305). He noted that if the assessment had tried to do both, it would have failed at both. Yet, NAPAP was created by Congress not as a scientific research program but in response to an issue relatively high on the political agenda (Blodgett 1992; Latin 1991).

It is perhaps easier to begin with a policy-oriented agenda than to stick to one. The assessment must be kept focused on the policy-relevant problems throughout the assessment, and those problems should be used as the basis for defining information and data collection needs. That is why it is probably useful for policy makers to limit an assessment up front with a deadline and cost ceiling. Cost boundaries are extremely important, since financially costly assessments can become politically costly issues. There is also a need to keep asking, either through debriefings or through periodic progress reports, whether the data and analyses are focused on the pertinent policy problems or whether they are veering off into the realm of scientific curiosity. This relates not only to what questions are asked but also to knowing when data may be sufficient for policy making, even if not for scientific closure. Policy action often cannot wait until comprehensive data sets are compiled, all possible variables tested, all relevant relationships specified, and scientific uncertainty reduced. While scientists may be uncomfortable moving forward with incomplete information, inaction may sometimes be politically untenable for policy makers. This does not mean, however, that scientists have no input to the policy agenda or that the agenda is necessarily inflexible. Periodic debriefings can also be useful as a mechanism for maintaining dialogue between scientists and policy makers so that the knowledge and information that the assessment team is compiling can be used to inform and, if necessary, revise the policy agenda.

Adaptive management (a core component of new ecological approaches to management, such as ecosystem management) addresses the problem of moving ahead in the face of uncertainty. Under adaptive management, innovations are implemented, but new research and monitoring results are routinely checked to determine whether there is a need for adjustment. Policies are viewed as large-scale experiments (Lee 1993). Policy makers and scientists jointly assess and learn from doing; decisions are continuously revisited and revised, never final. Considerable emphasis is put on the value of decentralized decision making to avoid the rigidities of highly centralized institutional arrangements with inflexible prescriptions.

For these reasons, we would question the efficacy of efforts to derive standardized, national protocols for regional assessments in order to achieve consistency. If assessments are truly placed in a political context, the major problem(s) at each local or regional level will define whether "best available information" is adequate for decision making or if new data collection is warranted, how much information needs to be collected and how it should be analyzed and acted upon. What is not needed is a "one-for-all" set of protocols. This results only in bigger and better rational-comprehensive processes that are centralizing in nature, expensive, inflexible, and ultimately may not address the problem(s).

Finally, assessments are only a starting point in a long process of policy formation, assessment, implementation, and reevaluation, and one in which budgetary resources are not unlimited. Costly and time-consuming scientific assessments come at the expense of implementation and evaluation; they can consume time and dollars that may just as productively be put toward taking action and monitoring in an adaptive management mode.

Take a Broad Perspective

A core principle of ecosystem or landscape-scale management emphasizes the need to investigate ecosystems at broader spatial and temporal scales; bioregional assessments are a consequence of this need to examine issues from a broader perspective. This principle applies to human dimensions as well as to biophysical aspects. A broad perspective means appreciating the historical antecedents of public problems and the historical context in which they have emerged. It may mean examining socioeconomic trends and conditions that are far outside the geographical bounds of the assessment (e.g., the vagaries of the Mexican economy in relation to bioregional assessments in the Southwest), or even issues that are beyond the immediate reach of the natural resource policy arena (e.g., the impacts of the growing disparity between rich and poor on future social values). No assessment can operate in a vacuum. Even though bioregions are spatially more encompassing units of analysis than those traditionally used, the social and political context cannot arbitrarily be bound by the biophysical boundaries.

Several years ago Iris Lloyd (1979) wrote a provocative article entitled "Don't Define the Problem." That may seem strange since our first criterion is to start and end with policy problem(s). Nonetheless, Lloyd makes some cogent points. She argues that "not defining the problem doesn't have to mean floundering in chaos. It can mean treating the problem as open-ended . . ." Noting that today's management problems are more complex and less static than they used to be, she argues for taking a broad perspective to problem definition. When problems are rigidly defined, says Lloyd, we set boundaries at the wrong time and for the wrong reasons. We become captives of mind-sets, more prone to mistake symptoms for problem causes, less capable of seeing the problem as part of the total situation, and unable to deal with a continuous state of change.

These are the points in Lloyd's article that are useful as we struggle with bounding assessments. While it is important to identify policy problems, it is equally necessary not to constrain an examination of the problem by looking at too few potential causes. This means recognizing that problem definitions often can't be agreed upon. It means continuing to define and redefine problems as assessment proceeds (Hjern 1987). It means being careful about letting scientists think that it is the province of experts via the scientific method to define the problem. Many times scientific problem definitions are sterile and simplified renditions of real-world conflicts and issues lived by decision makers and other stakeholders. Scientific and public definitions of problem priorities and risk are sometimes quite divergent (Slovic, Fischoff, and Lichtenstein 1980). Even though we need to start with the problem, there is also a need, as Lloyd cautions, for both policy makers and scientists to be open-ended as the real policy world swirls around an assessment as it is taking place. Analyses can therefore be updated and revised based on social, economic, and political changes.

Finally, in keeping a broad perspective, it is important to recognize that if we ask the question What sort of policy questions can be answered by scientific assessments, the answer is: very few. The range of policy questions for which science alone provides an answer is pretty narrow. However, if we pose the question another way, i.e., What questions of policy can be addressed by science, then the range is pretty broad. When asking what role assessments can and should play, we shouldn't confuse the word *addressed* with the word *answered*, or the word *discussed* with *decided*. Science information and analyses are just one part of an ongoing political discussion into which other considerations in addition to the information in an assessment need to be, and will be, factored before policy decisions are made.

Keep in Mind the Human Elements

While scientists can and do use their theories, methods, and perspectives to discuss almost every conceivable topic, there are respected and significant elements of the scientific community that have a less expansive definition of what constitutes science and the proper bounds of science. While some researchers have a disdain for anything less than "hard science" and view issues such as, for example, the distribution of costs and benefits as too "squishy" for scientific treatment, lessons from past assessments show that addressing such questions is critical (Lackey and Blair 1995; chapter 9, Hagenstein). As Blodgett (1992) notes, "Hard scientists may be able to restrict their research to hard science, but the policy-maker does not have that luxury."

Over the past twenty-some years, one of the biggest mistakes, if not the biggest mistake, policy makers have made is defining their political problems as technical problems to be solved by technical solutions. Whether it has been forest planning or public participation, this tendency has been repeated—despite the many articles and reports pointing out the pitfalls of this approach (e.g., Allen and Gould 1986; Baltic, Hof, and Kent 1989; Cortner and Richards 1983; Cortner and Shannon 1993; Larsen et al.1990; Wondolleck 1988). There is the decided possibility that those responsible for implementing ecosystem management on the ground will make the same mistake, despite the many instances where publics or policy makers have effectively derailed implementation of decisions that did not take them into account and despite a body of theory that attempts to embed human processes into ecosystems and deals explicitly with the need to address social and institutional change (e.g., Clark and Minta 1993; Cortner et al. 1996; Slocombe 1993).

Placing bioregional assessments in a political context means explicitly examining a broad range of questions about the human dimensions. It means beginning with an understanding of the social and political context and not treating it as an add-on. In addition to providing policy makers with information on how different policy options will differentially affect various segments of the community and how much those options will cost, assessments need to provide information concerning alternative institutional designs. Policy makers should demand this, and scientific teams should be structured so that examining these questions is an integral and integrated part of the assessment task. Implementation considerations, which are heavily people dependent, need to be factored into the assessment, not left for policy makers and managers to figure out after the scientists go home (Pressman and Wildavsky 1973; chapter 6, Nelson).

It is relatively easy to ask how the ecosystem is functioning but leave out the people. Bioregional assessments should ask how policies are working and how issues, problems, and policies impact people and vice versa. As the Northern

Forest Lands study shows, information about people's belief systems and core values can be critical (chapter 9, McCarthy). Having a public-involvement "plan" is not enough. It is easy to draw up a public-involvement plan, treating people as objects, and then leaving them behind for the rest of the assessment as the experts take over. Actually involving the public is considerably more complicated but essential to plan implementation.

Carefully Evaluate Analytical Approaches and Techniques

When science informs policy it is not truth informing decisions—it is understanding negotiated by democracy. Neither policy makers nor scientists should kid themselves—there is no such thing as "objective" science. "What passes as facts in science are matters of social construction, and the scientific process involves relationships between humans and among humanly constructed institutions. . . . Negotiation is fundamental to the elements, outcomes and procedures of fact production" (Schneider and Ingram 1997, 150, 157; Mitroff 1972). Science doesn't and can't deliver truth that is nonarguable (Cullen 1990). Assessment issues involve values and are therefore social and political issues. Subjective value interpretations are made by scientists all the time during an assessment process. Defining problems, selecting indicators, choosing criteria for analysis, and building and running computer models all involve value judgments (Brewer 1983; Cortner and Schweitzer 1983).

The "rational-analytical model that supposedly drives science is not shared by the rest of society, which operates in a bargaining-conflict containment mode" (Cullen 1990). This bargaining-conflict containment mode is a core feature of the kind of democratic governance model we as a society have chosen. Instead of objective truth waiting for the right scientists to pluck it from the sky, when science informs policy it is only providing understanding and perspective that will and should be negotiated by citizens in a democratic society. Scientists thus become negotiators. If they don't convince the publics and policy makers of the importance of their findings, the findings become irrelevant, no matter how important they may be to the issue at hand. Maintaining that assessment and evaluation are value free and apolitical ignores reality; by asserting such a notion, assessors as well as policy makers are engaging in politically value-laden behavior.

Quantitative techniques and computerized models often reinforce a perception that an endeavor is a rational analysis of quantifiable and therefore "unbiased" factual data. Relying exclusively on such approaches and techniques, or overemphasizing their importance, can limit meaningful participation to individuals or organizations who have the special knowledge and skills to question the assumptions and professional biases inherent in those methodologies. Additionally, experiential knowledge and other forms of available data that the public might contribute are likely to be excluded from analysis. People affected by policies can provide important insights about the physical landscape, physical interactions over time, and policy effectiveness on a daily basis. Further, because social values are often too complex to program into computer models, it is difficult to integrate such information into the scientific models. Consequently, social values and the analysts who examine them can frequently all but sit on the side. Assessment processes need to be capable of accommodating new forms of knowledge and multiple sources of information. A variety of analytical approaches, both qualitative and quantitative, need to be integral parts of assessment design from inception through execution. Because such choices are highly political, the choice of analytical approaches and techniques should not solely be the concern of scientists.

When the technical content of issues is emphasized, the criteria for decisions are often restricted to technical criteria and technical solutions are sought. Technology-based solutions are often preferred by the scientific community over solutions that call for modifications in human behavior, although solutions that avoid the human element may not prevent environmental degradation or ensure the protection of ecosystem integrity. There is a risk that decision choices will be narrowed to options that can be technically defined and measured rather than allowing open discussion of preferences (Schneider and Ingram 1997, 176–178). Science can become the framework for the debate—the surrogate for debate over values—and discourse can be confined to the specialized language, theories, and data of the expert. The political content of choice becomes obscured, and questions of morality, equity, and justice eclipsed. Because the fundamental issues of values are not resolved, effective long-term policy solutions become elusive (chapter 6, Nelson). It should be the role of the policy maker to prevent the assessment process from being limited to technical analyses that avoid "sticky" issues of morality, equity, and justice.

We should also "understand that the best scientific models alone are not necessarily the best tools for integrated assessments. There is a need for a hierarchy of models, data, and judgments with different levels of complexity and complementary strengths and capabilities that can help provide a range of insights for addressing complex environmental problems" (Rubin 1991). Complex science and the best models from the scientist's perspective may not be the best way to get answers to things that matter most to people. What counts for people often cannot be counted. What advances science or is good for science and the scientific community does not always serve the public interest.

In evaluating an assessment, much can be concluded by looking at the analytical approaches taken by the scientific team. Generally such approaches are either expert or discursive in design. In expert designs, scientists and professionals define policy problems, determine the means to attain goals, and use the scientific method to identify preferable choices among alternatives. Issues are constructed in such a way as to seem so complex that only the expert with specialized knowledge can solve them. Discursive designs promote democratic deliberation about the problems people, not just experts, see as important, give greater status to grassroots knowledge, and foster collaborative learning and deliberation about values (Schneider and Ingram 1997). In discursive design, policy makers and stakeholders work with scientists to design and interpret the assessment.

Define the Assessment Team's Role

Generally, when we think of the relationship of science to politics, we think of the perversion of science by politics. Scientists are vulnerable to being used and abused by government (Mukerji 1989). We know, for example, that in order to postpone facing problems, policy makers frequently call for research or form study commissions. Policy makers sometimes invoke science when it is in concert with their preferred policy preferences and ignore it or even repress when it is not (Blodgett 1992). Scientific methodologies, such as cost-benefit analysis, are distorted in order to cloak politically favored projects, e.g., federally financed water projects, with scientific respectability. Yet scientists are not simply unwitting pawns who despite their best efforts get swept up in politics. Scientists can also act as policy entrepreneurs to set the public agenda on topics in which they have a strong interest but that have not yet been recognized as important by the public, whether the topic is global warming, ozone depletion, or the consequences of the loss of spotted owl habitat. They can propose solutions that benefit the scientific community itself, e.g., by calling for solutions that require continued research or enhanced budgets. Scientists engaged in assessments have agendas and roles, some imposed upon them and some defined by themselves.

We also need to recognize that those roles are shaped by the very different cultures of the scientific and policy communities (Behn 1981; Blodgett 1992; Cullen 1990). Behn, for example, draws a distinction between the culture of the policy analyst and the culture of the policy politician. Part of the difference, he points out, can be attributed to the analyst's indifference to constituencies and the politician's devotion to them. Members of the scientific community often forget that public policy works through people and that most policies are directed toward "target populations" or specific institutions, groups, and individuals whose behaviors are relevant to creating and/or resolving issues. Obviously, assessments can't be successful if they ignore the human element and the target populations (including agencies) that will be called upon to help implement recommended solutions. Finally, while scientists tend to dismiss harmful effects not proven with a high degree of scientific certainty, much environmental policy operates under the "precautionary principle," i.e., that it is better to enact protection measures even in the absence of conclusive evidence of adverse environmental impacts. Scientists need to recognize also that policy makers deal with political uncertainty, focusing on how action or inaction affects their influence with key constituents and their election chances.

For their part, policy makers need more understanding of how science itself, although clothed with an aura of independence and objectivity that is fiercely promoted and defended by scientists, is very political. Science is socially constructed knowledge, and scientists are members of policy communities that react to the norms and conventions of their disciplines and the political incentives and competitions within their professional networks. While understanding the differences between the cultures of science and policy as they operate in a political context, we also need better attempts at bridging those differences through more discursive assessment designs.

Is it true that scientists can "no longer simply do the science and hope that someone else uses the information to make good laws that protect species and their ecosystems" (Meffe and Viederman 1995)? In bridging science and policy making, just how activist should scientists become? The trend toward activist science can be seen in the numerous advocacy groups with scientific staff whose job it is to serve an ideological agenda (Schneider and Ingram 1997).

Policy makers and scientists can and should inform each other. While scientists may not need to become activists, it is their responsibility to push for recognition and respect of the best scientific information available. This may even mean that in addition to recommending the policy alternatives that the assessment team believes will most effectively achieve ecological sustainability, members of the team might recommend the best strategy for adoption and implementation (Behn 1981). Scientists do not need to compromise their science, but they do need to recognize and work with political realities. By acknowledging the validity of each other's realities, scientists and policy makers can begin to collectively design assessments and examine scenarios that address both realities, are implementable, and are therefore of use.

It is critical for policy makers at the outset to define the role that the assessment team and its products will take in forming policy. Policy makers need to be explicit up front about what they want, keep the assessment bounded and directed, and be available to clarify their charge to the team throughout the process (chapter 6, Johnson et al.).

Treat Assessments as Tools for Learning

"Assessments are part of a learning process that pulls people together to develop information and insights that might not otherwise be developed" (Rubin 1991). Assessments assemble and organize information and provide insights about existing conditions. They can speculate about future potentials and be explicit about uncertainties. As such, they can serve as the focal point for discussion among all stakeholders about desirable futures, trade-offs, and potential paths to issue resolution. Adaptive management can then be designed to help inform areas of uncertainty. Both assessments and resulting management prescriptions can be adjusted as new information becomes available. This implies that policy makers, scientists, and publics have the capacity to identify and adapt to new knowledge and changing public attitudes and to learn lessons from new research and on-the-ground management experiences. The concept of assessments as tools for learning (Lee 1993; Shannon 1992a) also suggests the necessity of speaking in a language that is easily understood by a general audience, whether that audience is made up of scientists from other disciplines, policy makers, or members of the local "friends of the river" group. Both policy makers and scientists need to develop better ways of communicating with the public, recognizing that if it is worth making public decisions about, it can be understood by the public (chapter 8, Hurchalla). Distributing responsibility and authority widely among citizens can encourage people to engage in critical thinking and build the capacity of citizens to solve problems.

Thus, again we revisit the concept of science as an input to understanding negotiated by democracy. By treating bioregional assessments as tools for learning for all participants, we can talk about how their design and execution serve policy and social goals. In discursive policy designs, there are enhanced roles for the public as well as professionals (Shannon 1992b), and assessments become part of an integrative, interactive process. Thus, while a most important outcome of the Northern Forest Lands study is the model for involving people in contentious issues and diffusing the tensions to bring about realistic solutions (chapter 9, McCarthy), FEMAT was more technocratic than democratic (chapter 6, Johnson et al.). Because FEMAT was not a public process, there have been difficulties in getting public ownership of the solution (chapter 6, Nelson). Implementing recommendations from bioregional assessments will require widespread support among policy makers and the general public. To succeed, bioregional assessment must not be viewed solely as a scientific enterprise; it must be a politically democratic process.

Conclusion

In sum, the principal measurement of success in a political context model is not whether a bioregional assessment has amassed an impressive collection of scientific reports, using the latest state-of-the-art techniques and models in carefully conducted tests reported under the most rigorous of scientific review standards, although the "best available science" is always desirable. The bioregional assessment is not an end in itself and will not provide ultimate "answers." The bioregional assessment is an important tool in an ongoing, collective process of learning and evaluation. As participants in bioregional assessments have shown, there is no silver science bullet (chapter 6, Nelson; chapter 7, Jones; chapter 8, Hurchalla). The most policy makers can hope for is a reasoned assessment of environmental and social needs and conditions that takes into account the effect society and environment have on each other.

The search for scientific certainty is never ending, since often so little is known about the linkages between cause and effect in complex ecosystems. Yet policy makers often must make decisions with less than complete information and lots of uncertainty about the impact of changing policy direction. The policy-making process must necessarily be ongoing and adaptive, incorporating new information as it becomes available. It is the policy maker's role to keep the assessment on track, focus on key policy issues, and integrate social and scientific information in a directed dialogue. We cannot allow assessments to become the end product, consuming the resources that are needed for influencing action in the political and economic arena and impacting human behavior. Assessments should be policy relevant, done in a timely manner, and with clear restraints on costs. Assessments are one small part of a larger political process of debating and deciding about resource issues. They must be put into political perspective and kept in perspective.

LITERATURE CITED

Allen, G. M., and Gould, E. M. 1986. Complexity, wickedness, and public forests. *Journal of Forestry* 84(4):20–23.

Baltic, Tony J., John G. Hof, and Brian M. Kent. 1989. *Review of critiques of the USDA forest land management planning process.* General Technical Report RM-170. Fort Collins, CO: USDA Forest Service, Rocky Mountain Forest and Range Experiment Station.

Behn, Robert D. 1981. Policy analysis and policy politics. *Policy Analysis* 7(2):199–226.

Blodgett, John E. 1992. Six grim tales of science and public policy. In *Water resource challenges and opportunities for the 21st century. Proceedings of the first USDA water resource research and technology transfer workshop,* edited by Wilbur H. Blackburn and John G. King. Report ARS-101. Washington, DC: USDA Agricultural Research Service.

Brewer, Gary D. 1983. Some costs and consequences of large-scale social systems modeling. *Behavioral Science* 28:166–185.

Caldwell, Lynton. 1987. *Biocracy: Public policy and the life sciences.* Boulder, CO: Westview Press.

Clark, Tim W., and Steven Minta. 1993. *Greater Yellowstone's future: Prospects for ecosystem science, management, and policy.* Moose, WY: Homestead Publishing.

Cortner, Hanna J., and Merton T. Richards. 1983. Institutional limits and legal implications of quantitative models in forest planning. *Journal of Soil and Water Conservation* 38 (March–April):79–81.

Cortner, Hanna J., and Dennis L. Schweitzer. 1983. Institutional limits and legal implications of quantitative models in forest planning. *Environmental Law* 13(2):493–516.

Cortner, Hanna J., and Margaret A. Shannon. 1993. Embedding public participation in its political context. *Journal of Forestry* 91(7):14–16.

Cortner, Hanna J., Margaret A. Shannon, Mary G. Wallace, Sabrina Burke, and Margaret A. Moote. 1996. *Institutional barriers and incentives for ecosystem management: A problem analysis.* General Technical Report PNW-GTR 354. Portland, OR: USDA Forest Service Pacific Northwest Research Station.

Cullen, Peter. 1990. The turbulent boundary between water science and water management. *Freshwater Biology* 24:201–209.

Fischer, Fred. 1990. *Technology and the politics of expertise.* Newbury Park, CA: Sage Publications.

Hjern, Benny. 1987. Implementation and intermediation: Implications for theory and practice. In *Helping small firms grow: An implementation approach,* edited by B. Hjern and H. Hull. London: Croom-Helm.

Jasanoff, Sheila. 1990. *The fifth branch: Science advisors as policy-makers.* Cambridge, MA: Harvard University Press.

Lackey, Robert T., and Roger L. Blair. 1995. *Science, policy, and acid rain: Lessons learned.* Paper presented at the workshop "At the Crossroads of Science, Management, and Policy," November 6–8, 1995, Portland, OR.

Larsen, Gary, Arnold Holden, Dave Kapaldo, John Leasure, Jerry Mason, Hal Salwasser, Susan Yonts-Shepard, and William E. Shands. 1990. *Synthesis of the critique of land management planning, vol 1.* Washington, DC: USDA Forest Service.

Latin, Howard. 1991. Regulatory failure, administrative incentives, and the new Clean Air Act. *Environmental Law Review* 21:1647–1720.

Lee, Kai N. 1993. *Compass and gyroscope: Integrating science and the environment.* Washington, DC: Island Press.

Lloyd, Iris. 1979. Don't define the problem. *Public Administration Review* 38 (May/June): 283–286.

Meffe, Gary K., and Stephen Viederman. 1995. Combining science and policy in conservation biology. *Renewable Resources Journal* 13(3):15–18.

Mitroff, Ian I. 1972. The myth of objectivity or why science needs a new psychology of science. *Journal of Management Science* 18(10):B613–B618.

Mukerji, Chandra. 1989. *A fragile power: Scientists and the state.* Princeton: Princeton University Press.

Pressman, Jeffrey, and Aaron Wildavsky. 1973. *Implementation.* Berkeley: University of California Press.

Price, Don K. 1965. *The scientific estate.* Cambridge, MA: Belknap Press of Harvard University.

Roberts, Leslie. 1991. Learning from an acid rain program. *Science* 251(4999):1302–1305.

Rubin, Edward S. 1991. Benefit-cost implications of acid rain control: An evaluation of the NAPAP integrated assessment. *Journal of the Air and Waste Management Association* 914–921.

Schneider, Anne Larason, and Helen M. Ingram. 1997. *Policy design for democracy.* Lawrence: University Press of Kansas.

Schooler, Dean, Jr. 1971. Science, scientists, and public policy. New York: Free Press.

Shannon, Margaret A. 1992a. Building public decisions: Learning through planning: An evaluation of the NFMA planning process. In *Forest Service planning: accommodating uses, producing outputs, and sustaining ecosystems,* vol. 2: Contractor Papers. Washington, DC: Office of Technology and Assessment.

Shannon, Margaret A. 1992b. Foresters as strategic thinkers, facilitators, and citizens. *Journal of Forestry* 90(10):24–27.

Slocombe, D. Scott. 1993. Implementing ecosystem-based management: Development of theory, practice, and research for planning and managing in a region. *BioScience* 4(9): 612–622.

Slovic, Paul, Baruch Fischoff, and Sarah Lichtenstein. 1980. Facts and fears: Understanding perceived risk. In *Societal risk assessment: How safe is safe enough,* edited by R. Schwing and W. A. Albers, Jr. New York: Plenum.

Wallace, Mary G., Hanna J. Cortner, and Sabrina Burke. 1995. A review of policy evaluation in natural resources. *Society and Natural Resources* 8:35–47.

Wondolleck, Julia M. 1988. *Public lands conflict and resolution: Managing national forest disputes.* New York: Plenum Press.

PART THREE

Case Studies
and Reviews

CHAPTER 6

Forest Ecosystem Management Assessment Team Assessments

Case Study

K. Norman Johnson, Richard Holthausen,
Margaret A. Shannon, and James Sedell

Science Review

Logan A. Norris

Management Review

Judy E. Nelson

Policy Review

Paula Burgess and Kristin Aldred Cheek

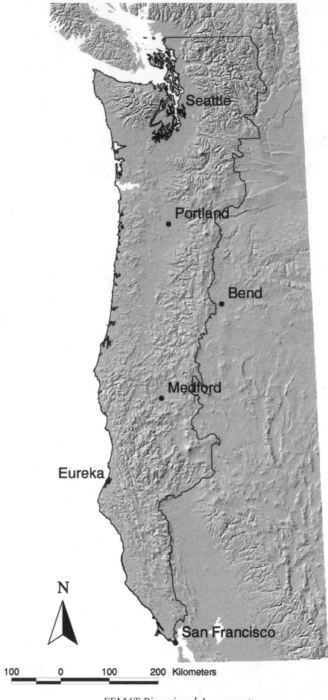

N

100   0   100   200   Kilometers

FEMAT Bioregional Assessment

Case Study

K. Norman Johnson, Richard Holthausen, Margaret A. Shannon, and James Sedell

If you believe that national forests should work as a smoothly running and efficiently operated wood fiber machine, you will be disappointed by both the framework of options and the policy outcomes of the Forest Ecosystem Management
Assessment Team report (1993). As Richard White eloquently teaches us in his
slim book *The Organic Machine: The Remaking of the Columbia River* (1995), the
development of the West resulted in transformation of its forests, rivers, lands,
animals, and people into cogs of "organic machines," machines in which nature is
a constituent but wild component. Viewed from the metaphor of the machine,
the value of any one element is embodied by its contribution to the smooth functioning of a machine that links human and natural energy through work. Thus,
whatever humanity individuals gain from their work is invisible from the perspective of the machinery.

The story of FEMAT did not begin on the day President Clinton convened the
Forest Conference, but long before when the organic wood machine was per-
fected in the fertile and fabulously valuable forests of the Pacific Northwest early
in the twentieth century. White's evocative metaphor of the organic machine is a
useful one for integrating the policies, organizations, agencies, disciplines, natural
resources, capital, and labor that were combined in the Northwest to "get the cut
out and the wood to people's use." Each law, each agency, each idea, each relationship, and all the other parts of the social-economic-political-biological-phys-
ical conglomeration fit together to create a machine all the more wondrous
because it works so smoothly no one realizes it exists. Until it breaks down.

This is the story of FEMAT and its role in developing a forest plan for the federal forests of the Northwest. But the conclusion of the story has yet to be imag-
ined. The working out of ideas and changes to policy and management implied
in the FEMAT report will shape natural resource policies into the next century, so
we must learn its lessons and work from them in building the future.

Creating the Need for a Comprehensive Plan

The northern spotted owl had been a concern in the scientific community since
the early 1970s. After passage of the Endangered Species Act in 1973, biologists
from state and federal agencies formed the Oregon Endangered Species Task
Force and identified the spotted owl as an indicator species pursuant to wildlife
diversity requirements of the National Forest Management Act. They recom-

mended that three hundred acres of old-growth forest be retained around every known nest site.

In 1986, the owl flew into public view when the Forest Service released proposed management guidelines that called for hundreds of spotted owl habitat areas scattered across the landscape, each containing up to twenty-two hundred acres of federal forest. Logging would be allowed to continue in owl habitat outside these reserves. In 1988, the Forest Service released the final spotted owl guidelines, which called for an even more extensive system of spotted owl habitat areas; again, logging would be permitted to continue between them.

Convinced that this proposal was inadequate, Seattle Audubon Society, along with other environmental groups, sued the Forest Service for failing to adopt a credible conservation strategy that would comply simultaneously with the National Environmental Policy Act, the National Forest Management Act, and the Endangered Species Act. In spring of 1989, Federal District Judge William Dwyer granted a preliminary injunction on 135 timber sales.

To save the timber sale program, Congress added a rider to the annual appropriations bill (Section 318) that declared proposed Forest Service and Bureau of Land Management plans for spotted owls "sufficient" in preparing timber sales for two years after mandating a slight temporary expansion in the number and size of the spotted owl habitat areas. Both agencies were to minimize fragmentation of "ecologically significant" stands of old-growth forest. Citizen advisory boards were to be created to assist in preparing and modifying the sales. Further, the agencies were directed to reevaluate and improve their plans for owl protection.

In response, Forest Service Chief Dale Robertson called for the creation of an Interagency Scientific Committee (ISC) to "develop a scientifically credible conservation strategy for the northern spotted owl." Dr. Jack Ward Thomas, at the time chief research biologist with the Forest Service, was named as the chair of the seventeen-member committee. The ISC recommended the creation of large habitat conservation areas distributed throughout the range of the owl in which no further timber harvesting would occur, along with other restrictions on timber harvest.

The Forest Service declared that it would operate in a manner "not inconsistent with" the ISC Conservation Strategy without following normal rule-making procedures. Seattle Audubon returned to court charging that the Forest Service had complied with neither NEPA nor NFMA. In Seattle *Audubon v. Evans* (771 Supp. 1081), Judge Dwyer again ruled against the Forest Service in a scathing opinion, finding that the Forest Service had violated NEPA by not completing an acceptable spotted owl management plan. Judge Dwyer wrote, "The problem here has not been any shortcomings in the laws, but simply a refusal of administrative agencies to comply with them. . . . This invokes a public interest of the highest order: the interest in having government officials act in accordance with the law" (771 Supp. 1089). Dwyer castigated the Forest Service for "deliberate and systematic refusal" to comply with federal wildlife laws and also placed blame on "decisions made by higher authorities in the executive branch." The opinion, upheld on appeal, prohibited any timber sales in spotted owl habitat until the Forest Service had complied with NEPA and developed suitable owl protection plans.

During this time, two agencies of the Department of Interior were also in court over their treatment of the northern spotted owl. The federal courts ruled that the U.S. Fish and Wildlife Service acted arbitrarily in denying the petition to list the owl as endangered and ordered the agency to reconsider its decision. After a reevaluation, the FWS listed the spotted owl as a threatened species in Washington, Oregon, and Northern California. Additionally, the courts ruled that the Bureau of Land Management had not complied with NEPA relative to the implications of its management for the northern spotted owl, with the result that its timber sales slowed to a trickle. In response, the director of the Bureau of Land Management challenged the ESA by invoking the creation of the special seven-member committee authorized under the ESA (commonly called the God Squad) to consider exempting forty-four timber sales in spotted owl habitat from compliance with the ESA. The God Squad voted to exempt some of the sales in exchange for an agreement from BLM that it would undertake long-range planning to protect the spotted owl. As the undersecretary of commerce said, "What's missing is a long-term solution. In the absence of any other instructions from this committee, we'll be here a year from now attempting to resolve the (next) sale program."

The ISC report, as powerful as it was, explicitly considered only one species, the northern spotted owl, albeit a wide-ranging species whose habitat requirements would have a significant effect on forest management. In spring 1991, two committees of the U.S. House of Representatives commissioned four scientists to provide an independent assessment of options for managing late-successional federal forests in the Pacific Northwest. The Scientific Panel on Late-Successional Forest Ecosystems (known as the Gang of Four) delivered a sobering, and from a political perspective, unwelcome message to Congress (Johnson et al. 1991). There is no "free lunch"—that is, no alternative provides abundant timber harvest and high levels of habitat protection for species associated with late-successional forests and threatened fish stocks. In order to give fish and wildlife species even a moderate chance of survival, timber production would have to drop much more than estimated for the ISC conservation strategy.

The Forest Service's immediate reaction, under the Bush administration, was to ignore the Gang of Four report. Repeated attempts by various congressmen to craft bills based on its findings went nowhere. Instead, the Forest Service proceeded with implementation of the earlier ISC report. The final EIS for the northern spotted owl, largely based on the ISC report, was successfully challenged in court, in part, for acknowledging (but not refuting) the conclusion in the Gang of Four report that protecting the northern spotted owl would not protect major elements of biodiversity (Caldwell, Wilkinson, and Shannon 1994).

The Forest Conference

With the 1992 presidential campaign and elections, decisions about management of federal forests were put on hold, except in Oregon, where management of these forests became a campaign issue. At a campaign stop in Oregon, candidate Bill Clinton promised that, if elected, he would hold a forest summit to find an equitable solution to the federal forest crisis in the Northwest.

To keep that commitment, in April 1993 President Clinton held a Forest Conference in Portland, Oregon, at which a diverse group of citizens, scientists, and experts from the region were asked to give their ideas about the importance of the forests and about how to break the gridlock. In his introduction to the conference, President Clinton posed a fundamental question to the participants: "How can we achieve a balanced and comprehensive policy that recognizes the importance of the forest and timber to the economy and jobs in this region, and how can we preserve our precious old-growth forests, which are part of our national heritage and that, once destroyed, can never be replaced?" And he said, "The most important thing we can do is to admit, all of us to each other, that there are no simple or easy answers. This is not about choosing between jobs and the environment, but about recognizing the importance of both and recognizing that virtually everyone here and everyone in this region cares about both."

At the conclusion of the Forest Conference, President Clinton committed himself to producing a plan in sixty days that reconciled legal requirements for species and forest protection with the contribution of forests to the economic and social well-being of the people in the Pacific Northwest. To accomplish that goal, he set up three task forces, one to help develop a forest management plan for federal forests of the spotted owl region, a second to organize labor and community assistance for that area, and a third to improve federal agency coordination in the region. The president said five principles should guide the task forces' work:

> First, we must never forget the human and the economic dimensions of these problems. Where sound management policies can preserve the health of forest lands, sales should go forward. Where this requirement cannot be met, we need to do our best to offer new economic opportunities for year-round, high-wage, high-skill jobs.
>
> Second, as we craft a plan, we need to protect the long-term health of our forests, our wildlife and our waterways. They are a . . . gift from God, and we hold them in trust for future generations.
>
> Third, our efforts must be, insofar as we are wise enough to know it, scientifically sound, ecologically credible, and legally responsible.
>
> Fourth, the plan should produce a predictable and sustainable level of timber sales and nontimber resources that will not degrade or destroy the environment.

Fifth, to achieve these goals, we will do our best, as I said, to make the federal government work together and work for you. We may make mistakes, but we will try to end the gridlock within the federal government and we will insist on collaboration not confrontation.

The FEMAT Process

Following the Forest Conference, senior scientists and managers from federal agencies and universities quickly assembled a team to begin work on the plan envisioned by President Clinton. Membership and organization of the team evolved over time; by the time the report was published, over one hundred scientists and managers were listed as substantial contributors. Members of the office of the secretary of interior represented the Clinton administration in the team's discussion, clarifying the mandate and probing, on occasion, the basis for different team proposals and approaches.

Jack Ward Thomas, the team leader, began the effort with a summary of the charge from the president and a rough outline of tasks and time lines. From the beginning, tension existed between the magnitude of the task and the time that was provided to complete it. This conflict was somewhat ameliorated when the team (by now calling itself the Forest Management Ecosystem Assessment Team) requested and was granted an extension of one month to complete its work.

FEMAT's understanding of its charge evolved over time from the production of a single conservation strategy to the production of options with the final choice left to the president. Thus, the focus of the analysis underwent a subtle shift. Initially, the focus was simply to allow the team to make an informed series of decisions leading to a single option. With the altered charge, it became clear that the analysis would have to be completed and presented in a way that would allow someone outside the team to fully understand the ramifications of all the options. The charge grew from developing a single plan to developing options because: (1) it became clear that the requirement to provide for viability of species was not one of black and white but rather one with many shades of gray; (2) some members of the team had participated in the Gang of Four study and were committed to presenting scientific information in an analytical framework that did not usurp the policy makers' decision prerogatives; (3) these results would be folded into an environmental impact statement and an EIS generally includes a range of alternatives; and (4) it quickly became clear that it would be difficult to get the team members to agree on a single plan.

The evolution of FEMAT from a single plan to a set of options reflects the way many policy decisions were made in FEMAT. The operation functioned for the first month without a signed charter. During this time many decisions about how to proceed were made and remade as needed. Finally, a charter was finished that matched the administration's expectations with what the scientists found to be possible and fruitful.

From the beginning of their assignment, FEMAT members knew that two or three months was not enough time to develop a full-scale ecosystem management plan. Therefore, they designed a continuing three-phase process. The first phase would be development and assessment of management options for establishing a network of late-successional and old-growth forest reserves and prescriptions for the management of the intervening forested land (the matrix). This phase also included selection of an option and the completion of the procedures required by the National Environmental Policy Act (i.e., the environmental impact statement) to meet the standards from Judge Dwyer's court. A separate interagency team was recruited to produce the environmental impact statement for the president's preferred option. The second phase involved a revitalized forest planning process that would integrate the FEMAT results with existing forest plans and address problems that FEMAT did not consider. The third phase would be implementation, monitoring, and adaptive management.

There were several key biological objectives in the first phase. A central goal was to assure adequate habitat on the federal lands to aid "recovery" of species such as northern spotted owls and marbled murrelets, listed as threatened under the Endangered Species Act and associated with late-successional forests. In addition, the team assessed the "viability" of a wide variety of species of plants and animals under each suggested management option, in keeping with federal responsibilities to prevent species from being listed as threatened and endangered under the Endangered Species Act and with the regulations issued pursuant to the National Forest Management Act. The team developed riparian management options for habitat adjacent to streams, on the basis of the understanding that aquatic and riparian habitats and wetlands on federal lands are key to numerous aquatic organisms, including some 13 species and approximately 260 runs of anadromous fishes considered to be "at risk" of extinction.

The Influence of Studies before FEMAT

Development of forest management options by FEMAT was strongly influenced by the recent scientific and management history of Pacific Northwest forests. Studies describing the ecological characteristics of old-growth Douglas fir forests (Franklin et al. 1981) and wildlife and vegetation of unmanaged Douglas fir forests (Ruggiero et al. 1991) provided a scientific base for the studies that followed.

The Interagency Scientific Committee (Thomas et al. 1990) had proposed a management plan for the northern spotted owl that called for the establishment of a series of large habitat conservation areas (HCAs), totaling more than five million acres, across the federal forests. Forestland between the HCAs (the matrix) was to be managed to facilitate dispersal of the owls with at least 50 percent of each quarter township maintained in trees averaging eleven inches in diameter at breast height and at least a 40 percent canopy closure.

When the report was published, it was considered to be the most extensive application to date of the concepts of conservation biology to a large management problem. The idea of a system of reserves combined with management of the intervening land to facilitate dispersal strongly influenced the studies that followed, including FEMAT. The owl population data and the reserve map developed by the ISC influenced and constrained future reserve systems. The conclusion by the ISC that insufficient knowledge existed to prescribe thinning inside reserves to accelerate the development of owl habitat also influenced later studies. Finally, the process used by the ISC in which scientists developed a plan isolated from managers carried over into later studies, as well.

The third major study underlying the work of FEMAT was the report of Johnson et al. (1991) to the Agriculture Committee and the Merchant Marine and Fisheries Committee of the U.S. House of Representatives. This Gang of Four report analyzed the biological and timber production effects of thirty-six possible alternatives for management of late-successional federal forests in Oregon and Washington. The report based biological effects on the likelihood of maintaining viable populations of spotted owls, marbled murrelets, sensitive fish species, and an unspecified list of other species associated with late-successional forests. In addition, the report addressed the likelihood of maintaining a functional late-successional forest network, as well as the employment and income effects of timber harvest levels associated with each alternative.

This report was significant to the later FEMAT effort in three ways. First, the alternatives in the report were founded on a new mapping of late-successional and old-growth forests on federal land, developed with the help of resource specialists from the affected national forests and BLM districts, which gave it a high level of credibility. Second, the report added the issue of conservation of salmon stocks to the issue of conserving late-successional processes and species. Conservation strategies involving key watersheds and enhanced riparian buffers strongly influenced later studies. Third, in contrast to the ISC, this report included as a policy analysis of a series of options rather than a single plan, and these options considered both ecological and economic effects.

The fourth effort that played an important part in the development of the FEMAT options was the Final Draft of the Northern Spotted Owl Recovery Plan (USDI 1992), which included a refined mapping of the spotted owl conservation areas to those originally proposed by the ISC.

The recovery plan was soon followed by the report of the Scientific Analysis Team (SAT) (Thomas et al. 1993), published under direction of the court to provide adequate consideration to other species associated with late-successional and old-growth forests. The SAT report represented the first attempt by the government to look at broad ecosystem effects of late-successional forest management and riparian management. In addition to terrestrial and aquatic vertebrate species, which were the historic focus of such assessments, the report analyzed effects on plants and invertebrates.

The SAT report recommended that the Forest Service adopt various mitigation measures in conjunction with its proposed management of spotted owl habitat. These included large, interim reserves for marbled murrelets, riparian reserves to protect aquatic systems, and buffers established for certain terrestrial species that were thought not to be adequately protected by other measures. All of these recommendations resurfaced in the work of FEMAT. Central to SAT and later to FEMAT was the risk analysis process used to assess the likelihood that viable populations of species would be sustained by proposed management actions. This process had been under continual development by scientists with the Pacific Northwest Research Station and the Pacific Northwest Region of the Forest Service. It had been employed in a series of environmental impact statements and was a key test of Forest Service ability to implement provisions for diversity under the National Forest Management Act.

Choosing Strategies and Developing Options

With an extremely short time line, FEMAT set about identifying those strategies that would best satisfy the president's mandate for the plan. That mandate was to "attain the greatest economic and social contributions from the forests" while meeting "the requirements of the applicable laws and regulations, including the Endangered Species Act, the National Forest Management Act, the Federal Land Policy Management Act, and the National Environmental Policy Act."

The Clinton administration's mandate stated that a central component of meeting the laws was maintenance or restoration of habitat conditions adequate to support viable populations of all species associated with late-successional forests. Particular emphasis was placed on northern spotted owls and marbled murrelets, both of which were listed as threatened under the Endangered Species Act, and on at-risk species and stocks of anadromous fish. A final requirement was to establish conditions that would maintain a connected old-growth forest ecosystem on federal lands. In sum, the team's instructions from the Clinton administration were to investigate options that would provide for moderate to very high likelihoods of achieving these conditions while providing for the greatest possible economic contribution.

The option development process began with a coarse screen of forty-eight options that had originally been brought forward in the documents described above or that had been proposed as agency management plans. This screen identified options that could potentially meet the biological portion of the president's mandate. Options that passed this screen were brought forward for further consideration.

Ultimately, the team selected eight options to be fully evaluated. Major components of these options included late-successional reserves, taken from either the report of Johnson et al. (1991) or the Northern Spotted Owl Recovery Plan (USDI 1992); riparian reserves adopted from the SAT report (Thomas et al. 1993); and various prescriptions for management of the forests between the reserves. Notably, options for long-rotation forestry without provisions for reserves were not brought forward for full analysis. The objective of the reserves was to establish well-distributed areas where historic patterns of habitat fragmentation could be halted or reversed. The ISC, Gang of Four, SAT, and the northern spotted owl recovery team had all judged that the reserves were necessary for spotted owl recovery, at least over the short term, and FEMAT accepted that judgment.

It soon became clear, however, that the initial mix of options was insufficient. The options represented ideas generated from previous exercises and did not adequately reflect the definition of the problem that emerged from the Forest Conference and that constituted the instructions to FEMAT. Because the options were created from previous exercises, they did not necessarily have a core philosophy or vision. And a review of preliminary timber production analyses revealed that volumes projected under the various options were lower than expected by the Clinton administration.

The team became convinced that a new option, developed with all ecosystem concerns considered simultaneously rather than sequentially, would be more efficient than any of the existing options in terms of impact on timber harvest. Dr. Jerry Franklin led the development of Option 9, which was based on selection of reserves with the simultaneous consideration of spotted owl habitat, marbled murrelet habitat, old growth, key watersheds for protection of at-risk fish species and stocks, and the impact on timber production. As scientists from the many working groups and the field specialists worked on creating a map of reserves, managers from the administrative units in question came to review and comment on the work.

Toward the end of this process, the idea emerged of creating areas to test new management approaches. Ten adaptive management areas (AMAs) were mapped, ranging in size from 80,000 to 400,000 acres, with widely differing biological, administrative, and jurisdictional characteristics. Extensive enough to support large-scale experiments on ecosystem management, these areas were meant to provide natural laboratories for both the technical and the social challenges of ecosystem management.

Evaluation of Options from a Biological Perspective

Each option was analyzed relative to its success in protecting species and ecosystems. Two key criteria drove the biological analysis: providing habitat adequate to support viable populations of species associated with late-successional forests and anadromous fish; and providing for a late-successional forest ecosystem well distributed across the federal forests.

Panels of experts made these evaluations within a carefully-designed assessment process based on the level of habitat support for populations across a time frame of at least 100 years. Panelists expressed their judgment relative to the likelihood that each of four specified outcomes would occur for each species or group of species under each option. The most favorable outcome was a projection that habitat would be of sufficient quality, distribution, and abundance to allow the species population to stabilize, well distributed across federal lands; the least favorable was a projection that habitat conditions would result in species extirpation from federal land. The likelihood of maintaining a well-distributed late-successional and old-growth ecosystem was assessed in similar fashion using modified definitions of the outcomes.

Fourteen separate expert panels assessed outcomes for 82 terrestrial vertebrates, 21 groups of fish, 102 species of mollusks, 124 vascular plants, 157 species of lichens, 527 species of fungi, 106 species of bryophytes, and 15 functional groups of arthropods that may include 7,000 to 10,000 species. For experts assigned to the panels, the assessment process was an enormous task. For each option, the panelists had to mentally project the future of the landscape and then consider its interaction with the species. This was complicated by numerous significant issues that arose concerning the assessment process. While FEMAT scientists generally agreed that viability required habitat able to support well-distributed populations of vertebrates, there was not general agreement on the level of assurance needed to meet legal challenges. Appropriate outcomes for plant and invertebrate animal species were even less certain.

Due to the number of species being addressed, the effort required for viability analysis was far greater for the terrestrial working group than it was for the aquatic working group. The terrestrial group dealt with over one thousand species or groups of species, while the aquatic group worked with less than twenty. Consequently, the terrestrial group devoted virtually all of its attention to the viability analysis, while the aquatic group focused on refinement of its approach for managing and conserving aquatic systems. The product of this effort was a well-articulated Aquatic Conservation Strategy that formed the nucleus of the aquatic chapter in FEMAT. The terrestrial group, in contrast, intended that the results of the viability panels would be used for later modification of the options, but this loop was never fully closed. Mitigation measures responding to the viability analyses described in the FEMAT report were not fully incorporated in the options.

Despite technical difficulties and differences in approaches, strong patterns emerged from the assessment. The options had clearly been designed for relatively large-bodied vertebrate species with strong dispersal capabilities (spotted owls, marbled murrelets, and anadromous fish). Species with similar life history characteristics received favorable ratings under nearly all options. At the other extreme, species with limited distributions and poor mobility were rated as being at risk of extirpation under many options. Lack of knowledge often contributed to the level of risk assigned these species. Specific features of the options had predictable effects on the ratings assigned by the experts. Those options with larger riparian reserves were universally rated more favorably for anadromous fish and for terrestrial and aquatic species associated with riparian and aquatic systems. Options with more acreage in late-successional reserves were rated more favorably for all terrestrial species.

Evaluation of Options from an Economic and Social Perspective

The team assessed economic and social effects that were largely contingent on the timber harvest level and associated employment and county payments in the different options. The land allocations and management rules of each alternative were given to each national forest and the BLM state office to develop estimates of likely timber harvest levels associated with each alternative, levels they could be reasonably sure that they could deliver.

The economic assessment entailed a detailed analysis of timber markets, which included measuring market effects (including private timber harvests) resulting from federal harvest levels, estimating timber industry employment impact, and assessing economy-wide effects. The team attempted to consider many other outputs including special forest products (such as mushrooms), commercial fishing, grazing, minerals, and outdoor recreation, but it encountered numerous difficulties in relating changes in these outputs to changes in alternative design. In the end, only a few recreation categories, such as dispersed recreation, showed much variation between the alternatives.

FEMAT undertook an innovative social assessment of the alternatives. The social assessment team analyzed community-level effects of declining federal timber harvests, the social values of a wide variety of resources on federal lands, the history of the current problems, and the public definition of issues related to federal forest management. Given that systematic comparative studies did not exist, FEMAT gathered over fifty community experts who could assess the probable consequences for the three hundred rural communities affected by changes in federal land management policies. One important contribution of this assessment was to identify regions negatively affected by the changes implied by the options as well as to identify regions positively affected by those changes. The core concept of the community analysis was "community capacity" to absorb and respond to changes brought on by declining federal timber harvest. Since this was a driving policy issue for FEMAT, the role of the social assessment was to locate the areas and sectors in which the effects were most acute and to identify possible solutions. The obvious limits of doing a social assessment in the context of FEMAT led to the recommendation for community self-assessments as part of the implementation process.

Role of the Public, Managers, and Policy Makers

At the beginning of the FEMAT process, the Clinton administration decided not to involve the public in developing alternatives, beyond the discussion that occurred in the Forest Conference. There was concern that public involvement was a spigot that could not be turned off once it was turned on and that another step could not be added to a process that already would be stretched to deliver a product in the time allowed. Thus, the public had little opportunity to craft its own detailed proposals for Northwest forests beyond what it had already stated in the development of forest plans and at the Forest Conference. While the social assessment worked to make the linkage with the public comments and issues raised in the recent planning efforts, it was difficult to compensate for the loss of creativity from not having open, public discussion of the issues after completion of the Forest Conference.

The FEMAT analysis was not subject to formal public review and comment. Instead, the EIS team worked concurrently on the development of a formal EIS based on the options and analysis of FEMAT and published a draft EIS, which was open to comment and review. While the FEMAT alternatives were largely unmodified by this review, challenges to the legal adequacy of the species protection provisions in Option 9 helped ensure that these provisions would be strengthened in the record of decision. Also, challenges to the long-rotation strategy in Option 9 for federal forests in Northern California, based on the role of fire in shaping those forests, resulted in the dropping of that requirement.

Representatives of the Clinton administration were present throughout the FEMAT analysis to clarify the charge and contribute to the discussion. Otherwise, administration members stayed largely in the background, although they obviously were anxious about the possible outcomes. Congress, on the other hand, saw it as a presidential initiative, expressed appreciation that the president had assumed responsibility, and stayed three thousand miles away.

Managers were, by and large, not included in the FEMAT process, except for occasional, informal contacts between individual scientists and managers and a brief review of Option 9 near conclusion of the effort. The ISC had set the precedent of excluding managers by having a sequestered group of scientists develop a conservation plan. Additionally, managers were identified with the forest plans that had been judged as inadequate in the protection of species by a number of studies (Thomas et al. 1990; Johnson et al. 1991; Thomas et al. 1993). Some concerns existed that managers would have difficulty in taking an objective look at options that might eliminate or alter timber sales in which they had invested significant time and effort.

However, the exclusion of managers in FEMAT had costs. First, managers who would be in charge of implementing the President's Plan had little initial ownership in it or knowledge of it. The difficulty encountered by the Clinton administration in getting its plan off the ground in the Pacific Northwest reflects this problem. Second, a set of rules for forest management considered necessary by scientists, and compounded by the addition of more measures to protect species in the record of decision for the President's Plan, has proven very difficult to implement. Involving managers from the beginning could have improved the implementability of the plan. Third, creative solutions to the issues being faced that might have been suggested by managers were not heard.

FEMAT was composed primarily of federal researchers but also included a number of academics. The Federal Advisory Committee Act (FACA) requires that all groups "established or utilized" by the federal government "in the interest of obtaining advice or recommendations" must operate in accordance with a series of procedural requirements. The statute also indicates that advisory groups are to be "fairly balanced in terms of the points of view represented and the functions to be performed."

Almost a year after the FEMAT process ended, a federal judge held that FEMAT had been subject to FACA, essentially because the academics involved did not qualify as federal employees, and that FEMAT violated FACA's procedural requirements. The court refused to enjoin the use of the FEMAT study, however, because the plaintiffs failed to demonstrate that the study would have come out differently if FACA had been followed and also because the court was mindful of the president's constitutional prerogative to act upon advice from whomever he might wish.

The Practice of Science in FEMAT

Generally, classical science attempts to narrow uncertainty to the point where acceptance of an argument among scientific peers is essentially unanimous. It does this by being narrowly focused and conservative. From an ecosystem perspective, classical science is incomplete and fragmentary. It is a science of parts (Gunderson, Hollings, and Light 1995). Classical science is organized around developing and testing hypotheses and building a strong inference case. Murphy and Noon (1992) argue that the classical model was used in the ISC to develop its conservation strategy by testing the ability of alternative landscape designs to promote survival and growth of owl populations. Such an approach provides a clear model and testable hypotheses and helps focus subsequent discussion.

Another approach to science focuses on integration of parts. It uses the results and technology of classical science but attempts to integrate those results to answer questions about the functioning of entire systems. It identifies gaps and develops alternative hypotheses, often using multivariate models in the effort. Developing and testing hypotheses are often difficult to do in any comprehensive sense. Uncertainty is high; the premise of this science is that knowledge of the systems we deal with is always incomplete. Not only is the science incomplete, but the system itself is a moving target because of the impact of management, natural disturbance, and expansion of the scale of human influences (Gunderson, Hollings, and Light 1995).

The approach taken in FEMAT utilized both kinds of science but did not adopt either model as its own. The FEMAT approach might be best seen as a science assessment (Gordon 1993) in which scientists were asked by policy makers to address a policy question in a pragmatic, managerial context, not a question arising from the scientist's curiosity. The effort was not about "doing science" but rather was an effort by scientists to bring together existing knowledge and hypotheses to develop an informed judgment regarding potential outcomes of alternative policy choices.

Little formal hypothesis testing occurred. Rather, FEMAT undertook the complex process of assembling ideas and testing them against the views of research and agency resource scientists with experience in the field, to gradually build a working hypothesis about the functioning of late-successional forests and watersheds of the Northwest. The use of adaptive management is intended to continue this experimental approach in a managerial context.

Beyond the development and testing of hypotheses, another attribute of classical science is the peer review process, in which results are evaluated by experts somewhat distant from the process and without a personal stake in the outcome. In FEMAT, the amount and form of peer review varied among the working groups. Some working groups used professional societies to find reviewers for their work. Other working groups, such as the one estimating timber harvest levels, relied on the judgments of field specialists and did not use a formal peer review process. In most cases, it proved difficult to find reviewers who could address the strategic approaches in addition to the substantive details, and the reviewers had very limited time in which to respond. Referees were not employed in the usual sense of arbitrating the process of manuscript revision and making decisions on adequacy of the work. Rather, the scientists themselves decided how to react to review comments. With policy makers, including the president, waiting anxiously for the results, outright rejection of the results as technically inadequate was not a realistic possibility; at most, the scientists could patch up any weak parts that had been identified and push on.

The FEMAT Results

The options vary in four respects: the quality and location of land placed in reserve; the activities permitted within those reserves; the delineation of areas outside the reserves; and the activities allowed in areas outside reserves. Many on the team believed that some system of reserves was required to assure the viability of threatened and at-risk species. Consequently, each of the options contains reserve areas in which timber harvests are either not allowed at all or are limited, and areas outside of reserves (referred to as the "matrix") where most timber cutting would occur. The reserves are of two types: (1) late-successional reserves, encompassing older forest stands, and (2) riparian reserves, consisting of protected strips along the banks of rivers, streams, lakes, and wetlands.

Late-successional reserves contain a mix of late-successional and younger forest as a function of the disturbances that have occurred. In some options, design of late-successional reserves centered on the habitat needs of individual wildlife species. Other options developed late-successional reserves around remaining old growth. Option 9 developed late-successional reserves through an integration of these two approaches with protection of key watersheds.

Under most options, some thinning of stands less than eighty years of age would be allowed in the portion of the reserves that does not currently meet the definition of late-successional forest to accelerate the development of late-successional forest conditions. Also, under most options, trees with a small likelihood of existing when late-successional conditions are reestablished (eighty to one hundred years into the future) could be salvaged in late-successional reserves.

All options contained some form of riparian reserves (areas adjacent to streams) intended to address the habitat requirements of fish and other aquatic and riparian species and to protect water quality, maintain desired water temperatures, and reduce siltation and other degradation of aquatic habitat that results from timber cutting on adjacent land. Under different options, riparian reserves varied in width, depending on stream or lake characteristics. Some options designated key watersheds, where riparian protection might be greater than in other locations. Initially, under most options, no harvest would be allowed in riparian reserves until completion of watershed analysis. Following such analysis, timber harvest might occur within these reserves, if it helped achieve the objectives specified for them, and their shape and extent might be modified.

Under all options except Option 9, timber harvesting outside of reserve areas would have to meet, at a minimum, the specifications in the plans that existed for the national forests and the lands of the Bureau of Land Management. Option 9 allowed the national forests to modify special protections for late-successional species such as pileated woodpeckers and pine martens, where those species exist outside of the established reserves. Most options incorporated additional guidelines for timber harvests in the matrix, including provisions of dispersal habitat for the northern spotted owl and green tree retention at time of regeneration harvest.

Land remaining in the matrix, after withdrawal of various reserves, varied from 10 to 30 percent among the different options (figure 6.1). The proportion of the remaining late-successional and old-growth forest available for harvest varied from 1 to 30 percent of the forests on matrix lands, by the measures employed in this study.

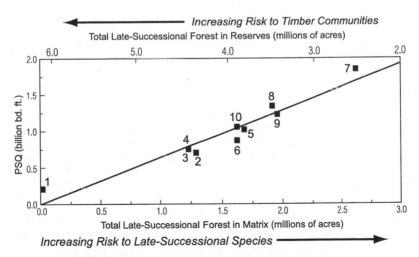

Figure 6.1. *Area of late-successional forest in reserves and matrix for each option. No data available for Option 3. Reserves include late-successional and riparian reserves; additional late-successional forest occurs within congressionally and administratively withdrawn areas. (Read up from an option point to derive the acres in reserves. Read down to derive the acres in the matrix. Read left to derive probable sale quantity, PSQ.)*

Option 9 included the concept of adaptive management areas in ten relatively large areas that would be allowed more flexibility in achieving its conservation objectives. These areas would rely more on the experience and ingenuity of resource managers and communities than on the prescriptive approaches that are applied in other areas. A full-scale monitoring program would be particularly important in these areas to ensure adherence to plans that clearly spell out the goals to be achieved through management.

Most options provided a high likelihood of providing well-distributed habitat for the northern spotted owl. Many options also provided a high likelihood of well-distributed habitat for most vertebrates, but the options were much less successful in providing well-distributed habitat for invertebrates, nonvascular plants, and fungi. Much of the effort in writing the record of decision focused on mitigation measures to improve the likelihood of providing well-distributed habitat for these latter species. The options varied considerably in the likelihood of providing well-distributed habitat for salmonids; those with extensive buffer systems scored the highest. Adoption of these extensive buffers in the record of decision raised the species-viability-likelihood estimates for Option 9.

Timber harvests under the different options varied from 2 to 30 percent of the 1980s levels; Option 9 provided about 25 percent. Federal timber harvest in the region during the 1980s averaged about one-third of the total harvest on all ownerships; by the early 1990s, federal harvest had fallen to approximately 25 percent. Under the options, federal harvest was projected to equal from 2 to 20 percent of the regional total, with Option 9's contribution equaling about 10 percent.

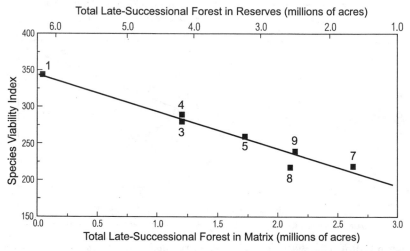

Figure 6.2. *Expected numbers of viable species in relation to acres in reserves and in the matrix. (Read up from an option to determine acres in reserve. Read down to determine acres in matrix. Read left to derive the number of viable species.)*

Figure 6.2 shows that increasing the probable sale quantity requires increasing the acres of late-successional forest in the matrix, which in turn increases the number of species at risk. While this result may seem simple and obvious, it summarizes the federal forest management dilemma in the Northwest: timber harvest from federal forests of the owl region in the near future will come at the cost of increasing the risk to species dependent on those forests.

"Option 9 Shall Be My Forest Plan"

President Clinton announced the selection of Option 9 as his forest plan on July 2, 1993. In an interview with the *Oregonian* newspaper, he made the following comments:

> We believe that this plan is, given the scientific and legal realities, the best and most balanced approach available. It contains some real innovation in its approach to forest management based on protecting watersheds and valuable old growth, and it really is a fundamental shift away from land management based on political boundaries to planning based on ecological boundaries . . .
>
> I think that the most important thing I want to emphasize about that is that the scientists dealt with the reality . . . that the volume of available trees for harvest was much lower than had previously been the case.
>
> Secondly, . . . the effort drew not only upon scientific studies from the spotted owl recovery plan or the scientific assessment team or the Gang-of-Four, all of which have been criticized by different people at different times, but also involved a hundred of the best scientists from all the agencies involved—from the Forest Service, Fish and Wildlife, BLM, Marine Fisheries, EPA, as well as from universities. So, it seems to me that there was a real effort to get people from diverse backgrounds who would have diverse points of view.
>
> I was under a legal obligation under the present law and, given the lawsuit that now exists, to present a plan that was clearly scientifically defensible.

The Draft Environmental Impact Statement

The work done by FEMAT was simply a report to the president and did not in itself produce changes in federal land management. Implementation of any of the options developed by FEMAT required production of an environmental impact statement (USDA and USDI 1994a) and a record of decision (USDA and USDI

1994b) detailing the options considered, their anticipated environmental effects, the option selected as the preferred alternative, and the reasons for its selection.

The EIS team was given two months for preparation and release of the draft supplemental environmental impact statement (SEIS), so it was clear from the outset that the team would largely depend on the work of FEMAT in production of the draft. The ten alternatives in the draft SEIS were identical to the options developed by FEMAT, with some clarifications provided by the EIS team, which had greater experience with agency planning documents.

The ninety-day comment period resulted in approximately 102,000 public comments, indicating the level of interest in management of the Northwest's federal forests. Most of these comments came from people in the three involved states, and almost all opposed Option 9 in part or whole. Issues that might legally block implementation of the plan were of the highest concern, and two of these stood out. First were potential violations of the Federal Advisory Committee Act in the way FEMAT had done its work. Second were serious questions about whether the selected alternative would withstand legal tests surrounding the provision of habitat for maintenance of viable populations of species associated with late-successional and old-growth forests. These questions were to have a significant influence in shaping the final SEIS.

Questions about species viability sat at the intersection of the FEMAT results, the Forest Service regulations implementing the National Forest Management Act, and the history of successful legal challenges to a succession of Forest Service plans for spotted owl management. The existing Forest Service regulation (36 CFR 219.19) states, in part, that "fish and wildlife habitat shall be managed to maintain viable populations of existing native and desired non-native vertebrate species in the planning area. . . . In order to insure that viable populations will be maintained, habitat must be provided to support, at least, a minimum number of reproductive individuals and that habitat must be well distributed so that those individuals can interact with others in the planning area." Interpretation of this regulation raises many questions, including the level of certainty required to "insure" viability and the appropriate definition of "well distributed."

In previous court cases regarding spotted owl plans (*Seattle Audubon Society v. Mosley*), the court had stipulated that "what is required is that the plan adopted not be one which the agency knows or believes will cause the extirpation of other native vertebrate species in the planning area." While the application of this provision to nonvertebrate species had not been similarly tested in the courts, interpretation of the Endangered Species Act would suggest that a similar standard would apply to plants and invertebrates. FEMAT and draft SEIS analysis of Option (now Alternative) 9 suggested that there was some risk of extirpation for a significant number of the species that had been analyzed. In addition, there were challenges to the credibility of the analysis of northern spotted owls and questions about whether the most recent information had been used in those analyses. Because of the legal risk posed by these challenges, the Department of

Justice provided strong encouragement to the interagency EIS team to deal with the issues in preparation of the final SEIS.

The Final SEIS and ROD

A team of federal scientists reviewed the outcomes that had been projected for late-successional associated species under Option 9. An initial screening identified species whose projected outcomes might present legal barriers to implementation of the forest plan. This initial screening resulted in a list of almost five hundred species or groups of species that required further attention. After significant analysis, the secretaries of interior and agriculture decided to modify Option 9 prior to adopting it in the record of decision (ROD) to improve protection for those species. They adopted larger riparian reserves along intermittent streams and added protection for known sites and/or newly discovered sites of more than three hundred species or groups of species. Adoption of these mitigations decreased projected timber outputs somewhat but may have been instrumental in later court decisions in favor of the plan.

These provisions have been controversial with some of the former FEMAT members who argue that they shift the focus of the plan from ecosystems to species. The mitigation measures had been described in FEMAT's original report but not adopted in Option 9. Rather than viewing them as shifting the focus to a species plan, they might be viewed as resolution of the need to deal with locally distributed species in a regional plan.

Another potential controversy stemmed from the procedures used to estimate the timber harvest impact of the added provisions. In development of the FEMAT options, the national forests and the BLM state office were heavily involved in estimating the likely timber harvest level under Option 9. That process was not continued in the revisions to Option 9 adopted in the record of decision. Rather, the EIS team made estimates of the likely effects based on trade-off analysis done by FEMAT scientists (Johnson et al. 1993). Thus, the reality check that the national forests and the BLM state office had provided in the previous estimates was lost, as was the potential for a sense of ownership of those estimates by the federal managers.

The President's Plan was challenged on two sides in Judge Dwyer's Court. The Seattle Audubon Society (SAS) and several other environmental groups contended that compliance with environmental laws was still inadequate and sought a remand of the plan to the agencies and an injunction against timber sales in the meantime. The Northwest Forest Resource Council (NFRC) challenged the plan under a number of laws and sought a remand to the agencies (although they did not want their case heard in Judge Dwyer's court). Both sides moved for a summary judgment. Judge Dwyer's overall ruling: For the reasons given in his decision, he found that the federal defendants had acted within the lawful scope of their discretion in adopting the 1994 Forest Plan.

Looking at Dwyer's reasoning on one of the many points he addressed will give the flavor of his decision. One of the suits claimed that inadequate consideration was given to opposing scientific views on the protection needed for the northern spotted owl. He found as follows: There is reputable scientific opinion supporting the secretaries' view. The plan must be designed by the agencies not the courts; the question for judicial review is whether NEPA's requirements have been met. A disagreement among scientists does not in itself make agency action arbitrary or capricious, nor is the government held to a degree of certainty that is ultimately illusory. The final SEIS does take a "hard look" at the available data and opposing opinions. It has a reasoned discussion of a myriad of factors. Careful monitoring will be needed to assure that the plan maintains owl viability. New information may require that timber sales be ended or curtailed. Dwyer ruled that on the present record, the SEIS adequately disclosed the risks and confronted the criticisms as required by NEPA.

In his decision, Dwyer made three points that might influence implementation of the President's Plan. First, he emphasized the key role that the agencies had given to monitoring to help with the uncertainties of the plan. If this monitoring does not occur, the agencies may end up back in Dwyer's court. Second, he ruled that the cumulative effects analysis, necessary for site-specific actions, was adequate for purposes of programmatic EIS. Whether the procedures that the agencies develop for this analysis are adequate is still an open question. Third, he noted that the secretaries indicated that the President's Plan would provide the highest sustainable timber levels of all the alternatives likely to satisfy law and policy. In other words, according to Judge Dwyer, any more logging sales than the plan contemplates would probably violate the laws. Thus, attempts to produce more timber than indicated under the plan might result in scrutiny from Dwyer.

Implementation of the President's Forest Plan

Implementation of the plan began with the signing of the record or decision, in winter 1994. While FEMAT was preparing its ecological, economic, and social assessment, an Agency Coordination Team discussed ways of achieving the interagency coordination needed for successful implementation of the plan. Recommendations of this group were eventually embodied in a Memorandum of Understanding for Forest Ecosystem Management. Under terms of that agreement, four interconnected committees containing members of eight federal agencies have been formed to facilitate and oversee implementation.

Success of this interagency structure has been mixed. The structure has been very successful in increasing the level of communication and understanding among the agencies, which may seem a trivial accomplishment to those outside the government, but in bureaucratic circles it is a significant achievement and lays the groundwork for further success with interagency collaboration. The interagency structure has also improved the ability of the agencies to share data, and efforts are ongoing to develop common data standards. On the downside, the interagency structure has had only limited success in developing tools and processes needed to implement the plan. Agency decision makers may not accept some of the priorities of the new forest plan. For example, funding to accumulate information on "survey and manage" species has been slow in coming. Also, much of the work is being done by ad hoc committees, whose members have less than full-time assignments and meet only periodically, which may have difficulty in developing quality products. Needed technical skills are in short supply, causing unavoidable delays in development of technical tools. And the regulatory agencies and land management agencies often come to problems with different worldviews.

Measuring Success

The first major test of the success of the plan was its ability to withstand legal challenges. The importance of this hurdle, and of the plan's success in overcoming it, can hardly be overstated. With success in Dwyer's court, the government overcame five years of legal setbacks surrounding its management of late-successional and old-growth forests. Subsequent to this initial legal victory, however, other measures of success emerged.

Biological Results

Of all the measures of success of FEMAT, the biological indicators may be the most difficult to assess. Important measures of biological results would include stable or upward trends in populations of threatened or endangered species, with their eventual delisting; stable or upward trends in species with high social importance, particularly anadromous fish; stable or upward trends in populations of other species of concern; improvements in the amount and distribution of late-successional habitat, particularly within late-successional reserves; improvement in water conditions and other indicators of the health of riparian systems; and improvement in the condition of forests, particularly those that had deteriorated due to the disruption of natural disturbance cycles. It is much too early to judge the success of the plan in accomplishing any of these objectives. Still, there is little doubt that the management of these forests is on firmer footing relative to protecting late-successional species and ecosystems and at-risk fish stocks with the President's Plan than with previous plans.

Outputs

Forest outputs that could be used to gauge success of the plan would certainly involve measures of timber harvest. However, other measurable forest outputs could include recreation, hunting and fishing use, and products from a variety of plants and animals.

Attainment of timber harvest goals has been difficult at best. The harvest goal in the plan was slightly over one billion board feet. When President Clinton and his aides announced Option 9 in July 1993, Secretary of the Interior Bruce Babbitt estimated that two billion board feet would be available over the following year. While that estimate apparently included the timber under contract, it established expectations that proved to be unrealistic. In the first two years after announcement of the plan, approximately one-quarter of a billion board feet was sold. The following few years have been closer to their targets, but the sternest test may not come until the turn of the decade, when many of the sales of relatively low controversy will be complete, leaving some very tough decisions for future programs.

Adaptive Management

If the adaptive management framework of the plan is successful, implementation of the plan will result in new information about forest resources and new knowledge about the interaction of those resources with management activities, and the plan will adapt significantly over time based on that new knowledge. The plan will also be adjusted in an ongoing way based on the use of local information to refine the initial standards and guidelines.

The designated adaptive management areas are a major component of the adaptive management strategy. They are intended to support development of innovative, collaborative techniques for achieving objectives of the plan. Early efforts in the AMAs were frustrated by concerns about the Federal Advisory Committee Act and the need to invent AMA procedures on the fly. The FACA concerns have now been largely resolved through establishment of advisory committees in compliance with the law, AMA procedures have been developed, and activities in the AMAs are underway.

To date, the AMAs have often provided an effective forum for discussion and collaboration. Some have also generated innovative approaches and solutions, but progress has been slow. One limit to the potential for innovation comes from the conditions under which they were established. The Clinton administration's interest in them was in part contingent on whether they would maintain their contribution to regional timber harvest goals. It was agreed that such areas could explore innovative ways to meet the harvest goals but that establishment of the AMAs would not result in reduction of the goal.

The perilous situation of many fish and wildlife species is another challenge to the use of innovative measures in AMAs. Learning often involves experimentation and taking risks, yet it is difficult to justify risk taking when these areas contain a number of species whose survival is in jeopardy. Most AMAs were located adjacent to late-successional reserves, thus lessening the potential risk to some species due to AMA experiments. Still, there are major risks to species from experimenting in riparian reserves, and those reserves cover over half of many AMAs.

The second component of adaptive management, the overall commitment to use new information to refine the plan, has yet to be translated into specific actions. The interagency committees have made several attempts to define a process for identifying, analyzing, and using new information but have not reached agreement. The slow progress results, in part, from concerns that new information could produce changes that would hurt the interests of specific forest users.

The third component of adaptive management is the use of local information to improve the plan. The standards and guidelines for management in the record of decision were intended as defaults that could be refined through more detailed analysis at the local level. Moving from those default standards to more site-specific standards has been difficult at best. For example, interim widths and standards for riparian reserves are specified in the record of decision. Watershed analysis was to assess whether some other configuration of buffers would enable each particular watershed to more efficiently meet the goals in the President's Plan. Those analyses have been complicated by the large number of species associated with riparian systems. While the riparian reserve system was proposed by aquatic scientists to protect aquatic biodiversity, it also provides habitat for a large number of terrestrial species for whom risk to survival would increase without this protection. The expertise necessary to consider this diversity of species is generally not present in any individual ranger district, so it is not surprising that managers have been reluctant to suggest changes in the interim standards for riparian reserves.

A further impediment to the use of watershed analysis for improving resource management has been the decision made early in the process that watershed analysis would not foray into planning but would focus exclusively on analysis. Thus, watershed analysis is not a decision document and does not come under the requirements of the National Environmental Policy Act. While such an approach ensures that the analysis does not get caught up in the sometimes lengthy NEPA process, it also creates an artificial barrier between analysis and planning that makes it difficult to direct the analysis toward decisions that are needed.

Integrating the President's Plan into Operations of the Federal Agencies

Forest plans are the major guiding document for national forest management under the National Forest Management Act, and District Resource Plans guide management of BLM districts. A long-term goal of the FEMAT effort was to incorporate the President's Plan into forest and district plans. In FEMAT, the President's Plan was placed, by and large, over the forest and district plans without reevaluation as to whether the forest plans could be modified to make a better fit with the President's Plan. That reanalysis was left to the national forests and BLM districts and has proven to be very difficult because of the lack of accepted techniques for measuring improvement in meeting the goals of the President's Plan.

Also, there is a lack of fit between the planning hierarchy currently in use in the national forests and that envisioned in FEMAT. The Forest Service over time has evolved into a three-tier planning system, including regional guides, forest plans, and project plans. Recently proposed revisions of the regulations implementing NFMA attempt to formally codify forest plans and project plans into a two-tier approach. FEMAT, on the other hand, proposed a planning hierarchy that uses ecological boundaries instead of political boundaries: (1) province plans (plans for large ecological units such as the Oregon Coast Range), (2) watershed plans, and (3) project plans. It is unclear which of the approaches to planning will be used in the future.

One key to the integration of the President's Plan into agency operations is the availability of adequate budgets. Early in this century, Congress designed a system in which federal forest management would be funded from congressional appropriations and financial returns from the sale or lease of resources would be returned to the general fund. Studies continue to confirm, though, that much of the gross revenues from national forest receipts is retained by the Forest Service for management activities or is paid to counties in lieu of taxes (GAO 1995). In the case of the northwest forests, which return more to the treasury than most, reductions in receipts, as are predicted to occur under Option 9, will be broadly felt by the agency, the counties, and the treasury.

Throughout the FEMAT process, and in the general shift toward ecosystem management on federal land, it has been repeatedly noted that attending to the multiple values of forest ecosystems and managing for diversity will likely increase the cost to the taxpayer. Current and proposed appropriations, though, are actually lower (in real dollars) than past funding levels, in part due to Congress's desire to reduce the federal deficit. While it is difficult to predict the future, it seems likely that the kind of analysis, monitoring, experimentation, research, and restoration envisioned in FEMAT is unlikely to be fully funded. Whether that will cause the plan to reappear in Dwyer's court, given his emphasis on agency commitments to monitoring in his decision, is anyone's guess.

Economic, Social, and Political Indicators

The somewhat mixed results of new interagency structures have been discussed above. Within single agencies, some adaptation to new ways of doing business and new priorities has occurred. The U.S. Fish and Wildlife Service, most notably, has established new branches with newly hired employees to help discharge its responsibilities under the plan. The land-managing agencies (Forest Service and BLM), appear, on the surface, to have made relatively few structural changes to adapt to the plan. They have, though, undergone significant contraction of their workforce in many places as timber sales targets have declined.

As predicted in FEMAT, the economies of the Northwest continue to grow, even with sharp declines in federal timber harvests. Interestingly, newspaper reports suggest that many communities in the heart of the area that suffered employment declines from reduction of federal harvest have survived better than expected. Still, the source of family-wage jobs in areas outside of metropolitan areas remains an issue.

It is too early to read many of the social and political indicators. However, it is fair to say that federal land management is under attack throughout the western United States, and Congress has passed legislation that would override federal land management laws. It is unclear how much of this might be attributed to the President's Forest Plan, but that plan has been the subject of specific congressional actions, including legislation that limits appeal of actions under the plan and legislation that allows harvesting of some sales that would have been off-limits under the President's Plan. While these sales have been welcomed by the timber industry, they have also resulted in renewed environmental protest of federal timber harvesting practices.

The continuing controversy over forest management on federal lands in the Pacific Northwest should not be surprising. The President's Plan restricted timber harvest to a smaller area of the federal land, but it did not fundamentally change the focus on old forest that created the controversy. Most of the harvest will still come from old-growth timber for many years into the future; the harvests still detract from protection of species associated with these forests (as figure 6.2 shows). These activities, though, will occur on a much reduced scale. Even without congressional action, some protest might have been expected.

Some Lessons from FEMAT

We learned many lessons about conducting bioregional assessments from the FEMAT experience. Here are a few with general application to other assessments:

- *It was harder than we thought it would be to construct integrative looks at the implications of policy options for ecosystems.*

Much of the work of scientists has a strong disciplinary focus with accepted protocols and procedures. Cooperative, cross-disciplinary work runs across the grain of these normal efforts and proved very difficult to do in FEMAT. As we tried to examine the implications of the options from different perspectives, we encountered significant cultural differences among the resource specialties. Differing worldviews and differing protocols for determining the validity of evidence hampered attempts at integration, as did a lack of trust among the different specialties that occasionally surfaced. This might be viewed as the biological scientists versus the social scientists, but it is more complicated than that. As an example, the worldviews of landscape ecologists, wildlife biologists, and silviculturists appear to differ significantly, especially in terms of how to consider the dynamics of forest ecosystems and the effects of natural disturbances such as damage resulting from insect attacks. Part of the difficulty also may stem from the desire of each discipline to play a preeminent role in the evolving realm of ecosystem management.

• *It is difficult to gain an integrative review of the science assessment.*

Generally, the FEMAT work was not about advancing knowledge of particular subjects; rather, it was about organizing what was known to answer the policy questions being asked. In our case, much of this work involved the assessment of ecological, economic, and social implications of different ways of managing the forests of the Northwest. Obtaining a review of the details of each piece of the puzzle was possible; obtaining a review of how the pieces fit together was much more difficult. Generally, the science community appears to be a poor reviewer of integrative science-policy approaches.

• *More attention needs to be given to how to portray the level of uncertainty surrounding estimates of implications of alternatives.*

FEMAT attempted to make estimates of the ecological, economic, and social implications of policy options for meeting the president's five goals. And a sophisticated system of expressing likelihoods of outcomes was developed for the ecological assessment. Still FEMAT was only partially successful in expressing the uncertainty surrounding its estimates.

FEMAT made progress dealing with the uncertainty of estimates on at least two fronts. The likelihood voting methodology used in the species viability panels allowed expression of uncertainty about the likelihood of each of the alternative outcomes. However, FEMAT could not find a way to express this uncertainty in a form that could be readily understood for the thousands of judgments made by the viability panels. Consequently, it remained buried in Team documents and did not appear in the final report. Similarly, significant uncertainty surrounds the estimates of timber yield. This was partially expressed by changing the description of timber outputs from "allowable" to "probable," but the more detailed discussion of the uncertainty surrounding timber harvests was contained in a background document rather than in the main report.

• *The degree of trust in agency implementation can greatly influence the results; this aspect of an assessment needs explicit recognition.*

In FEMAT, we encountered two worldviews of the effectiveness of agency implementation of any prescriptions that might be developed. Some people assumed that agency personnel would generally (or always) faithfully implement the prescriptions developed and that they could be trusted to use their common sense in doing that; others assumed that the inertia of the old culture and old objectives would retard implementation. Those views affected the development and evaluation of options in two ways. First, those who assumed faithful implementation of the prescriptions tended to specify the prescriptions as goals and let the agency determine how to achieve them; those who had little trust wanted to detail what actions were permitted. (It must be acknowledged, though, that some who advocated specifying permitted actions did so out of concern that the goals themselves were not sufficiently instructive.) Second, those who had little trust were tempted to adjust their estimate of implications for the degree to which they thought the agency might wander from the prescriptions—like a Kentucky windage adjustment to a rifle sighting—and to penalize active management.

In FEMAT, we were not completely successful, for example, in excluding the trust issue from the species viability assessment or in identifying its contribution to the results there. Future assessments might address the trust issue more directly from the beginning.

• *With species on the brink, it is difficult to advocate experimentation or actions that improve the ecosystem in the long run at some short-run cost.*

A number of creative policy options in FEMAT, including more aggressive forest management strategies, were eliminated from detailed consideration because of the precarious position in which many late-successional species seem to be. It is difficult to justify actions that have any significant chance of harming a threatened species no matter what the long-run benefits might be.

• *The information needs of ecosystem management and adaptive management are significantly greater than the needs of extractive management.*

The management changes that culminated in FEMAT display an evolutionary change in management from a focus on sustaining outputs to a focus on sustaining ecosystems. Past policy frameworks that supported extractive management ignored many parts of the ecosystem, thus simplifying management. At least over the short term, this lack of knowledge did not impede management success.

By contrast, adaptive management that is intended to achieve ecosystem-based goals appears to require knowledge of a multitude of system components. The watershed analysis called for in the FEMAT report and the significant "survey and manage" requirements in the ROD implementing Option 9 illustrate this need for knowledge.

The "hidden cost" of adaptive, ecosystem management needs recognition by policy makers, and it must serve as input to realistic projections of both resource outputs and rates of adaptation. If the costs are not recognized, the expectations of ecosystem management plans will not be met.

• *The paradigms for ecosystem management are still quite rudimentary, and truly new ideas are rare. Even the definition of ecosystem management continues to evolve.*

The federal forests of the Pacific Northwest were largely managed under a single dominant paradigm for many decades—clear-cutting in staggered settings and planting. Transition to a completely new paradigm—under the banner of "ecosystem management"—in a single step is a huge undertaking and is probably not realistic. To be successful, this leap requires significant scientific breakthroughs. Yet, those breakthroughs are rare.

Definitions of ecosystem management continue to be fuzzy. We know that we want to "sustain" the ecosystem, but we have only the most rudimentary concept of how to take its pulse. In trying to develop these new ideas, we are constrained by past practices. A landscape shaped by decades of clear-cutting practices provides a poor laboratory to aid understanding of more creative forestry practices. Thus, the practices proposed to implement ecosystem management, such as in FEMAT, look suspiciously like the practices of past management paradigms. They may be somewhat reduced in extent and intensity but they are not truly new. More creative practices are cautiously added to the mix, but only in small amounts because we do not have the information to evaluate them. Thus, evolution toward ecosystem management is likely to be a slow, continuous process, especially in the world of forest management, where the results of our actions often take decades to be revealed.

- *Achieving a blend of technocratic and democratic approaches to science-based assessments is essential.*

Jasanoff (1990) suggests that science assessments can be classified as to whether they are "technocratic" or "democratic" processes. "Technocratic" assessments convey enormous responsibility on scientists to formulate and evaluate alternative approaches to solving some problem for policy makers. This work is done far removed from the rest of the world and is based on the notions that the organization of knowledge is a key element in solving the policy problem, that facts and values can be separated in the policy-making process, and that scientists should be best left alone to organize these facts. Then, after the scientists are done, values can be applied in the policy-making process. "Democratic" assessments, on the other hand, emphasize the need to incorporate a full range of values in the development of options for policy making, the difficulty of separating facts and values, the need to have a broad and balanced membership of technical advisory groups, and the need for open review and decision making.

FEMAT was much more technocratic than democratic. While this can be partially explained by the perceived need to insulate scientists from political pressure, and the history of communication with Judge Dwyer, the potential costs of this approach in terms of developing effective, implementable policies must be acknowledged.

The best procedure for working with policy makers, managers, and citizens remains to be developed. It should not be overlooked, though, that the scientific studies that have recently resulted in major changes in forest management in the Northwest—the Thomas report, SAT report, Gang of Four report, FEMAT, CASPO (Verner et al. 1992), and the East-Side Screens (USDA FS 1994, 1995)—have been done with almost no public interaction during the critical stages of developing and evaluating options. Other processes that have extensive public involvement, such as national forest planning, often seem mired in the status quo and have not always proven to be effective vehicles for solving ecosystem problems.

• *Field examples can help test ideas and ease implementation.*

FEMAT set up generic prescriptions to be applied over millions of acres of federal land. These prescriptions are described in terms of "Standards and Guidelines" in the FEMAT report. For several years, though, field examples of what is meant by these prescriptions were (and still may be) unavailable to the people who must try their best to implement the plans. Future assessments, which evaluate alternatives, should consider developing field examples of management actions that meet the goals of the different policy options—at least for the option chosen for implementation. This will help the assessors test their ideas and ease implementation—perhaps an obvious idea but one that often seems to be forgotten.

• *Context is important.*

There is a tendency for bioregional assessments to concentrate on their own part of the world with little reference to the broader ecological, economic, and social environment in which they sit.

FEMAT was greatly strengthened by efforts, such as those of the economic subgroup, that viewed the economic implications of FEMAT as marginal changes in a broader economic environment. By helping us understand the role of federal timber harvest in overall regional harvest and the role of timber employment in overall regional employment and growth, they provided a context for evaluation that greatly broadened and strengthened our understanding of the options being evaluated.

• *Leadership is everything.*

FEMAT was held together by an exceedingly strong and visionary leader named Jack Ward Thomas. Amid the chaos and confusion that often surrounded the FEMAT effort, his steady focus on the president's goals and his willingness to make decisions made it possible for FEMAT to function and to complete its mission in a short period of time. Without a leader like Jack Thomas to hold things together, projects like FEMAT tend to fly apart. If we had only one bit of advice to give future science assessments, it would be: first pick a strong, credible leader.

LITERATURE CITED

Caldwell, L. K., C. F. Wilkinson, and M. A. Shannon. 1994. Making ecosystem policy: Three decades of change. *Journal of Forestry* 92(4):7–11.

Forest Ecosystem Management Assessment Team (FEMAT). 1993. *Forest ecosystem management: An ecological, economic, and social assessment.* Portland, OR: USDA Forest Service; U.S. Department of Commerce, National Oceanic and Atmospheric Administration, National Marine Fisheries Service; USDI Bureau of Land Management, Fish and Wildlife Service, National Park Service, Environmental Protection Agency. 1033 pp.

Franklin, J. F., et al. 1981. *Ecological characteristics of old-growth Douglas-fir forests.* Pacific Northwest. For. and Rang. Exp. Sta. Gen. Tech. Report PNW-118.

GAO. 1995. *Budget trends—Obligations by item of expense, fiscal years 1971–1994.* GAO/AIMD-95-227, September. Washington, DC: Government Accounting Office.

Gordon, J. 1993. Assessments I have known: Toward science-based forest policy. In *1993*

Starker Lectures: Communications, natural resources, and policy. College of Forestry, Oregon State University. 49 pp.

Gunderson, L. H., Hollings, C. S., and Stephen S. Light. 1995. *Bridges and barriers to the renewal of ecosystems and institutions.* New York: Columbia University Press.

Jasanoff, S. 1990. *The fifth branch: Science advisors as policy-makers.* Cambridge: Harvard University Press.

Johnson, K. N., S. Crim, K. Barber, M. Howell, and C. Cadwell. 1993. *Sustainable harvest levels and short-term timber sales for options considered in the report of the Forest Ecosystem management Assessment Team: Methods, results, and interpretations.* Corvallis: Oregon State University. 93 pp. Mimeo.

Johnson, K. N., J. F. Franklin, J. W. Thomas, and J. Gordon. 1991. *Alternatives for management of late-successional forests of the Pacific Northwest. A report to the Agricultural Committee and the Merchant Marine Committee of the U.S. House of Representatives.* Corvallis: College of Forestry, Oregon State University.

Murphy D. D., and B. R. Noon. 1992. Integrating scientific methods with habitat conservation planning: Reserve design for northern spotted owls. *Ecological Applications* 2(1):3–17.

Ruggiero, L. F., et al. 1991. *Wildlife and vegetation of unmanaged Douglas-fir forests.* General Technical Report PNW-GTR-285. Portland, OR: USDA Forest Service, Pacific Northwest Research Station.

Sessions, John. 1994. Obstacles to implementation. *Journal of Forestry,* 92(4):42.

State of Oregon. 1997. *Coastal salmon restoration initiative: Executive summary and overview.* Portland, OR. 13.

Thomas, J. W., et al. 1993. *Viability assessments and management considerations for species associated with late-successional and old growth forests of the Pacific Northwest.* Washington, DC: USDA Forest Service.

Thomas, J. W., E. D. Forsman, J. B. Lint, E. C. Meslow, B. R. Noon, and J. Verner. 1990. *A conservation strategy for the northern spotted owl: A report of the Interagency Scientific Committee to address the conservation of the northern spotted owl.* Portland, OR: USDA Forest Service, USDI Bureau of Land Management, Fish and Wildlife Service, National Park Service.

USDA Forest Service Region 6. 1994. *Interim management direction establishing riparian, ecosystem, and wildlife standards for timber sales for east-side national forests.* Portland, OR: USDA Forest Service Regional Office.

———. June, 1995. Continuation of interim management direction establishing riparian, ecosystem, and wildlife standards for timber sales for east-side national forests. Portland, OR: USDA Forest Service Regional Office.

USDA and USDI. 1994a. *Final supplemental environmental impact statement on management of habitat for late successional species and old growth forest related species within the range of the northern spotted owl,* volumes 1 and 2. Portland, OR: USDA Forest Service Regional Office.

———. 1994b. *Record of Decision for amendments to Forest Service and Bureau of Land Management planning documents within the range of the northern spotted owl.* Portland, OR: USDA Forest Service Regional Office.

USDI. 1992. *Recovery plan for the northern spotted owl—Final draft.* Portland, OR: U.S. Fish and Wildlife Service, Regional Office.

Verner, J., et al. 1992. *The California spotted owl: A technical assessment of its current status.* USDA Forest Service General Technical Report PSW-GTR-133. Albanu, CA: USDA Forest Service Pacific Southwest Research Station.

White, R. 1995. *The organic machine: The remaking of the Columbia River.* New York: Hill and Wang.

Wilderness Society, National Wildlife Federation, and Natural Resources Defense Council. 1994. *Review of the Final EIS for the Clinton Forest Plan.* Washington, DC.

Science Review

Logan A. Norris

FEMAT was an undertaking of unparalleled complexity, proportion, and importance in the Pacific Northwest. It included key elements of management, science, and policy, focusing almost exclusively on management strategies for federal forests. The April 1994 *Journal of Forestry* has a series of articles about this assessment and is a helpful reference. The following are my own perspectives about the science.

Science is about knowledge—often defined as the framework within which we obtain, catalog, analyze, and use knowledge. Research is one process by which new knowledge is gained (through the application of the scientific method), but new knowledge is also formed in other ways. FEMAT was not research. It was the synthesis and integration of existing knowledge to provide a new level and type of knowledge. It took bits and pieces of knowledge, making a larger whole from them, creating new knowledge. FEMAT science was shaped by the *context* of the problem, the *assignment*, and the *composition* of the team.

Conflict about the management of forest-related resources on federal lands led to FEMAT. The conflict is rooted in the changes occurring in social, physical, and biological systems. Increasing population, education, and affluence lead to greater consumption of resources, higher demand for services, and aspiration to include a broader array of values in natural resource management. This is not to the exclusion of commodity values, but to complement them and to have a better balance among these values.

The management of forest resources in the Pacific Northwest in the 1970s was dominated by utilization and the drive to create a "fully regulated forest" that would provide a sustainable flow of commodities (and, later, amenities) indefinitely. But times changed, especially what people wanted. Humans initiate or respond to change by developing or changing the policies intended to promote the welfare of one system or another. However, because of the piecemeal approach to policy formulation historically, the framework for solving the conflicts that led to FEMAT was inadequate. Conflict of enormous proportion was the result and was the trademark of the management of federal forest resources in the early 1990s in the Pacific Northwest.

The charge to the team was not to optimize the benefits, i.e., What is the "best mix" of timber and old growth we can have? In fact it was hierarchical, i.e., First meet the law, then provide the timber to the extent consistent with meeting the law. While several laws were involved (ESA, FLPMA, NFMA, and NEPA), for FEMAT the essence of each focused on wildlife (broadly defined) at the species level and (more broadly) in terms of biodiversity. This distinction between an optimization approach and a hierarchical one is fundamental because it shaped the science and therefore the outcome. FEMAT identified management options and assessed their likely effect—*first* on the forest and then on people.

FEMAT was performed almost exclusively by scientists and technical experts (managers were not included) drawn from the federal sector. Many of those who were not in federal employment at the time of FEMAT had a long formal or informal association with the federal community. The composition of the team established the culture of the group—influencing the interpretation of the assignment, understanding the context, identifying alternatives (options), shaping the analysis, and interpreting results.

Crucial Scientific Questions

The most challenging science questions involved the "biological" assessment of wildlife habitat condition and distribution and population viability. Other questions are important, but they are more analytical and are driven by the results of options identified in the biological assessment. The specific questions were: (1) What is necessary to maintain and/or restore habitat conditions for the northern spotted owl and the marbled murrelet? (2) What is necessary to maintain and/or restore habitat conditions for viable populations of species associated with old-growth forest conditions, well distributed across the range of these species? (3) What is necessary to maintain and/or restore habitat for spawning and rearing of viable populations of anadromous fish (and other aquatic species) considered to be sensitive or at risk on federal lands?

The greatest challenges involved in accomplishing the assignment were the combined factors of limited time, limited information, and secrecy. Limited time meant having to do this job on 25 million acres in sixty days, later expanded to ninety days. This caused the team to rely heavily on the use of panels and "expert opinion." It underutilized the power of peer review, especially from individuals who may have had a different perspective. It also limited the interaction between disciplinary groups within the team and with management and technical experts outside the team. Finally, it provided little time for introspection, consideration of review or other comments, or thoughtful analysis.

Limited information hampered the attempt to assess habitat needs, condition and distribution, and population viability for more than one thousand species (not counting arthropods). This caused the team to lump geographic areas and species into large groups and to utilize information sources that had not been the subject of independent peer review and referee evaluation. It also caused the team to be very conservative in the face of uncertainties and to reach conclusions with a significant (but poorly articulated) uncertainty.

Maintaining secrecy meant not telling what was going on, especially what options were being considered. This challenge severely limited input from outside the team, heightened anxiety in the community, and created an impression of "federal" science elitism.

Strengths of FEMAT Science

FEMAT science accomplished some things at an unprecedented level. These include achieving integration across disciplines—for instance, it tied the aquatic and terrestrial systems together in ways that promoted an integrated approach to their management as parts of one system. It included an enormous scope of coverage—geographic, temporal, and disciplinary. It covered all federal forests within the range of the northern spotted owl. This coverage extended through time, as reflected by the concept of initial stream and riparian strategies that might be changed after watershed analysis. The coverage included a wide array of social as well as biological and physical sciences. FEMAT science also provided or clarified newer resource management concepts such as adaptive management, adaptive management areas, and watershed analysis. And it provided the basis for decision making and implementation through, for instance, the development of integrated databases of unprecedented scope for spatial analysis and planning. Finally, it identified some limitations in knowledge—for instance, about the dispersal ability of terrestrial amphibians in managed forests.

Weaknesses of FEMAT Science

FEMAT science also had some weaknesses. For instance, the team shared (largely) a common culture (derived from members' employment histories), limiting the range of perspectives included in the assessment. The use of panels and opinion reduced the "quantitative" flavor of the assessment. Additionally, the uncertainty associated with various analyses and estimates was poorly articulated, and federal lands were treated as if they were isolated from the larger ecosystem. Finally, the peer review process was uneven among groups and did not usually include those most likely to be critical of the analyses. I don't believe this was intentional, but it occurred. Often the most helpful reviews come from those who may not agree with the premise.

Lessons from FEMAT

Complex regional assessments take an enormous amount of time, personal energy, and other resources. These resources are usually diverted from other programs and purposes, increasing the importance of designing and conducting such assessments carefully to ensure that the products are worthy of the investment.

The challenges, strengths, and weakness of FEMAT science provide some lessons in this regard. For instance, it is important to articulate the policy questions involved and maintain strong linkages between the policy level and the assessment level. This is crucial for ensuring understanding, but it is dangerous because of the potential for compromising the integrity of the assessment. Also, it is important to ensure that the team composition and process include the full array of perspectives on issues and the science embedded within them. This will increase the difficulty of the task but produce a "better product" and one more likely to be accepted. It is important to deal explicitly with uncertainty and the possible consequences of it. Not doing so will mislead those dependent on the analyses, causing poor decisions and enormous loss of credibility of the science. Finally, it is important to provide sufficient time to accomplish what is needed and to compensate for the reallocation of resources, especially people, in the conduct of assessments. This is crucial in research, where the products of research may be jeopardized by lack of attention by the scientists assigned to assessment teams. This reallocation of personnel resources continues long after the assessment is released, due to the need to share information about and defend the assessment.

Conclusions

FEMAT was, at the time, unique in the forestry experience in the Pacific Northwest. It is easy to be critical of such a large and complex effort, especially one that is so important and that touches on subjects about which there is no unanimity of opinion. Science was a crucial part of this effort, and while it had its weaknesses, by and large it was successful.

Many people in both the forestry and environmental communities do not like the outcome. Some believe it overemphasized environmental issues and underemphasized social and economic ones. However, criticism of the outcome should not be automatically extended to the science. The outcome was largely determined by the context and the particulars of the assignment. While the science and the outcome are obviously related, each must stand on its own merit.

FEMAT was an intense and exhausting exercise—for team members and their families, and for the many others who waited for and then studied the product. I give credit to the science team for its integrity and commitment, and especially its courage for undertaking this assessment.

Management Review

Judy E. Nelson

In 1992, federal forest management in the Pacific Northwest was in gridlock as a result of a series of lawsuits. The Forest Ecosystem Management Team opened the way for the renewed management of the public forests by developing an ecosystem management plan that protected the species, northern spotted owls and marbled murrelets, that had been the cornerstone of the litigation. The Northwest Forest Plan (NFP), which resulted from the FEMAT process and agency processes including environmental impact statements, the record of decision, and the amendment of agency planning documents, provided the framework for answering the legal challenges.

Mechanisms created by the NFP for different agencies to develop common goals and work together to solve problems were a major accomplishment of the plan and include the following examples: (1) The Regional Ecosystem Office, which is made up of representatives of eight participating agencies, provides clarification and consistency of NFP interpretation, guides monitoring across the landscape, and coordinates research. (2) Regulatory and land management agencies work together in project design and consultation to minimize impacts on sensitive and endangered species. (3) Provincial advisory committees provide a way to accomplish ecosystem management by including private landowners, tribal governments, local and state governments, and citizen representatives in the decision framework with the federal agencies.

Through the NFP, the potential to accomplish ecosystem management has improved; however, the NFP faces barriers that may threaten its long-term survival and its vision of ecosystem management. The following needs, based on almost three years of implementation, are what I see as some of the key lessons learned from the FEMAT and NFP processes.

The Need for Public Debate

To get public ownership of a plan, a public process is necessary. By the 1990s, public uses of the forests, and values associated with those uses, had shifted dramatically from the 1960s, when the previous forest management policies were developed. By the 1990s, a wide spectrum of the public did not like clear-cuts and valued old growth. Instead of responding to the shift in public values, land management agencies focused on scientific principles such as maximizing sustainable timber harvests with appropriate mitigation for wildlife. Not being listened to, segments of the public started framing their concerns in legal and scientific terms; they argued that land management agencies were not following the

Endangered Species Act in providing proper protection for spotted owls. So "science" became the basis for all discussion.

The FEMAT process tried to resolve the legal and scientific debate, but it did not address the underlying value debate. Its decision framework became a simplistic trade-off between timber production and species protection. FEMAT itself was not a public process. Following the president's acceptance of Option 9 as his preferred alternative, the agencies used a traditional public process (written comments on the EIS) as public input to the NFP. Although the agencies considered over 100,000 comments to the EIS, it was hard to convince the public that the agencies had not already chosen the president's alternative. Changes between FEMAT and the ROD tightened environmental standards but did little to address the emotional issues on either side by explicitly saving additional old growth or protecting timber jobs. In the response to public comments, the final EIS states that "there were many personal, emotional letters," but agency responses addressed no emotional or value questions.

In July 1995, Congress, frustrated by the agencies' lack of progress in implementing the NFP, as measured by the volume of timber sold, included a provision temporarily removing the NFP sales from legal challenges as a provision of Public Law 104-19. What became known as the salvage rider also released disputed timber sales for harvest (or required the agencies to offer replacement timber) and gave direction on expediting the harvest of salvage timber. Over one thousand people were arrested in 1996 trying to stop the cutting of old-growth timber, both in the NFP and in congressionally released sales. What was viewed by a segment of the public as "lawless logging" destroyed the very fragile trust many environmentalists had in the NFP process. The renewed controversy has again focused national attention on the forests of the Pacific Northwest, a signal that the debate on forest policy is still highly polarized.

Given the underlying value issues associated with forest management, the debate over management will most likely remain polarized. However, the NFP still suffers from the initial lack of a public forum and the feeling that many issues surrounding forest management have not been addressed. The NFP will not be a long-range solution until the majority of the public understands and has some ownership in the solution.

Science by itself cannot solve public policy issues. Future planning should avoid a top-down approach that excludes stakeholders; once you have a "preferred alternative," it is hard to make people feel that they are a part of the solution. Additional time in the FEMAT process to involve stakeholders both in the development of alternatives and in the selection of the preferred alternative might have provided more ownership in the solution. Thoughtful inclusion of stakeholders' ideas would have revealed value questions and provided a forum for hearing those issues.

Attempts to expedite implementation of the NFP by excluding public review through sufficiency language hurt the plan's credibility. Once a plan is in place, it is important to allow normal legal processes to proceed. The salvage rider, by preventing resolution of the issues surrounding the plan through the normal legal process, created unnecessary distrust of the plan and the motives of those implementing the plan.

The Need for a Funding–Planning Link

Without broad public ownership of the solution, the NFP has few funding advocates. Congressional funding decisions are based largely on program "advocates." Full implementation of the NFP is inherently expensive. In addition to the production of traditional forest outputs, the NFP adds additional planning steps; ecosystem restoration goals; survey-and-manage requirements; and new monitoring, research, and adaptive management goals. Neither the FEMAT report nor the NFP deals with what elements could be eliminated if funding is not available. Implementation of the plan has been caught among disillusioned environmentalists, a cost-cutting Congress, and constituents who are interested in assuring the forest continues to produce commodity outputs. Lack of full funding has forced the agencies to emphasize funding those portions of the plan that produce the outputs directed by Congress or are required for legal sufficiency. Most of the more creative elements of the plan—e.g., work on river basin and provincial analysis, research, and adaptive management area work—have suffered from lack of funding. Indeed, funding restrictions have slowed progress on implementation of the entire NFP and have created an atmosphere in which plan critics can point to failures in meeting public expectations.

Long term, the lack of funding for the plan components necessary for adaptive management will create a rigid plan. More problematic is lack of funding for the components that are important to the assessment of legal sufficiency—e.g., endangered species assessments and effectiveness and validation monitoring. These, in turn, may impact the plan's ability to survive future legal challenges. Future bioregional plans, such as FEMAT, should include costs of implementing alternatives as an explicit part of the plan, so trade-offs resulting from lack of funding are considered in the selection of the preferred alternative. There might be some very different solutions if budgets are considered as part of the planning process. Explicit budget-output links might serve as a very effective way to build public support to obtain additional funding.

A more radical way to stretch funding is to look at changing not only forest management but also the institutions that manage the forest. Instead of just requiring the agencies to coordinate, combine them. Do away with old systems and regulations that are not applicable to the new plan. Combining agencies would allow streamlining of many processes, although there would be high upfront costs—e.g., monetary costs, inefficiencies as jobs are combined and people relocated, and costs in rebuilding public trust that different agencies have gained with constituents. However, long-term savings could be substantial. Agency unification may also allow for a more efficient process in establishing and implementing NFP priorities.

The Need to Recognize Risk

As a result of changes between FEMAT and the NFP, the agencies are implementing "Option 9+," which requires additional protection measures, without a corresponding decrease in public expectation of timber production. FEMAT recognized the risk to species in managed ecosystems (see the case study). To reduce the risk to species viability, the ROD increased the size of the riparian reserves and identified species for future emphasis with special survey-and-manage guidelines. These guidelines reduce risk by requiring that we manage known sites of occurrence; survey for species prior to ground-disturbing activities; complete extensive regional surveys within specified time lines; and develop management plans for the survey-and-manage species. We hope to have funding to accomplish this ambitious mandate. Yet, ironically, lack of funding may require us to protect and manage more than an optimum number of sites until we have enough information to remove species from the survey and manage lists. The trade-off between a reduced risk management strategy and timber production is unknown. Limited funding and the ROD mandate to develop protocols, survey, and develop management plans for the hundreds of species in the NFP also divert significant time and resources from broader ecosystem management efforts.

Adoption of the NFP also did not prevent additional species listings. The Umpqua cutthroat trout and the Columbia steelhead have been listed, many varieties of coastal salmon are undergoing listing evaluations and are expected to be listed, and a court has ordered the U.S. Fish and Wildlife Service to review its listing decision for bull trout. The agencies are currently evaluating the adequacy of the NFP in protecting these fish species, and additional protections have been required in some instances. The issue is similar to that of the survey-and-manage species, with risk assessments again moving the implemented alternative away from the commodity production goals of Option 9.

The Need for Flexibility

Expectations were unrealistic that the standards and guidelines could be extensively changed at the local level. For example, watershed analysis was expected to allow local determination of the riparian reserve widths. However, the standards and guidelines contain the statement: "Watershed analysis should take into account all species that were intended to be benefitted by the prescribed Riparian Reserve widths." So in order to reduce riparian reserves, local biologists with generalized knowledge have to certify that the change will not be detrimental for hundreds of rare species. With very limited information on these species, few biologists are willing to take that risk, and few riparian buffers have been changed.

Adaptive management is the process envisioned to provide long-term management flexibility. To create confidence in adaptive management, formal adaptive management mechanisms should be included in the plan. However, the plan has neither those mechanisms nor the funding to develop them. In addition, any major plan adaptations include costly amendments to existing agency planning documents and process supervision by the courts.

The role assigned to the adaptive management areas is ambitious and may not be realistic. AMAs carry most of the burden of implementing adaptive management in FEMAT and the NFP. They are the primary place to do both biological and social experimentation, yet these areas have land-use constraints (late-successional and riparian reserves) and management constraints (target timber-harvest levels). Extensive public involvement is part of the design of AMAs, however, they were chosen without public input, and there is little public interest in some of them. Also, some AMA areas already have an extensive planning infrastructure, to which the AMA is an awkward addition. In addition, much of the public likes quick answers to questions. This does not fit well with the research process for proposals (which often includes large time delays between proposal and funding), experimental design, peer review, and publication, all of which slow down the development, evaluation, and dissemination of results. Creating meaningful partnerships between the research community and the public is proving difficult.

Future planning can improve the AMA process by soliciting public input into the location and purpose of the AMAs and by not burdening them with extensive rules and requirements that limit social experimentation. Clearly separating social experimentation from biological experimentation may also reduce some of the frustrations with process delays until scientific results are available. It is very important to resist the strong bureaucratic desire to prevent failure by standardizing AMA processes, which would destroy some of the experimental value of the AMAs.

The Need to Deal with Different Land Uses and Patterns

While encompassing in scope, FEMAT and the NFP did little to consider and examine other forest uses, such as recreation, urban/rural land interfaces, harvest of special forest products, etc. Without additional planning, the NFP is interpreted to mean that goals of species protection are paramount to other uses. For example, you remove established uses such as campgrounds if they detract from meeting plan goals. This element of the plan received little attention in the plan development, and it has received little public attention to date. However, removing uses or preventing expansion as the population and demand grow could well be one of the most contentious issues in the NFPs.

Most of the land that BLM manages in the NFP results from the government taking back the land grants made to railroads to finance railroad construction. This results in BLM managing a "checkerboard" land pattern made up of alternating sections. The NFP does not examine BLM's checkerboard management in any rigorous fashion. The same standards and guidelines that make sense in blocked ownership intuitively do not make sense in checkerboard ownership. The fragmentation of the riparian reserves on checkerboard ownership complicates management actions. For example, roads that traverse multiple-ownership lands can seldom be closed. Ecosystem management must make sense across the landscape, and we must find ways to ensure meaningful connectivity in a variety of land patterns.

The Future of Regional Planning

FEMAT provided the regional analysis that was essential to answer the scientific questions on forest management to move the public forests out of legal gridlock. The FEMAT vision was fossilized by the agencies' process for translating FEMAT to the NFP, by reducing the risk to species through the survey-and-manage process, and by lack of full funding to carry out the more creative elements. Lack of funding has not been accompanied by public understanding or acceptance of the reduction in NFP accomplishments. Nor has the NFP stopped endangered species listings or reduced the polarization of the forest debate. Future regional planning can avoid some of these problems by: including a wide variety of public in the planning process, explicitly addressing all important issues including human values, explicitly addressing risk and plan outputs that are associated with different levels of risk, and including funding as a planning element. It is very important that the public be engaged in a debate about the underlying value issues of the management plan, which may result in broad public ownership of the plan's outcome and produce the advocates who would ensure both the budget necessary for implementation and the flexibility necessary for experimentation. Only then could we realize the long-term goals of the FEMAT plan.

Policy Review

Paula Burgess and Kristin Aldred Cheek

FEMAT was a remarkable turning point in the way development and assessment of options for the management of federal forest lands is done, and in the way those lands are managed. Let us review some of what has been described previously: The scientific assessment expanded from consideration of a single species to include an unprecedented number of species and their habitats. The FEMAT process included social scientists as well as biophysical scientists. The plan embraced ecosystem management and included the new concept of adaptive management areas. Federal agencies made strides toward better interagency collaboration; the BLM and the Forest Service adopted a common management strategy. And, although the FEMAT process itself was closed, federal agencies are now, for the first time, sitting at tables with states, local governments, and tribes—all providing input into decisions about federal lands.

It is important to recognize these significant changes that the Forest Plan has brought about, but it is also important to look carefully at how far the plan went in meeting its stated goals. As described in the case study, President Clinton outlined five principles at the Forest Conference that provided guidance for the three working groups whose findings became incorporated in the plan. These principles reflected the President's desire to achieve a balanced solution to the forest problem.

- *We must never forget the human and the economic dimensions of these problems.*

The FEMAT options and the Forest Plan were driven primarily by the desire to protect the species and ecosystems of old-growth forests. Given that this protection could be achieved, there was a secondary attempt to contribute as much as possible to economic and social well-being. This goal hierarchy, established in the authorizing letter to FEMAT, did not call for a "balance" of the different considerations, as had been done in previous efforts. While it can be debated whether sufficient consideration was given to the human and economic dimensions, it cannot be debated that some people, especially families economically dependent on federal timber, were disproportionately impacted. To help lessen this blow, the Economic Adjustment Initiative has provided grants and loans to communities for such things as improving physical infrastructure, employing timber workers to restore watersheds, and building community capacity for economic diversification.

- *We need to protect the long-term health of our forests, our wildlife, and our waterways.*

One continuing criticism from the environmental community has been that the plan does not adequately protect old-growth forests but rather continues to rely on harvest from old growth for timber volume (Wilderness Society et al., 1994). The reality that policy makers face, however, is that old growth is what is available for harvesting on federal lands. If timber harvest is to continue in the short term, they are faced with limited choices.

There is risk associated with the plan, as there would be with almost any management plan. Ecosystem management seems to provide adequate protection for the forests and its inhabitants, and the plan provides for continual monitoring and evaluation of progress. It contains protective measures for hundreds of species, and a significant aim is to avoid future environmental problems. Funding for the plan to date has been adequate. If Congress continues to provide adequate funding for implementation, the plan can provide this protection. Yet long-term protection requires long-term funding, which is uncertain.

- *Our efforts must be scientifically sound, ecologically credible, and legally responsible.*

Much criticism has revolved around the FEMAT process; in particular, concerning which scientists were involved and whether they represented different scientific approaches, how quickly the assessment was completed and whether information fell through the cracks, and the credibility of the viability assessments that relied in part on professional judgment. This last criticism continues to be a source of disagreement, in the Northwest as well as in other regions where the analysis has been adapted. Some critics claim that the assessment was professional opinion misrepresented as a scientific study, while many scientists are quick to point out that professional judgment is a scientifically credible source for building understanding. It seems many of these concerns could have been avoided if the process had been more open, if not in participation, at least in communication.

The plan has proven defensible in court. The plan has also been defensible to the Fish and Wildlife Service and the National Marine Fisheries Service. Both agencies were consulted on the plan, and both concluded that the plan would not jeopardize or affect species listed as threatened or endangered (USDA and USDI 1994, appendix G).

- *The plan should produce a predictable and sustainable level of resources.*

Much debate has related to the level of timber harvest in the federal forests. The predictability of timber supply is of particular concern to rural communities and those closest to the forest products industry. The magnitude of changes called for in the Forest Plan created a need for a transition period before timber sales could reach their estimated levels. Indeed, initial harvest levels in the transition period have been below what was expected, primarily because federal agencies have to redesign timber sales to comply with the plan and must complete watershed analyses in dozens of areas critical for salmon habitat before timber sales can proceed in those areas.

A predictable and sustainable level of harvest will prove difficult to achieve beyond this transition period for four main reasons. First, the plan is complex. It created an intricate blend of different land allocations, with management guidelines for each, and the land allocations are not distinct. For example, riparian reserves are also located in matrix lands. The result is what some have called "unmanageable slivers" and has left many with questions about the practicality of timber harvest in some areas (Sessions 1994). Second, there is great uncertainty regarding the eventual impact of many of the guidelines. In looking at the Record of Decision, it is evident that many questions remain, that there are many chances for mitigation measures for species to impact harvest levels. Third, there was a failure to accurately adjust probable sale quantity (PSQ) estimates after changes were made to the preferred option following the FEMAT assessment. Fourth, conflict over timber harvesting continues in the region. Some see the difficulty with the PSQs not as a function of the details of the plan but as a question of how people in the region will react—that is, whether timber sales will be appealed or litigated.

- *We will insist on collaboration.*

One accomplishment of the plan is that federal agencies worked together in new ways. Yet managers were largely excluded from the FEMAT process. Their exclusion from the assessment process has been noted as contributing to misunderstandings and lack of commitment, further distrust between management and researchers, and difficulty in modifying the plan. Although this goal focused on collaboration within the federal government, improved collaboration among federal, state, and tribal governments was also important. The FEMAT assessment itself was not collaborative. But state and tribal governments have since been given a voice in advisory committees, and the agencies that have participated in those cross-governmental efforts since FEMAT deserve credit for opening up their decision-making processes.

Reduce Divisiveness

The reduction of divisiveness regarding forest management issues in the region was not an explicit goal of the president, but perhaps was an implied goal of the Forest Plan, which had the potential to reduce divisiveness. The president spoke with a certain conviction that the efforts coming out of the Forest Conference could break the gridlock. In one important way the efforts did that—they passed the initial legal challenges. For the first several years after implementation, the plan seems to have been successful in reducing divisiveness among the federal agencies. It also provided people with a definitive plan in place. In other ways, the plan provided ingredients to reduce divisiveness: it gave the public a clear picture of the problem; it gave people a clear cost of resolving the problem; and it gave a picture of the role of federal timber in the regional and national timber supply.

The plan that resulted from the FEMAT assessment seems to have focused the discussion but not reduced its intensity. One measure of this intensity is the comments received from public review of the draft and final environmental impact statements indicating that feelings were "still intense and still reflect all sides of the issue" (USDA and USDI 1994, 60). Why was the plan not successful at reducing this intensity over the issue? There are three possible reasons. The first is that the continued divisiveness surrounding the issue is not necessarily a reflection of the specifics of the plan itself but is a reflection of the difficulty in addressing the fundamental issues underlying the divisiveness. As mentioned above, some continued harvesting of old growth seems unavoidable if a timber program is to continue in the short term. On the other side of the issue, industry and community interests continue to be dissatisfied with the level of harvest from the federal lands. The two sides have fundamentally different ideas about the use of federal land.

The second reason that the intensity of the debate has not lessened is that the process by which the plan was put together exacerbated the conflict over the issue. The Forest Conference set high expectations for people and their potential to contribute to resolving the problem. They were let down when the doors were closed. The counterargument about the process says that people were given ample opportunity after the FEMAT process to comment. However, at that point, the range of options had already been determined, and people felt that their comments could only affect incremental, not broad changes. Another argument is that public involvement in other efforts has not necessarily led to resolution or less conflict, as Johnson et al. point out in the case study. However, it seems that the Forest Conference was presented as a preface to creating a solution that would welcome participation by many. The personal involvement by the president signaled that this situation would be somehow different. It raised the hopes of many people, but the opportunity to take advantage of that mood was lost.

Finally, the fact that a "no reserve" option was not fully analyzed by the FEMAT scientists has haunted the plan. Documentation of the plan shows that scientists felt that reserves were necessary, at least in the foreseeable future, and that they felt justified in eliminating a no-reserve option from full analysis. This left the plan susceptible to criticism about the range and creativity of options explored; critics argue that there is no definitive evidence that a no-reserve option is not viable. If a full analysis had been done, we might have a definitive answer.

The State's Involvement

The President's Forest Plan for Northwest Forests was announced in July 1993. After some deliberation, Oregon Governor John Kitzhaber endorsed the plan and became one of the few politicians to do so. He called the Forest Plan "a scientifically credible strategy that is regional in scope and carefully crafted as a compromise for protecting threatened and endangered species throughout western Oregon and Washington, while at the same time providing a stable supply of timber to local communities" (News release, Oregon Governor's office, January 22, 1998).

Governor Kitzhaber has acted on his belief in the plan and its commitments. His staff have participated in the advisory councils set up to allow the states and tribes a voice in implementation. When communities and mill operators pointed out that the projected harvest levels were not being achieved, the governor wrote the president requesting redoubled effort to meet the timber harvest targets. When it appeared that the salvage rider would open national forest lands to logging not envisioned in the Forest Plan that could threaten the integrity of the plan, the governor spoke out against that provision. When groups recently argued for a ban on road building in the roadless areas under the plan, as part of a larger national moratorium, he opposed the ban because it would "violate the commitments of the plan, and would likely threaten its very survival" (News release, Oregon Governor's office, January 22, 1998).

Regional Conservation Strategies

In the last few years, the governor, the state agencies, and the state legislature have embarked on a program "to restore our coastal salmon populations and fisheries to productive and sustainable levels that will provide substantial environmental, cultural, and economic benefits" (State of Oregon 1997). In contrast to many endangered species recovery plans that rely primarily on regulatory approaches, this plan represents a new way of restoring natural systems . . . the "Oregon Approach." It is built on both regulatory and voluntary actions. This new approach—called the Oregon Plan for Salmon and Watersheds—meshes scientifically sound actions with local watershed-based public support. When the National Marine Fisheries Service recently considered listing two groups of coho salmon as threatened under the Endangered Species Act, it agreed not to list the group on the Oregon coast and, instead, agreed to give the Oregon Plan a chance to work.

The protection of streams and watersheds in the Northwest Forest Plan form the backbone of protection of coastal salmon habitat under the Oregon Plan. Without such a strong aquatic strategy as that found in the Forest Plan, it would be exceedingly difficult to craft a salmon habitat protection strategy acceptable to both National Marine Fisheries and the landowners of the Coast Range.

Conclusion

The Forest Plan has been defensible on many grounds, but the assessment process by which it was created provided critics with much ammunition. There are many possibilities for the future of the plan. If it cannot deliver timber harvests as predicted, or if it cannot continue to provide adequate species protection (which diminishes with intervention by Congress), then it faces a questionable future. Keep in mind it was only intended to be an interim solution. Nonetheless, it is the only substantive plan on the table at this time, though, and the conceptual and political difficulties of replacing it should not be underestimated. In addition, specific components of the plan are becoming embedded in regional conservation strategies such as the Oregon Plan, which will tend to give them longevity long after the original sponsors have retired from the scene.

CHAPTER 7

Great Lakes–St. Lawrence River Basin Assessments

Case Study
Henry A. Regier

Science Review
Michael L. Jones

Management Review
James Addis

Policy Review
Michael Donahue

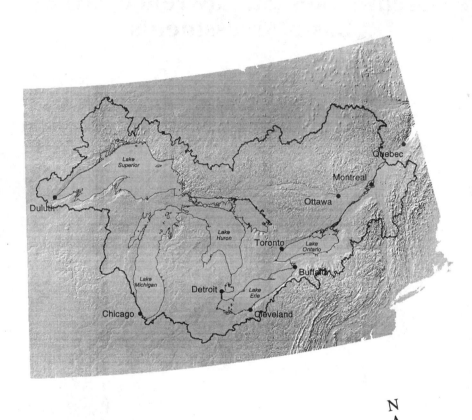

200   0   200   400   600   800 Kilometers

N

Great Lakes–St. Lawrence River Basin Bioregional Assessment

Case Study

Henry A. Regier

That the Great Lakes–St. Lawrence River basin might be meaningfully considered to be a "bioregion" is a notion that has been entertained casually by people inside and outside the basin for some time. In this chapter I explore the meaning of "bioregional assessment" and relate it, as an example of a scientific information service, to a wide range of such services as they have evolved in the basin. I then turn to the possible role of bioregional assessment in post-normal science and participatory governance, using the Great Lakes case to sketch a particular example of bioregional assessment.

What Is Bioregional Assessment?

Bioregion is defined by R. F. Dasmann (1995), one of the early proponents and users of the concept. He states:

> The value of the bioregional concept is primarily educational. If people can become familiar with the conditions for life and the other living species in the area in which they live, presumably they will be equipped to exercise greater care in the use they make of their bioregion, and may take greater interest in protecting its natural diversity. A knowledge of ecological constraints . . . could prevent the kind of misuse of land, water, and resources which has caused serious ecological damage in many parts of the world. . . .
>
> There would be obvious advantages to bringing a unified political control over the management of a single ecosystem [or bioregion], such as the Pacific Coastal Forest ecosystem. . . . In the latter part of the twentieth century it appears more feasible to seek close cooperation among the agencies involved in management of a bioregion than to attempt the redrawing of political maps.

Turning to *assessment*, an environmental encyclopedia edited by Paehlke (1995) offers definitions of environmental impact assessment, benefit-cost analysis, risk analysis, and other concepts. Most of these terms focus on unwanted adverse consequences of some human intervention in nature. Within benefit-cost analysis, both desirable and undesirable consequences are to be treated fairly, but there is usually a bias to overestimate the desirable and underestimate the undesirable from the perspective of socioeconomic development (Andrews 1995). With respect to bioregional assessment, the following description of environmental assessment in a Canadian context (Gibson 1995) may be particularly relevant: "New methodological challenges are arising from efforts to focus assessment work more effectively at the higher decision levels of planning and policy where overall guidance is given to many individual undertakings and where attention can be given to larger cumulative effects and sustainability considerations."

Based on the exegesis sketched above, *bioregional assessment* may belong to a family of terms including *sustainable development* and *ecosystem management*, superficially oxymoronic terms whose political purpose appears to be to create an opportunity for synthesis of two apparently contradictory commitments. In each of the three cases the first element—bioregional, sustainable, ecosystem—implicitly relates to the John Muir end of a policy spectrum, while the second element—assessment, development, management—relates to the Gifford Pinchot end.

So *bioregional assessment* may be the most recent manifestation of a need to foster cooperation rather than conflict between those who would preserve the cake and those who would eat it. The focus of contention in the synthesis, however, is not really a cake but rather a relatively open, self-organizing, holarchic system with a canon or rules of integrative behavior (Koestler 1978) that can be impaired, subverted, or destroyed to the disadvantage of current users (Regier and Kay 1996) and perhaps the ultimate peril of all humans. To function properly, the cultural and natural polarities of such a system need to interact synergistically in a healthy caring way rather than in a pathological careless way. If humans understand such coevolving bioregional systems, with the help of assessment services, then pathological traps are not likely to be sprung and the bioregional system may evolve in a generally desirable way. What is understood to be desirable is ultimately a political decision.

Scientific Information Services Related to Bioregional Assessment in the Great Lakes Basin

Several centuries ago, the Great Lakes Basin was dominated by the Iroquois culture (Hodenoshone) and the Chippewa/Ojibwa culture (Anishinabe). Conflicts with European Americans rearranged settlement patterns, but native cultural knowledge of the basin has endured. Many explorers and Christian missionaries provided early assessments that were widely used in attempts to induce Europeans to migrate into the basin. In the nineteenth century, a binational commission of inquiry was established by the U.S. government and the government of Great Britain, then acting on behalf of the Canadian colony. The charge of the commission was to consider and report on a wide range of topics, including "the limitation or prevention of exhaustive or destructive methods of taking fish and shell-fish in the territorial and contiguous waters of the United States and [Canada] respectively, and also in the waters of the open seas outside of the territorial limits of either country to which the inhabitants of the respective countries may habitually resort for the purpose of such fishing. The prevention of the polluting or obstructing of such contiguous waters to the detriment of the fisheries or of navigation" (Rathbun and Wakeham 1896).

This early example of interjurisdictional inquiry was among the first of numerous kinds of environmental assessments of a bioregional scale that have been conducted in the Great Lakes Basin over the past century. Assessments have been undertaken for different purposes and can be described as six different types as follows:

1. To foster immigration and development, early explorers conducted surveys of various kinds. Beginning in the nineteenth century, geological prospectors and geographic surveyors collaborated to provide regional assessments, which grew during the present century into quite detailed assessments of watershed basins in New York in the 1920s, in Michigan in the 1930s, and in Ontario in the 1940s to 1960s. In Ontario, about equal emphasis was placed on soils, forests, waters, fish, and recreational resources, with explicit efforts at synthesis to inform the separate conservation authorities subsequently created for these basins. Strong governance initiatives have not followed from these efforts, perhaps because they would have had to resolve conflicts in land-use rights. And no jurisdiction in the basin has, as yet, addressed such conflicts at all effectively. Updated versions of these watershed assessments are playing a useful role in some remedial action planning processes, as described below.

2. To develop a consensus or at least an acceptable compendium of information, interpretation, and advice on a specific issue, objective expert inquiries, such as the Rathbun and Wakeham Commission, have been established in which duly accredited experts interacted with a variety of regional experts, often in the setting of public hearings.

3. To facilitate the convening of expert inquiries and related interjurisdictional governance services, standing commissions have been created to which particular and relatively noncontentious issues were formally delegated and to which contentious issues could be referred for "general assessment" and advice to the national parties. Both the International Joint Commission (IJC) and the Great Lakes Fishery Commission (GLFC) serve such a function. The IJC was established to serve the purposes of the Boundary Waters Treaty of 1909 and has since received numerous formal "references" or requests for disinterested expert assessments and advice. Until the 1970s, assessments by these commissions were usually narrowly focused on interests and concerns of a particular socioeconomic sector. Since the 1970s, the references have become progressively more pluralistic, encompassing interacting interests.

4. To serve the interests of broader regional planning, the Great Lakes Basin Commission, an interstate compact, produced a fourteen-volume *Basin Framework Study* in the 1970s, whose framework was extended to some extent into the Canadian part of the basin.

5. To address fisheries rehabilitation in the Great Lakes, six comparative symposia—on salmonid communities in oligotrophic lakes, on percid communities in mesotrophic lakes, on sea lamprey, etc.—have provided

assessment-like compendia of information and have served as a basis for a Strategic Great Lakes Fishery Management Plan and subsequent revisions (Christie 1995).

6. Since the mid-1980s, bioregional assessment activities have gradually organized in a three-level nested holarchy, which integrates governance and assessment at the local level, subregionally, and basinwide. Fisheries issues have been addressed at three levels of governance for several decades. Environmental governance emerged later at the three nested levels, and now interactive governance is in place to address many different issues, including waterfowl, water levels, and flows, in an "ecosystemic" approach. The IJC is generally considered to be the lead agency (under the Great Lakes Water Quality Agreements, GLWQA, of 1972, 1978, and 1987), but initiatives spearheaded by others—notably by the GLFC, the interstate Great Lakes Commission, and the nongovernmental Great Lakes United—have contributed as much to emerging shared governance as has the IJC.

Nested Levels of Assessment and Governance

The six types of bioregional assessment services sketched earlier illustrate a gradual attempt to transcend the polarization between a narrow analytical focus and a broad synthetic emphasis. As information and interests have become more complex, the bioregional assessment process has self-organized into an implicit three-level holarchy. There is no deliberate, planned direction of this evolutionary process.

The three levels of governance and relevant assessment activities are the:

- *local level,* where some forty-three areas of concern (degraded inshore areas) have been formally designated under the Great Lakes Water Quality Agreements; each has a remedial action plan (RAP) under development or in implementation, with the goal of complying with water-quality standards and subsequent delisting;
- *subregional level,* where each of the five Great Lakes has a lakewide management plan under development with leadership from environmental and fisheries agencies; and
- *regional basin level,* for which the comprehensive ecosystemic state-of-the-lakes reports are being produced biennially by the two national governments, consistent with their responsibilities in the Great Lakes Water Quality Agreements.

Many people of the basin have long been conscious of their interconnectedness within the basin. Federal bureaucracies, which are centered in the national capitals of Ottawa and Washington, have often interfered with this emerging basin awareness. Federal civil servants, collaborating under the aegis of binational commissions within the basin, have suffered from the conflict of divided loyalties.

Nevertheless, many of these people have played key roles in sponsoring and conducting bioregional assessments.

As native peoples clarify their unceded rights to natural resources in many parts of the Great Lakes Basin (Satz 1991; Spangler 1997), they are developing intertribal institutions with scientific information services relevant to their interests. These services usually complete some version of a bioregional assessment as a basis for negotiations on treaty rights. For example, the Chippewas on the U.S. side have created a Great Lakes Fish and Wildlife Commission and a Chippewa-Ottawa Treaty Fishery Management Authority, with powers roughly equivalent to those of the states in three transboundary Lake Fish Committees. Other tribes or first nations are developing similar alliances to share information and assessment services for mutual benefit.

Bioregional Assessments in a Broader Cultural Context

The culture and politics of North America, and apparently of the whole globe, are well into a major transformation. Increased awareness of the massive degradation of nature is part of the motivation for this transformation. George Francis of the University of Waterloo and I picked as pivotal the year 1968 (Regier 1995), a point in time when the conventions of the older "modern" era began to give way to a revisioned "emerging" era (EE). The evolution of bioregional assessment contributes certain science-related innovations to this Emerging Era.

Some of the conventions of the Modern Era are disintegrating while new conventions are self-organizing (Drucker 1989). Simplistically, the Modern Era can be characterized by systems that are "formally-organized, hierarchic, and closed" and the Emerging Era as "self-organizing, holarchic, and open" (Koestler 1978). From one perspective, this vast transformation involves redemocratization and devolution of centralized government functions and strengthens the integration of features at regional and global levels. The influence of specialized national sectors disintegrates, at least partially, under federal cost-cutting and local democratization policies. Regional features may cross formal boundaries between states and nations, and their integration may call for informal, extraconstitutional arrangements that preclude the need for intervention by a nation's central government. For example, the Great Lakes Fisheries Commission has quietly facilitated a Law Enforcement Committee, sponsored informally by both federal governments and all the state and provincial jurisdictions, to conduct low-key joint surveillance and occasional sting operations. Of course, individuals and agencies that still adhere to the old conventions find such a transformation to be subversive of their ideals and threatening to their security. And some of these new efforts may turn out to have been ill-advised and will need to be corrected or mitigated eventually.

A similar devolution of responsibilities has been evident in the process to develop remedial action plans (RAPs) (Hartig and Zarull 1992). The international protocol that designated the areas of concern did not specify in detail what a RAP would entail, except to put a strong emphasis on cleaning up hazardous contaminants. A rather noisy evolutionary process then ensued. Some agencies assumed that an old-fashioned technocratic approach was appropriate and hired contractors to provide remedial action plans in quick order. However, the Great Lakes Science Advisory Board, a junior advisor to the IJC, was adamant that the RAP process should be one of participatory democracy involving all willing stakeholders within an ecosystem approach. The technocrats initially had better control of available funds and thus briskly moved ahead, until the IJC recommended more participatory democracy in the planning process. Eventually, however, with lower levels of funding and reduced staff, the more adaptive technocrats and the more sober democrats began actively cooperating.

One of the themes of the Modern Era in North America that relates particularly to the governance of human interactions with the natural environment is that of progressivism. Nelson (1995) aptly describes the emergence of the Progressive movement in the early twentieth century, when officials turned away from policies favoring disposal of the public lands and resolved to keep the lands under federal control, with a strict separation of administration and politics. An early champion of progressivism, Gifford Pinchot believed that the government would do much better than the private sector in achieving the scientific management of natural resources. Science and technology would reveal the one correct answer that the experts would apply. The Progressive program for expanding the scope of government did not reflect a high level of confidence in democratic decision making.

However, Nelson points out that the government role of command-and-control envisioned by Progressive theorists of scientific management was much more difficult to accomplish than they ever really understood. Progressives tended to assume that all necessary economic information would be available to government decision makers. In practice, however, important details were almost always missing. Information was costly to obtain, and it was not always consistent across an increasingly diverse society. Nelson concludes that eventually the idea of one set of values and one administrative design for an entire nation seemed not only flawed but increasingly an outright impossibility.

Devolution of Governance in the Great Lakes–St. Lawrence River Basin

Almost all of the surface waters of lakes and rivers in the Great Lakes–St. Lawrence River basin are in the public domain, yet a relatively small proportion is under federal control. National and international transboundary issues were once a jealously guarded prerogative of the federal governments, but extraconstitutional governance has replaced some of the earlier federal control. "Land-use planning" at a municipal level, under guidelines from higher levels of governance, has haltingly evolved to relate more responsibly to ecosystemic values. Optimistically, this could lead to better governance at the municipal level to complement the strengthening roles of states and provinces.

For example, municipalities are increasingly taking responsibility for land-use planning in flood zones. Water levels and flows rise and fall unpredictably with the amount of precipitation over several years. During periods of low water, uninformed people may purchase property that would be subject to flooding or storm damage during a subsequent period of high water. For decades, the U.S. Army Corps of Engineers and its Canadian counterpart were funded by their federal governments to do something about such unfortunate events after they occurred. Eventually, after several references to the IJC, an inference emerged that municipal land-use zoning should clearly identify at-risk lands, and any prospective owner should be advised that no compensation for losses would be available from federal, state, or provincial governments. This inference has not been welcomed by those who oppose land-use planning as a matter of higher principle than the acceptance of implicit government subsidies. However, scientific information services are increasingly helping municipalities identify at-risk lands and develop ecosystem-based zoning.

There have been major changes with respect to science information service in the basin during recent decades. Until some twenty years ago, a handful of federal agencies in each country dominated science information in the region. Many universities in the basin made strong contributions to scientific understanding, with much of the research—as opposed to educational training—funded by the federal governments. Recently, the states and provinces have expanded their roles through development of expert staff, and private enterprises have funded increasing proportions of the science-based assessments through a rapidly evolving network of private consulting firms.

Generally, the lower the level of jurisdiction, the less clearly identifiable is the role of science. For example, much of the science mobilized for the RAPs may be participatory in nature and conducted "in real time" as a need for the information becomes clear to the stakeholders. This assumes the preexistence of much relevant information among stakeholders that can be mobilized and tested quickly. Native people are demanding that cultural knowledge, especially relevant to their traditional interests, be accepted respectfully in such a sharing process.

The devolution of responsibility for Scientific Information Services (SIS) from the federal to lower levels of governance has been facilitated more or less deliberately by a series of interstate and binational commissions:

- *The Great Lakes Basin Commission (GLBC)*, a U.S. federally funded basin commission with informal participation by Ontario, served a devolutionary function in the 1970s by involving people from all levels of government; it was a casualty of Reagan administration decisions in the early 1980s.
- *The Great Lakes Commission (GLC)*, a U.S. interstate compact now with informal participation by Ontario and Quebec, evolved further to continue some of the initiatives of the defunct GLBC together with its own emphases on

marine transport, on redevelopment of degraded urban shore zones, and on rehabilitating the Rust Belt's economy.

- *The International Joint Commission,* a treaty-based binational agency to which some responsibilities related to water quantity and water quality were delegated, has played a particularly active role in the emergence of SIS at state and provincial, municipal, and private corporation levels and thus has fostered the prerequisite of appropriate expertise for the devolution of governance responsibility from the federal level to lower levels. Formal "references," the activities of several advisory boards, and the Great Lakes Regional Office (see Donahue 1995) have all contributed to implementation of binational agreements and indirectly to the devolution of governance.
- *The Great Lakes Fishery Commission,* a convention-based binational agency to which responsibility for control of the exotic sea lamprey was delegated, also had a formal responsibility to coordinate and undertake research related to sustainable fisheries on common stocks. It has been effective in facilitating collaboration in SIS among the states and province, as well as in extraconstitutional cooperative decision making, e.g., by lake committees with respect to shared stocks. A lake committee includes formal delegates from all the relevant state, provincial, and tribal jurisdictions but not from the federal levels. Advisory boards and an office with appropriate experts have been active here too.
- *The Great Lakes Indian Fish and Wildlife Commission,* created jointly by a number of tribes of Chippewa with usufructuary rights related to many species over large areas in the Lakes Huron, Michigan, and Superior watersheds of the United States, is increasingly collaborating in initiatives of the other commissions. Much of its effort has been devoted heretofore to assisting tribes in their litigation to have nonextinguished treaty rights clarified and enforced.

Many individuals have had personal professional involvement with more than one of these interjurisdictional commissions, and a few individuals have been connected with all of them. In effect, these people serve as a largely invisible informal network between the commissions and other institutions. They are usually encouraged to serve such a function by their employing institutions. The efficacy of such a self-organizing capability must reflect a satisfactory level of trust among the many stakeholder groups of these jurisdictions (Fukuyama 1995; Hartig et al. 1995; Hartig et al. 1996).

When federal governments dominated science information services in the Great Lakes, decision making was strongly sectoral, with each sector (water quantity, water quality, air quality, sustainable fishing, sea lamprey control, waterfowl) largely isolated from the others. Such isolation was considered to be a good thing under progressivism, providing a sense of utilitarian efficiency in the delivery of government services, reminiscent of the early factory model designed by Frederick Winslow Taylor. Ecosystem connections between the different sectors were ignored or even dismissed indignantly and self-righteously by many progressivist scientists. But with devolution to local levels of governance, economies of scale did not justify permanent staffs with all the types of expertise included at higher levels of government.

Part of the weakening of the federal sector-based approach to binational issues in the Great Lakes Basin was perhaps due in part to a degree of mismatch in how the two federal systems were organized. The Washington way of defining the central purpose for an agency was not identical to the Ottawa way. There was a further mismatch between the sectors of federal government and the sectors of states or provinces of the basin. Still further, there were mismatches among the states of the basin as well as among the provinces. The story goes on to the level of the municipality, and in a different cultural dimension to the tribes of the native people. Thus, the emergence of interjurisdictional involvement inevitably put a strain on sectoral agencies and triggered their reorganization. In some ways, bureaucratic manipulators gained ascendency over science-based experts. The necessary "energy of activation" for scientific and technical cooperation often came in the form of new funds for a particular interjurisdictional purpose, sometimes quite large amounts of funds. Collaborators were scorned as "biostitutes" by fundamentalistic purists of the old conventions who tried to remain aloof.

The Ecosystem Bioregional Approach in the Great Lakes Basin

The devolution of governance has been occurring in part because federal governments have decided to reduce their levels of expenditures on Great Lakes environmental issues. It also occurred because the transition from the Modern Era to the Emerging Era brought an awareness that top-down sectorally specialized federal programs, each with good intentions, sometimes threatened to do more harm than good at local and regional levels of the basin. Within the basin a political commitment to an "ecosystem approach," beginning in about 1968 (Caldwell 1970), expanded and intensified at all levels of governance (Hartig and Zarull 1992; Regier 1992; Christie 1995). This approach created major problems, even for some federal agencies that were created in the early 1970s with a broader focus. Agencies related to environmental protection and coastal zone management were constrained to work within the old conventions by traditionalists within these ostensibly reform-oriented new agencies.

These internal contradictions led to a number of bizarre events. At one time traditionalists within the federal agencies responsible for water quality attempted to close down the Great Lakes Regional Office of the International Joint Commission, which was responsible for facilitating the implementation of the Great Lakes Water Quality Agreement. The regional office was apparently too energetic in implementing the ecosystem approach, to the consternation of the diehards of the federal agencies who were combating the ecosystem approach, partly through inertia and partly through their continuing commitment to the old ideals of progressivism. They had hoped, I suppose, to slow down the evolution of the Emerging Era, to turn back the clock to the traditions of the 1950s, say, when the strongly and narrowly disciplined scientific expert would once again have executive responsibilities, formally delegated.

Scientific Information Services in the New Civics

Until 1987, the Great Lakes Water Quality Agreement directed the federal governments to take the lead in implementing remediative programs and report publicly to the IJC, which would then issue its own version of what had been achieved. In 1987, the federal governments undertook to issue their own joint report on the state-of-the-lake ecosystems to the public, independently of the IJC. The intent was to document not the progress of particular programs but rather the ecosystemic consequences of all the corrective work and of uncontrolled cultural stresses such as introduction of exotic species. Since 1994, State of the Lake Ecosystems Conferences (SOLECs) have been convened every two years and have issued reports that are clearly more comprehensively ecosystemic than the concurrent biennial IJC reports, which have emphasized hazardous contaminants and their teratogenic and carcinogenic effects. In 1997, the IJC requested authority to prepare environmental reports on the state of several watersheds straddling the Canada–U.S. border, including the Great Lakes–St. Lawrence River basin, which may lead to a yet broader State of the Basin Ecosystem Conference in the future.

A schema sometimes termed "comprehensive rational action" was a favorite of planners, researchers, decision makers, and managers under progressivism. It was a kind of generalization of the old way of doing science with its rote linear sequence of objective, methods, analysis, findings, and conclusions. The broader context in which the hypothesis or problem emerged was irrelevant to the science-based technical expert, and the question was considered resolved when a predetermined criterion for closure, however artificial, was met. With comprehensive rational action, the roles of the different types of experts were clearly specified and arrayed linearly from the mysterious first appearance of a problem to its miraculous full resolution. Such rational action employed some kind of comprehensive quantitative model—e.g., regression relationship, benefit-cost, input-output, ecological simulation—to coerce coherence among the collaborators. Toward the end of the Modern Era, such schemata included linear feedbacks so that the increasing number of pluralistic concerns could be fed into the process in an ad hoc way, sometimes through an advisory body with weak responsibilities and with little expectation of influencing any pending decision.

Numerous versions of strategic planning emerged in the 1950s and 1960s, usually based on some system concepts in cybernetics. Most cyberneticians treated the system as sufficiently closed and with predominantly linear behavior to make it possible to model a comprehensive, rational strategic plan. However, few such plans were ever implemented, and it is now generally accepted that such strategic planning was a failure (Westley 1995).

With the Emerging Era has come the broadening of computer-based simulation models to attempt participatory real-time research and decision making. These are used as a heuristic or self-learning device, rather than to provide definitive numerical measures on which to base management decisions directly. A version of "adaptive environmental assessment and management" (Holling 1978) was applied this way beginning in the early 1980s, with the help of Carl Walters, Michael Jones, and Joseph Koonce, to the problems of integrated management of sea lampreys. The first generation of simulations, built from information from many stakeholder groups, was followed by several revisions and extensions, providing consensus information toward lamprey management through a collaborative process.

Reciprocally Responsible Action

Innovations such as the work with sea lamprey management may serve as a bridge from the old comprehensive rational action of the Modern Era to the "reciprocally responsible action" of the Emerging Era (Westley 1995). Reciprocally responsible action (RRA) emphasizes collaboration within a context of action and consequences. The RRA process lacks the clear beginning and ending characteristic of comprehensive rational action (CRA). Activity focused on a particular issue may be transferred sequentially through three stations, something like a meal passing through the gut due to peristaltic action; or the issue may bounce back and forth, or spiral a number of times. Some big, bad, but relatively simple issues, like Godzilla, may be transferred sequentially, similarly to CRA perhaps. In the Great Lakes Basin, the phosphate issue was something like the problem with Godzilla. Other insidious, complex issues, like the problem with Andromeda Strain, may spiral erratically through several cycles (Bulkley, Donahue, and Regier 1989). The hazardous contaminant issue has been like the Andromeda Strain.

In figure 7.1, I have elaborated the somewhat simpler schema by Westley so that each of the three "stations" has three "substations." This schema may be used to sketch the dynamics of a recent IJC reference concerning fluctuating levels of lakes and flows of rivers in the basin. The many individuals who became involved in this latest reference can, with apologies about oversimplification, be aggregated into three "actor groups":

- *nature stewards*—the people who valued nature and coastal wetlands that would be harmed by artificial shoreworks;
- *riparian owners*—the people who owned built structures and things along shore that suffered harm at excessively high or low levels and wanted more protective works; and
- *consensus builders (or conflict resolvers)*—the people in many organizations of governance who were caught in between the other two actor groups.

Nature stewards, at the signifying station, declaim from high moral ground the need for preserving natural wetlands. At the legitimizing station, they use

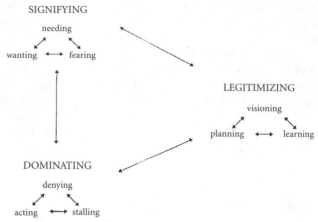

Figure 7.1. *A schematic classification of governance within the new inquiry relevant to the ecosystem approach in the Great Lakes–St. Lawrence River basin. Governance may be perceived as involving stations on a path, along which some movement, forward and backward, occurs. Each station necessarily involves some interaction between stakeholders for a particular issue. This figure is an elaboration of some concepts presented by Westley (1995).*

visioning to portray the benefits of wetlands and the consequences to humans of degrading the biosphere on which humans are ultimately dependent. At the dominating station, they use zoning processes to protect heritage natural areas from harm that would follow efforts to artificially control levels and flows. Thus, they deny all such efforts, or shift responsibility onto the proponents of such measures, to demonstrate a priori that no adverse consequences would result and to make legally binding commitments to correct adverse consequences if they were to occur.

Riparian owners signify their desire for security in locations where they had not taken adequate precautions for major fluctuations in levels and flows. They legitimize their demands for protective works on the basis of many precedents, in part because no agency of governance had provided legally binding warnings or had posted exclusionary boundaries of the floodplain through land-use zoning. They dominate by mobilizing political support through the conventional democratic process to achieve their desires. They are supported by the construction industry that had benefited from largely government-funded activities in earlier years and was now facing cutbacks due to the changing priorities of the Emerging Era.

Conflict resolvers, mostly in government agencies, signify their fears of economic losses, social disharmony, and political disruption. Politicians become increasingly displeased with the prospect that they will need to earn their pay and would sooner see that delegated "managers" of the civil service take the heat. They legitimize by conducting various kinds of "assessments"—such as environmental impact, benefit-cost, social impact, hazardous risk, human health, and political feasibility—all as an interminable learning process that has been variously caricatured as "paralysis through analysis" or "inaction through abstraction" or "perspiration through participation." They dominate by stalling direct action at higher levels of government, with hopes that lower levels will solve the problem by proactively zoning floodplains to exclude buildings and commercial activities that would be seriously harmed by large fluctuations in levels and flows and by retroactively providing assistance in relocating structures from hazardous sites.

Admittedly, this may all be too simple and cartoonish. But if it has some heuristic utility, then the question may be asked about the role of science in such reciprocally responsible actions of the new governance. The wanting-planning-acting sequence, as sketched earlier for the utilitarian riparian owners, can make some use of the old scientific offerings of comprehensive rational action under progressivism. How is science serving the two other sets of actors within the ecosystem approach as a theme of the Emerging Era?

Incidentally, in recent bioregional assessments within the Emerging Era in our basin, women have been most active in the needing-visioning-denying sequence and least active in the wanting-planning-acting sequence, as these terms are used above.

RAPs and Their Prospects

Hartig and Zarull (1992) and Hartig et al. (1995) have provided much information on institution building with respect to RAPs. It is apparent that the RAPs are not evolving in the same way across the whole set of forty-three. Whether many of them will in fact lead to the thorough rehabilitation of their degraded areas of concern is still in doubt (MacKenzie 1996). However, after careful consideration of current prospects, local people have not written off these waters; a sense of ownership and commitment to rehabilitation is still apparent in most of these areas of concern.

In the 1960s, the waterfront areas of most cities of the basin were in horrid condition. People and politicians turned their backs to the shores and the lakes. With remediation action plans and more attention to industrial slums, municipal leaders became more involved. This has led to the creation of the International Great Lakes–St. Lawrence Mayors' Conference and its recognition of the link between ecological condition and economic development of the region's urban harbors and waterfronts.

The Ecosystem Approach and the New Inquiry

Versions of an ecosystem approach to managing the watery parts of the Great Lakes–St. Lawrence River basin have been self-organizing in recent decades. Of the score or more organizations and institutions that have played lead roles in this evolutionary process, three are highlighted below.

The International Association of Great Lakes Research (IAGLR) emerged from annual scientific conferences starting in the 1950s and convened alternately at the University of Michigan and the University of Toronto. David Chandler

(University of Michigan) was for years the leader, in a friendly and cooperative way. Instead of publishing conference proceedings, the association now sponsors a journal for primary scientific literature. Collectively, the contents of the conferences and journal over the years illustrate the sequential dominance of different disciplines in addressing the problems of the basin. Physical sciences and engineering were prominent early in the series, followed by chemists, biologists, ecologists, and, into the twenty-first century, social scientists. With increasing frequency, women are leading the association.

Great Lakes United (GLU) emerged in the 1980s as a loose federation to lobby senior governments on environmental issues. The first two leaders of the GLU were Thomas Washington, subsequently president of the U.S. National Rifle Association, and Abbie Hoffman, then a fugitive from justice because of events in Chicago in 1968 and working as "Barry Freed" in a St. Lawrence River activist group. In a putsch led by women, leadership of GLU was wrested from Washington and Hoffman (who loathed each other), and it became an innovative, if rather well-behaved, federation of activists within IJC circles. In the mid-1980s, the GLU organized a series of public hearings in preparation for revision of the Great Lakes Water Quality Agreement. GLU's public success led to an invitation to directly participate in the negotiation of that agreement. The resulting Protocol of 1987 reflects quite an evolutionary advance over the 1978 version in moving the international agreement into the Emerging Era.

The Canada–U.S. University Seminars (CUSIS) began in 1971 as a cooperative initiative of academics interested in emerging governance in the basin. Each has been a three-year process, generally following the course sketched for reciprocally responsible action. For example, the formal process for remedial action plans that followed the 1987 Protocol to the Great Lakes Water Quality resembles a process that first emerged from work some years previously at Green Bay, Wisconsin, by CUSIS leaders. Later, the presence of CUSIS leaders on the Great Lakes Science Advisory Board may have facilitated the process of participatory democracy developed from the Green Bay approach. Incidentally, most CUSIS funding comes from private foundations. Academics "volunteer" their time, and graduate students receive some financial support for thesis-related research. Government agencies permit their policy-oriented experts to participate but do not interpret their mandate to include direct funding.

Numerous other organizations with interest in an ecosystem approach may be found at work in the basin (Hartig and Zarull 1992; Regier 1992; Lerner 1993; Francis and Regier 1995; MacKenzie 1996: Shimizu et al. 1997).

Mindscapes Competing and Collaborating in Bioregional Assessment

The Emerging Era is a result of evolution rather than revolution. This can be demonstrated conceptually with and influence by of Magoroh Maruyama (Caley and Sawada 1994), that four "mindscapes" occur prominently in many cultures in the world, including ours. In science, some people have preferred one mindscape, or a hybrid of two, over the others. I have sketched these mindscapes, comparing their notion of causality, giving them the designations of: hierarchism, independence, stasis, and genesis.

Hierarchical mindscapes (H) choose nonreciprocal notions, in which causality can only act deterministically and linearly without causal loops. Such science is applied in building dams, canals, and harbors, always incorporating a safety factor to cover those features of reality that do not conform to this causal mindscape.

Independent mindscapes (I) choose independent event models in which the most probable states are random distributions of independent events, and nonrandom features are only temporary and subject to decay. Such science performs risk assessments of the effects of human activities on nature, assuming that any linkages between effects are transient, without any systemic synergism. In this mindscape, the term "ecosystem" has little meaning.

Stasis mindscapes (S) choose homeostatic causal models, in which causality occurs deterministically with preordained organization maintained in causal loops in closed systems. Such cybernetic science perceives natural areas to be preserved as sealed ecological systems that can be locked up to remain in a particular wild state in perpetuity.

Genesis mindscapes (G) choose a morphogenetic causal loop model in which probabilistic or deterministic causal loops in an open system increase their patterned heterogeneity toward more organized complexity, with features that are not fully predetermined or predictable. This mindscape sees life as a nonequilibrium phenomenon in open self-organizing systems. Such science facilitates rehabilitation of a degraded system by working with nature to transform it into a desirable ecosystem with robust self-organizational and survival capabilities, as well as some surprising emergent features.

The H and I mindscapes are compatible with reductionism, and the S mindscape with old-fashioned holism. The G mindscape transcends these toward a contextual narratology, which uses a case study, a dynamic simulation, or an evolutionary or revolutionary story. From the G mindscape, conventional analysis, such as with hierarchical or independence mindscapes, may be useful as a particular kind of simplification. Narratives are widely used in all cultures, particularly in traditional cultures (Hanna, Folke, and Maeler 1996).

Technocratic managers of natural resources during the Modern Era preferred the deterministic hierarchic mindscape and used the independence mindscape to apply "error statistics" and risk analysis. Parks managers preferred the stasis mindscape. The Emerging Era seems to be dominated by the genesis mindscape, by those, for example, who take an ecosystem approach to bioregional assessments. Again, the other three mindscapes may be perceived as useful simplifications of the G mindscape narratology, depending on which dimensional features are simplified for study. In such cases, the onus is on the investigator to declare why and how the simplification was done (or so a proponent of the genesis mindscape might argue).

If pluralistic democracy is the dominant ideology in the Emerging Era, then a tolerant commitment to a G mindscape would imply continuing roles for the other mindscapes. They are likely to persist indefinitely anyway. Yet one should always be on guard to prevent domination by the fundamentals of any one mindscape. In a way, a G mindscape transcends some excessively constraining fundamentals of the Modern Era and so provides a better base for balanced reciprocity with respect to linked rights and responsibilities in our interactions with each other and with other creatures and things. In other words, American democracy continues to evolve.

Closing Comments

This chapter has been an exercise to describe some features of bioregional assessments in the Great Lakes–St. Lawrence River basin using concepts and schema of a new inquiry. It is an attempt to see into the present. It may add stimulus for networking among many people who have long been at home in the mindscape of narratological context and may not yet recognize their fellows in this Emerging Era.

Finally, an admonition from C. West Churchman (1979, 26): "One last remark on my learning journey, which is a reflection of older but not necessarily wiser years. The mood of the journey was serious, deadly serious. Now there is one thing we all must learn about seriousness—namely, that it is ridiculous. None of this story can be told without recognizing the underlying humor of the systems approach. It forever runs the risk of being caught with its pants down in [the] Park."

Surprise happens. So bioregional assessments should emphasize perceptions of turbulence with some unpredictable inevitabilities.

LITERATURE CITED

Addis, J., R. Eckstein, A. Forbes, D. Gebken, R. Henderson, J. Kotar, B. Les, P. Matthiae, W. McCown, S. Miller, B. Moss, D. Sample, M. Staggs, and K. Visser. 1995. *Wisconsin's biodiversity as a management issue: A report to Department of Natural Resources managers.* Madison: Wisconsin Department of Natural Resources.

Andrews, R. N. L. 1995. Benefit-cost analysis. In *Conservation and environmentalism: An encyclopedia,* edited by R. Paehlke. New York: Garland Publishing.

Bulkley, J. W., M. L. Donahue, and H. A. Regier. 1989. The Great Lakes Water Quality Agreement: How to assess progress toward a goal of ecosystem integrity. In *Post-audits of environmental programs and projects,* edited by C. G. Gunnerson. New York: Environmental Impact Assessment Research Council, American Society of Civil Engineers.

Caldwell, L. K. 1970. The ecosystem as a criterion for public land policy. *Natural Resources Journal* 10:203–221.

Caley, M. T., and D. Sawada. 1994. *Mindscapes: The epistemology of Magoroh Maruyama.* London: Gordon and Breach.

Christie, W. J. 1995. The ecosystem approach to managing the Great Lakes: The new ideas and problems associated with implementing them. *University of Toledo Law Review* 26: 279–304.

Churchman, C. W. 1979. *The systems approach and its enemies.* New York: Basic Books.

Coleman, J. S. 1988. Social capital in the creation of human capital. *American Journal of Sociology* 94:95–120.

Dasmann, R. F. 1995. Bioregion. In *Conservation and environmentalism: An encyclopedia*, edited by R. Paehlke. New York: Garland Publishing.

Deming, W. E. 1982. *Out of the crisis.* Cambridge: Massachusetts Institute of Technology, Center for Advanced Engineering Study.

Dillon, P. J., and F. H. Rigler. 1974. A test of a simple nutrient budget model predicting the phosphorus concentration in lake water. *J. Fish. Res. Bd. Can.* 1771–1778.

Donahue, M. J. 1995. The IJC and its advisory boards: Setting the record straight. *Journal of Great Lakes Research* 21:405–407.

Drucker, P. F. 1989. *The new realities.* New York: Harper and Row.

Francis, G. R., and H. A. Regier. 1995. Barriers and bridges to the restoration of the Great Lakes Basin ecosystem. In *Barriers and bridges to the renewal of ecosystems and institutions*, edited by L. H. Gunderson, C. S. Holling, and S. S. Light. New York: Columbia University Press.

Fukuyama, F. 1995. *Trust: The social virtues and the creation of prosperity.* London: Hamish Hamilton.

Gibson, R. B. 1995. Environmental assessment (Canada). In *Conservation and environmentalism: An encyclopedia,* edited by R. Paehlke. New York: Garland Publishing.

Gunderson, L., C. S. Holling, and S. S. Light, eds. 1995. *Barriers and bridges to the renewal of ecosystems and institutions.* New York: Columbia University Press.

Hanna, S. S., C. Folke, and K. -G. Maeler, eds. 1996. *Rights to nature: Ecological, economic, cultural, and political principles of institutions for the environment.* Washington, DC: Island Press.

Hartig, J. H., N. L. Law, D. Epstein, K. Fuller, J. Letterhos, and G. Krantzberg. 1995. Capacity-building for restoring degraded areas in the Great Lakes. *International Journal of Sustainable Development and World Ecology* 2:1–10.

Hartig, J. H. and M. A. Zarull, eds. 1992. *Under RAPs: Towards grassroots ecological democracy in the Great Lakes Basin.* Ann Arbor: University of Michigan Press.

Hartig, P. D., J. H. Hartig, D. R. Lesh, D. G. Lowrie, and G. H. Wever. 1996. Practical application of sustainable development in decision-making processes in the Great Lakes Basin. *International Journal of Sustainable Development and World Ecology* 3: 31–46.

Holling, C. S., ed. 1978. *Adaptive environmental assessment and management.* London: John Wiley and Sons.

Koestler, A. 1978. *Janus: A summing up.* London: Hutchinson.

Koonce, J. F., and M. L. Jones. 1994. *Sustainability of intensively managed fisheries of Lake Michigan and Lake Ontario: Final report of the SIMPLE task group.* Ann Arbor, MI: Great Lakes Fishery Commission.

Lee, K. N. 1993. *Compass and gyroscope: Integrating science and politics for the environment.* Washington, DC: Island Press.

Lee, K. N. 1995. Deliberately seeking sustainability in the Columbia River basin. In *Barriers and bridges to the renewal of ecosystems and institutions,* edited by L. H. Gunderson, C. S. Holling, and S. S. Light. New York: Columbia University Press.

Lerner, S., ed. 1993. *Environmental stewardship: Studies in active earthkeeping.* Department of Geography Publications Series no. 39, Ontario, Canada: University of Waterloo.

MacKenzie, S. H. 1996. *Integrated resource planning and management: The ecosystem approach in the Great Lakes basin.* Washington, DC: Island Press.

Naisbitt, J. 1982. *Megatrends: Ten new directions transforming our lives.* New York: Warner Books.

Naisbitt, J., and P. Alburdene. 1990. *Megatrends 2000: The next 10 years . . . Major changes in your life and world.* London: Sidgwick & Jackson LTD.

Nelson, R. H. 1995. *Public lands and private rights: The failure of scientific management.* Lanham, MD: Rowman and Littlefield Publishers.

Paehlke, R., ed. 1995. *Conservation and environmentalism: An encyclopedia.* New York: Garland Publishing.

Rathbun, R., and W. Wakeham. 1896. *Report of the Joint Commission relative to the preservation of the fisheries in waters contiguous to Canada and the United States.* Washington, DC: U.S. Government.

Regier, H. A. 1992. Ecosystem integrity in the Great Lakes basin: An historical sketch of ideas and actions. *Journal of Aquatic Ecosystem Health* 1:25–37.

————. 1995. Ecosystem integrity in a context of ecostudies as related to the Great Lakes Region. In *Perspectives on ecological integrity,* edited by L. Westra and J. Lemons. Dordrecht, The Netherlands: Kluwer Academic Publishers.

Regier, H. A., and J. J . Kay. 1996. An heuristic model of transformations of the aquatic ecosystems of the Great Lakes–St. Lawrence River basin. (Kluwer Academic Publishers). *Journal of Aquatic Ecosystem Health* 5:1–19.

Satz, R. N. 1991. Chippewa treaty rights. *Wisconsin Academy of Sciences, Arts and Letters, Transactions,* vol. 79, no. 1.

Schindler, D. W. 1974. Eutrophication and recovery in experimental lakes: Implication for lake management. *Science* 184:897–899.

Senge, P. 1994. *The fifth discipline: The art and practice of the learning organization.* New York: Currency Doubleday.

Shimizu, R., H. A. Regier, J. J. Kay, and R. E. Ulanowicz. 1997. Government organizations and Great Lakes rehabilitations. In *Saving the seas: Values, scientists, and international governance,* edited by L. A. Brooks and S. D. VanDeveet. College Park, MD: Maryland Sea Grant College.

Spangler, G.R. 1997. Treaty fisheries in the upper midwest. *Water Resources Update,* (Universities Council on Water) 1:54–64.

Toffler, A. 1980. *The third wave.* New York: William Morrow.

Trist, E. 1983. Referent organizations and the development of interorganizational domains. *Human Relations* 36:269–284.

Vollenweider, R. A. 1966. Advances in defining critical loading levels for phosphorus in lake eutrophication. *Mem. Ist. Ital. Idrobiol.* 33:53–83.

Westley, F. 1995. Governing design: The management of social systems and ecosystems management. In *Barriers and bridges to the renewal of ecosystems and institutions,* edited by L. H. Gunderson, C. S. Holling, S. S. Light. New York: Columbia University Press.

Woodley, S., J. Kay, and G. Francis, eds. 1993. *Ecological integrity and the management of ecosystems.* Delray Beach, FL: St. Lucie Press.

Science Review

Michael L. Jones

A comprehensive assessment of the issues that have preoccupied environmental scientists, managers, and policy makers in the Great Lakes basin over the past thirty to forty years, even from a purely scientific perspective, is an overwhelming task. The Great Lakes Basin includes more than 20 percent of the earth's surface freshwater resources, is home to millions of people, and is the concern of innumerable agencies and jurisdictions, national and sub-national. The resources are vast; lots of people want to use them, and lots of institutions attempt to manage them or to influence their use, often in ways that conflict with one another. The past forty years have seen many issues involving the interaction of science, management, and policy in the Great Lakes—far too many to examine one by one. I will focus on a few particularly prominent issues. My goal is to use these examples to develop a perspective about the evolving role of science in policy development.

Historical Perspective: Obvious Problems, Obvious Science

In the 1960s, the environmental problems in the Great Lakes were finally recognized as being severe. Much of this recognition had to do with the developing environmental movement, but there were some water-quality problems in the Great Lakes that had become so obvious that they could not be ignored any longer. Three of the most prominent water management problems of that time were: (1) eutrophication of shallow embayments throughout the Great Lakes and particularly the central and western basins of Lake Erie; (2) excessively high levels of persistent toxic contaminants found in the water, sediments, and, most notably, the biota of the Great Lakes; and (3) the collapse of lake trout fisheries throughout the basin, due significantly to the effects of the invading sea lamprey. These problems exhibited obvious symptoms. Eutrophication caused algal blooms leading to severe taste and odor problems in drinking water, unsightly accumulations on recreational beaches, and even severe oxygen depletion in some areas, leading to fish kills. Toxic contaminants led to sufficiently high residues in piscivorous fishes and birds to cause widespread fishery closures and obvious reproductive and developmental abnormalities in birds such as bald eagles and cormorants. The lamprey decimated lake trout populations that were probably already stressed by overfishing and ultimately led to the collapse of locally important commercial fisheries.

Remarkable progress has been made in addressing these problems, although it would be wrong to suggest that they no longer trouble us. Remediation required considerable policy and legislative commitment on the part of many institutions. In developing the policies and regulations that have combated eutrophication, toxic contamination, and lamprey, these institutions relied greatly on the scientific community to unequivocally link cause and effect. Science played a key role, providing the justification for and means of control.

The eutrophication example is particularly clear. Once the problem was identified, there was considerable debate about its cause. The superb empirical studies of David Schindler (1974) and the development of validated models by Vollenweider (1966), and Dillon and Rigler (1974), demonstrated "beyond reasonable doubt" that phosphorus loadings to lakes were the primary cause of eutrophication and provided a basis for judging what acceptable levels of phosphorus loadings might be.

These problems are examples of the kinds of environmental problems that the science of what Regier refers to as the Modern Era were able to successfully address. While not trivial, they are relatively uncomplicated. The chain of cause and effect is indeed a chain, rather than an intricate web of (nonlinearly) interacting factors. Because of this, traditional scientific research can provide answers, and policy-making institutions can boldly assert that "the research has provided a clear answer to this question." Perhaps when the problems are acute and the enemy is known to all, "progressivism" works.

Today: Complex Problems, Uncertain Science

Today's environmental issues on the Great Lakes are rather different. Most important, they are far more complex, for two related reasons. First, they tend to involve more subtle effects, often not so severe that all agree a problem really exists. Second, the linkages between cause and effect are far more indirect and thus uncertain than was the case for the phosphorus-eutrophication connection. Many of these more recent problems involve ecosystemic interactions—the interplay among the numerous species and communities, including humans, that the Great Lakes ecosystem comprises. Examples include the management of offshore fish populations, including an intensively managed mix of native and non-native salmonine, the impacts of myriad nearshore habitat alterations on the aquatic communities that depend on those habitats, and the recurring threats to Great Lakes native species from accidental introductions of exotic species.

These problems do not have simple solutions. This is in part because of the devolution of top-down authority and the legitimization of a multitude of stakeholders which Regier talks about. The institutional milieu within which decisions have to be made (multiple governments, influential private interests, numerous private interest groups) demands recognition of a diverse array of interests, many of which are (or at least are perceived to be) profoundly conflicting. As a result, the scientific questions being asked are technically more complex. Science cannot provide the answers; the complexity means that it is frequently not possible to predict the outcome of even major shifts in management direction. The most pervasive feature of these problems is the profound uncertainty associated with the web of cause and effect that governs their behavior in response to management intervention. Progress in scientific understanding of the Great Lakes bioregion and its constituent ecosystems has been far outpaced by the evolution of complexity in both the institutions with a stake in Great Lakes management and in the issues for which they seek scientific guidance. As a result, science cannot hope to play the role it has played in the past.

A New Role for Science

Where, then, does science fit in? I believe science still has a crucial role to play in resource management policy development, in the Great Lakes and elsewhere. The role demands a rather fundamental shift, however, in how scientists bring their expertise to the policy table. Scientists must convey to decision makers and stakeholders alike just how uncertain we are about these ecosystems we are trying to manage. We have to get over the hurdle that would have us believe that to admit uncertainty is to admit weakness, a hurdle that has deep roots in our adversarial system of decision making. Scientists should be at least as aware of uncertainty as anyone else. Environmental decision making must acknowledge uncertainty so as to allow for risks (i.e., fail-safe strategies) and for learning (e.g., adaptive management). It is within the context of recognizing, indeed probing, uncertainty that science can continue to play a pivotal role in environmental policy development. An exploration of our knowledge, and lack thereof, of an environmental issue provides a level playing field upon which options, risks, and opportunities can be explored by all stakeholders without being dragged into unproductive, adversarial debates surrounding conflicting objectives.

A methodology for using science in this context exists. Perhaps several do. The one with which I am familiar is the workshop approach described in Holling (1978) and referred to as adaptive assessment by Gunderson in this volume. We recently used this approach to examine one of the issues I mentioned earlier—the sustainable management of the mix of native and non-native salmonines in the offshore waters of the Great Lakes (Koonce and Jones 1994). Through a series of workshops, attended by scientists, decision makers, and stakeholder representatives, during which we discussed our knowledge of the relevant ecological and socioeconomic issues, we were able to develop a common sense of uncertainty and risk among the many people who have a stake in the issue. The workshops had an important influence on the nature and public acceptance of a decision to substantially reduce stocking of Pacific salmonines into Lake Ontario. We did not achieve consensus—it would have been naive to expect that—but we did allow stakeholders to see the issue from other points of view and to better appreciate the difficulties faced by those responsible for the ultimate decisions.

A more interactive, consensual role for science in bioregional assessments comes at a price. Bioregional assessment science must be responsive to a plurality of voices, often with widely varying priorities. Scientists accustomed to developing and testing ideas within their own technical peer group, with relatively little external influence, will feel understandably threatened by such a model. There is little doubt that the influence of scientific opinion will be weakened, relative to the early years in the Great Lakes bioregion. But this was occurring in any event. In their new roles, science and scientists can contribute to bioregional assessment activities, such as lakewide assessments and more localized remedial action plans, by ensuring that uncertainties are not ignored and by helping to design remedial strategies that both accommodate and address those uncertainties.

As stated by Regier, the world has changed, and the progressivist model of the Modern Era will not always work in the Emerging Era, to use his vernacular. Because of this, science, as it relates to policy development, must change as well. No longer can scientific research provide the "silver bullet" for most resource management problems. Instead, scientists can bring to the table a common currency for the many players now involved in the debate—the currency of knowledge, risk, and uncertainty. This is true not only for the Great Lakes Basin but also for all problems of managing ecosystems. Scientists can do this, but most still need to learn how. Welcome to the Emerging Era.

Management Review

James Addis

The Great Lakes–St. Lawrence River System (GLSLRS) scientific, management, and governance sectors have all undergone a significant shift in recent years in the way they operate. In the policy review that follows this chapter, Donahue provides us with a comprehensive list of specific transformational changes that have been occurring during the past three decades. The transformation from the Modern Era to the Emerging Era is not confined to the GLSLRS; it is a change in how people view the role of science, professional management, and governance across many sectors of society. These postindustrial changes were widely predicted by Naisbitt (1982), Naisbitt and Alburdene (1990), and Toffler (1980). The Wisconsin Department of Natural Resources has tried to create an organizational environment that is committed to adaptive management strategies, openly debating issues, and seeking to understand the needs of our customers and to learn along with them.

Recently, Holling (1978) and Gunderson, Holling, and Light (1995) have suggested that Emerging Era concepts were developed in an attempt to deal with the chaos-produced difficulty of predicting the outcomes of management actions in complicated, highly interconnected systems. Senge (1994) recommends creating a learning organization to manage in an uncertain or chaotic environment. W. Edward Deming (1982) defined knowledge as information that creates an ability to predict outcomes. His application of statistical process control and the scientific process is a key element of his management system.

The GLSLRS as a Learning Organization

The Emerging Era model described by Regier is now widely practiced in some form in most modern institutions (see Jones, this chapter). Senge (1994) describes the five disciplines essential to building a learning organization. They are: (1) systems thinking; (2) personal mastery of the discipline of deepening and clarifying our personal vision, focusing our energies, developing patience, and seeing reality objectively; (3) mental models examining our core beliefs; (4) building shared vision; and (5) team learning. He suggests that all learning follows a learning model that is encompassed in the scientific process, a model that continuously encourages refining understanding in the face of uncertainty. Nearly all modern organizations are seeking to define how to operate as learning organizations, and most are developing operational handbooks to implement notions that reflect the concept we call the ecosystem approach. To operate in the paradigm of the Emerging Era, the processes and knowledge developed in bioregional assessment provide the basis for assessing uncertainty, building mental models, and creating a shared vision for the community to address. In the GLSLRS this occurred in many places and times, such as the Great Lakes Fisheries Commission and the International Joint Commission. As Regier points out, this holistic process was at work as early as the nineteenth century, when the Rathbun and Wakeham Commission treated the Great Lakes as an entity. It was also at work during the 1970s, when the Great Lakes Basin Commission, along with the states, created the 27-volume *Great Lakes Basin Framework Study*. These institutions and others created knowledge, which when shared by those who lived and worked in the GLSLRS helped this region's leaders perform as a learning organization.

The evolution of Emerging Era systems in the GLSLRS has been erratic and has taken place over a period of more than ninety years. To a great extent the people in the region have seen the vast lake system as a single place and have expected agencies to respond accordingly. I conclude that in almost every case, regardless of the sector in which the system operates, a common set of circumstances must exist for the Emerging Era approach to become the operating culture.

Building Trust

First, in most cases clearly identifiable problems or crises occur that threaten the status quo or prevent things from happening that most people want to happen. In the GLSLRS, a series of nearly catastrophic environmental disasters and fishery collapses created public demands for government to reverse and repair the ecological and environmental damage to the system. Although the initial response to these crises was a lot of rhetoric and name-calling followed by throwing money at the ill-defined problems associated with these collapses, hindsight leads me to conclude that this somewhat directionless casting about led to connections between people, institutions, and government that built "social capital." Coleman (1988) defines the building of social capital as the creation of social networks that foster the development of norms, trust, and connectedness in communities, allowing people to work together to their mutual benefit. In the GLSLRS, this process led to an amalgam of interests, which created activism, starting at the local community level and culminating in Senator Stephan Young's famous (1992) speech announcing the death of Lake Erie.

Second, the results of applying the current management paradigm must fail or fall short of the expectations of the community served. The surprises resulting from the inability of scientists, managers, and public officials to accurately predict management outcomes and deal with cultural impacts undermine public confidence in reductionist scientific management and government. Management may be designed to occur at scales of time and space different from the real systems being managed, leading to the creation of inappropriate hypotheses that do not predict outcomes accurately, which further undermines public confidence.

Third, a force weakening the ability of federal government and other hierarchical institutions to dominate or control the process must arise. Donahue points out that the New Federalism weakened the ability of federal agencies to exert a dominant influence during the early 1980s. At the same time, a complex set of commissions, local initiatives, and strongly held convictions among the Great Lakes states that states' rights should dominate served to empower states and lower levels of government to increase their influence. It also created social capital, which allowed states and locals to use the new sense of community to create trust and public interest in participating in finding ways to solve regional problems. I believe a key factor in this successful building of trust was that leaders needed to and did talk to one another and to listen to demands from the constituents of other leaders, which led to a level of multi-interest understanding that created trustful working relationships in local communities. Thus, trust building and a shift into the emerging paradigm of ecosystem management were likely due largely to the inability of any single entity to dominantly influence the outcome. Fukuyama (1995) provides a convincing explanation of the role that trust plays in building the relationships required to bring widely different interest groups together to create a shared vision for managing very complex systems and for responding to the surprises that arise by adapting rather than finding blame.

The size and complexity of the region and the strong cultural identification of its residents with the Great Lakes created enormous public interest and political involvement. I recall how, as a teenager in Ohio, I watched a grandmother in Cleveland create a major environmental movement when her grandchildren couldn't go swimming because beaches were contaminated with fecal coliform bacteria.

Public Demand for Action

I believe an equally important force leading to collaboration among the jurisdictional players was the fact that the public was increasingly angry about the degradation of the systems and blamed industry, fishermen, government, and scientists. The public raised those concerns in the media, in local community meetings, and in the context of regional and local political debates. As the intensity of the outcry increased in the mid-1960s and the likelihood of repercussion rose, it was a natural survival behavior for various jurisdictions to pull together to find solutions and share the heat. Collaboration spread the risk associated with managing in the face of uncertainty and resulted in the unique bioregional, whole-system approach followed in rehabilitating the GLSLRS.

Hanna Cortner points out (chapter 5) that people count. And nowhere else have they made such a difference in the outcome of major cultural change as in this shift toward managing the GLSLRS as an ecosystem. Without a demanding, interested, and participatory public, the countervailing forces needed to begin the management transformation from the Modern Era paradigm to the Emerging Era paradigm in the GLSLRS would not have forced us to move as far as we have.

Leading from Behind

Agencies must learn to follow the people as well as to lead. Successful Emerging Era projects usually come from the community and are implemented within the community. Scientists and managers need to move from being experts to becoming facilitators for creating agreement, fostering the creation of social capital, and building partnerships. This can occur only if we include lay people at the table when we begin to define the issues associated with the problems we face at all levels. It can happen only if all of us look at the "whole" and describe our place and issues within the context of the whole.

We need to build systems that are flexible and adaptable. We need to incorporate the principles of lifelong learning and team learning into our agency training programs and agency culture. We must practice systems-thinking, using critical thinking skills to test the hidden assumptions of our culture.

As mentioned, the Wisconsin Department of Natural Resources is experimenting to create an effective Emerging Era organizational culture. We have defined an ecosystem management decision model that incorporates the scientific process and adaptive management and recognizes the complex interaction and interconnectedness of ecological, social, and institutional contexts (Addis et al. 1995). We believe that by having all entities—government, local people, scientists, and industry—talking about their interests in the larger landscape—the whole—we greatly enhance the ability to identify where each of our roles and interests fits into the whole. This helps us and others managing and working in the systems that we influence and which influence us to identify areas where we need to communicate and share information and decision making.

The GLSLRS science, management, and governance systems continue to evolve into the Emerging Era paradigm. The fact that change is continuing to be evolutionary rather than revolutionary is testament to the success we are enjoying.

To paraphrase John Gordon's description in chapter 3 of where Emerging Era systems must operate, we must:

1. Manage where we are. But also recognize the importance of scale, especially including the resources, people, and institutions adjacent to us.
2. Keep people in mind. Each person and group brings a set of values, mores, and beliefs that fit into the whole. Most of our present institutions use a wide range of techniques for identifying the needs and values of people. We then acknowledge these needs and values through our newsletters, public releases, presentations, and daily contacts, but most importantly through our decisions, which reflect those needs and values.
3. Manage across boundaries. Always examine whom your actions impact and whose actions impact you from the start.
4. Manage using local information. Unique insight comes from hearing the stories and history of a region, from laypeople as well as specialists. In the GLSLRS, laypeople who use and understand the issues we are working on are invited to participate in the decisions. Commercial fishing operators

sit on panels recommending commercial fishing management policy, industry sits on panels looking at ways to prevent and clean up pollution.

5. Manage whole systems. Seek the scale at which the system operates and use that as your planning scale.

6. Manage across disciplines; ecosystem management is a nondisciplinary issue. Discuss issues at the system level to bring out cross-disciplinary issues, knowledge, and needs.

7. Ecosystem management (Emerging Era management) portends change; that is tough, but also inevitable.

Implications for the Future

There is no prescription for moving into what Regier describes as this Emerging Era of resource management. I cannot identify any single piece or body of information developed in the long history of bioregional assessment in the GLSLRS system that could be described as a keystone to the emergence of this new paradigm. To the contrary, I believe this paradigm has emerged through a long history of people finding that the information they collected to try to solve a local or what seemed to be a single-issue problem proved inadequate because others outside their sphere impacted their goals. This discovery forced people to network in ways that have led to the creation of social capital and the joint pursuit of common objectives. The shift was then more a result of the failure of various earlier approaches than a result of their effectiveness.

The lesson to be learned from the GLSLRS is that it is critical for managers and leaders to scan scales well beyond the perceived scope of their jurisdictions and to network with adjacent players. The concept of building social capital, the ability to work effectively with others outside your jurisdiction and control to identify and achieve jointly important outcomes, seems to be the key to dealing with transdisciplinary issues operating at large biological, social, and political scales.

The GLSLRS has not only espoused these principles, but we have also increasingly adopted them as part of our culture. Time will determine the extent of our success. The ecosystem approach is not a fad here, but we still have a lot to learn.

Policy Review

Michael Donahue

The Great Lakes–St. Lawrence Basin has long been blessed (or, some may argue, cursed) with an elaborate "institutional ecosystem." It features numerous regional, multijurisdictional entities that transcend the parochialism of traditional regional approaches and focus on hydrologic rather than political boundaries. Cases in point include the International Joint Commission, the Great Lakes Commission, and the Great Lakes Fishery Commission.

Ecosystem Approach

To varying degrees, all tend to embrace an ecosystem approach, participatory planning, consensus-based decision making, interdisciplinary team building, and an emphasis on the science–policy interface. Complementing those institutions is an equally elaborate configuration of regional treaties, conventions, agreements, policies, and programs that both foster and reflect characteristics of what Regier has termed the Emerging Era of Great Lakes governance. These characteristics— increasingly evident over the last twenty years—represent an evolution in which top-down mandates have evolved into bottom-up initiatives:

- a vertical management hierarchy has evolved into a horizontal management hierarchy;
- a command-and-control emphasis has given way to partnership-oriented approaches;
- a regulatory emphasis has evolved into an emphasis on voluntary compliance;
- informing the public has given way to involving the public;
- developing a legal and institutional infrastructure has evolved into enhancing the efficiency of that structure;
- balancing economic and environmental issues has evolved into integrating economic and environmental issues;
- nongovernmental organizations as "reactors" to public policy have evolved into nongovernmental organizations as partners in developing public policy;
- federal agency leadership and oversight have evolved into federal–state partnership, with a strong community role;
- acknowledgment of socioeconomic considerations and differing value systems in planning efforts and assessments has evolved into the inclusion of these considerations and value systems in planning efforts and assessments;
- the use of geopolitical boundaries as the basis for planning and assessment efforts has evolved into the use of hydrologic boundaries as the basis for planning and assessment efforts;
- a single-media emphasis has evolved into a multimedia, ecosystem approach.

Evidence that we are in the midst of the Emerging Era is pervasive and can be found in new approaches to management, interagency agreements, and institution building. For example, we are moving steadily away from a highly structured, vertical hierarchy where federal laws, mandates, and agencies are highly prescriptive, and state and local governments served more as implementors than partners in the management effort. "Place-based" management is an emergent philosophy, and the International Joint Commission's areas of concern (AOC) program is an outstanding example of its application. The remedial action plan process in each of forty-three designated AOCs is distinguished by intergovernmental and public- and private-sector partnerships, local "ownership" of the effort, and a pronounced emphasis on behavioral change and voluntary compliance to complement the historic emphasis on regulatory solutions to remediation and restoration.

With regard to interagency agreements, nonbinding, good-faith documents are increasingly preferred to the historic emphasis on painstakingly prepared, time-consuming, and legally binding agreements. A case in point is the Ecosystem Charter for the Great Lakes–St. Lawrence Basin, coordinated by the Great Lakes Commission in 1994. A nonbinding, good-faith document that gained almost 160 signatory agencies and organizations in its first twelve months of existence, the charter presents a common vision and set of principles that both reflect current approaches to basin management and offer guidance in their application. Many signatories indicate that charter principles and action items have already found their way into strategic plans, work programs, and policy priorities.

The philosophy of the Emerging Era is also reflected in institution-building exercises in the Great Lakes Basin. The preponderance of activity is now at the local watershed level, where creative public agency and private sector and citizen group councils and associations are being established with some regularity. In some instances, as with the Grand Traverse Bay Watershed Initiative in northwest Michigan, the effort was entirely voluntary and grassroots oriented. Inclusive, watershed-based and ecosystem-oriented institutions are increasingly valued as the appropriate vehicle to translate a growing ethic of stewardship into practical applications.

There are, of course, many other trends that one might reference as evidence of this transformation to an Emerging Era.

This transformation was not an orderly process brought on by the careful planning of enlightened individuals with the full blessing and concurrence of the existing, traditionally structured bureaucracy. Rather, the transformation is proceeding unevenly—in fits and bursts—and with much resistance from those that Dr. Regier would call progressivists. In fact, in my view, the transformation to the Emerging Era was greatly aided by the New Federalism philosophy ushered in by the Reagan administration in the early 1980s. This philosophy—which has seen a rebirth and reinvigoration in the current U.S. Congress—views bioregional issues largely as concerns of state governments that are placed—ready or not—into a position of leadership. As noted above, it opens the door—somewhat inadvertently—to a much more inclusive, partnership-oriented approach to planning and management. As a consequence, the challenge of bioregional assessment is found not only in the pursuit of good science but also in the ability to bring together many different communities in the region to work toward an ecologically sustainable future.

Lessons

What are the lessons to be learned from this transformation, and what are their implications for the bioregional assessment process? A few observations are in order—not a comprehensive list, but several items that may help prompt informed discussion and debate.

First, the bioregional assessment process is no longer the exclusive domain of federal or state government. It is more of a partnership approach and, increasingly, an activity that is taking place outside of as well as within government.

Recent years have seen a marked increase in assessment activity in the Great Lakes–St. Lawrence Basin. Examples include, beyond subbasin-specific initiatives (e.g., Green Bay, Saginaw Bay), efforts such as the State-of-the-Lakes Ecosystem Conference (U.S. EPA and Environment Canada); State-of-the-Environment Reporting (International Joint Commission); and an array of State-of-the-Great-Lakes reports prepared by state agencies, nongovernmental organizations, and other interests. Also, the IJC plans to methodically initiate "site assessments" in various AOCs.

"Bioregional assessment" is a very broad classification of activity that can range from highly scientific, fact-based, objective research to highly qualitative, subjective analysis with the intent to advance a viewpoint or philosophy of the sponsor. "Let the reader beware" is an appropriate motto in most instances.

The scientific and technical challenges of a credible assessment process are now equaled or exceeded by the challenges associated with accommodating socioeconomic considerations and generating consensus among a diverse and often contentious array of stakeholders.

As noted earlier, having a well-developed institutional framework at the regional level can be a curse as well as a blessing in conducting bioregional assessments of any kind. The law of diminishing marginal utility applies in a consensus-building process where myriad agencies and organizations are active participants; particularly when the expressed motives of some are to undermine or indefinitely delay a given process.

Finally, at least in the Great Lakes–St. Lawrence Basin, bioregional assessment is more appropriately defined as a long-term, continuing process than discrete, finite, product-oriented activities. While the latter are most certainly pursued, it can also be argued that implementation of the Great Lakes Water Quality Agreement has been an ongoing bioregional assessment of twenty-five years' duration.

Conclusions

As a concluding thought, I would agree that much can be learned from case studies (of successes and failures) with regard to the bioregional assessment process. In the Great Lakes–St. Lawrence Basin, for example, much can be learned from the Remedial Action Plan process, and on a broader scale the Lakewide Management Plan process also offers meaningful insights. The planning, assessment, institutional, and community-based characteristics of the RAP process, among others, offer evidence of the Emerging Era described by Dr. Regier. Further, they reflect the various trends I have articulated earlier.

In sum, the bioregional assessment process is establishing itself as an increasingly critical tool in informed, science-based decision making. By necessity, it is both an art and a science and, to be meaningful, must truly embrace an ecosystem approach. Lessons learned in one region are typically highly applicable to other regions. Such exchange is both timely and necessary.

CHAPTER 8

Everglades–South Florida Assessments

Case Study
John C. Ogden

Science Review
Frank J. Mazzotti

Management Review
Susan D. Jewell

Policy Review
Maggy Hurchalla

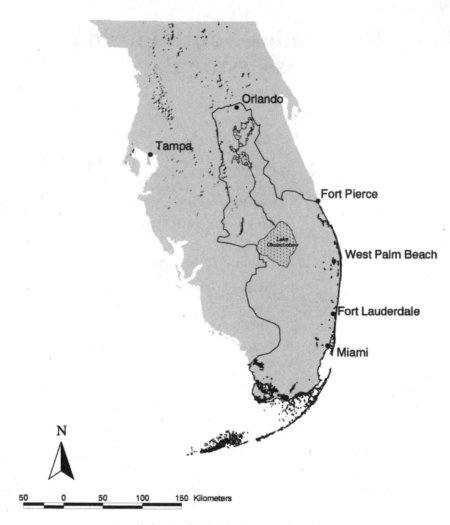

N

50   0   50   100   150 Kilometers

Everglades–South Florida Bioregional Assessment

Case Study

John C. Ogden

The task of recovering a sustainable, Everglades-type ecosystem in southern Florida is monumental (Gunderson, Holling, and Light 1995). The region has a human population of over four million, with high urban and agricultural demands for space and water; all of the currently considered conceptual plans for recovering this ecosystem are extremely expensive to implement; the ecological and societal targets for restoration are in need of better definition; ecological responses to the recovery of more natural hydrological patterns are uncertain; and the serious level of deterioration of the Everglades has created a crisis-driven, "hurry up with the solution" atmosphere. As the social and political elements of the Everglades restoration process increase in complexity, the role played by an array of land management–based science programs in South Florida needs to be reexamined. The current, organizational infrastructure by federal and state agencies, dominated by a hierarchical arrangement of task force (Washington), working group (Florida), and subgroups (advisory to the working group), has not adequately promoted the levels of integration of scientific information into the ecosystem restoration process as may be necessary if these efforts are to be successful. At the same time, science in South Florida needs to reexamine the steps it can take to become better synchronized and focused with policy makers and management.

The Pre-Drainage Everglades

Before drainage began over a century ago, the wetlands of southern Florida were a hydrologically interconnected mosaic of expansive freshwater sloughs, wet prairies, tree islands, cypress swamps, mangrove swamps, freshwater lakes and streams, and broad estuarine lagoons across an area of 3.6 million hectares. At the heart of these wetlands lies a broad "river of grass," the true Everglades, that covered 1.2 million hectares (10,520 sq km) between the southern shore of Lake Okeechobee and the estuaries of Florida Bay and the Gulf of Mexico (Gunderson and Loftus 1993). The imperceptible elevation gradient across the Everglades had an average slope of 2.8 centimeters per kilometer from north to south (Parker, Ferguson, and Love 1955). Deeper sloughs within the Everglades collected and stored heavy rainfall during a discrete summer and fall wet season. The flat terrain with its low hydrological gradient, and the resistance to flow caused by the marsh vegetation, resulted in a broad "sheet flow" from north to south at velocities that averaged only 0–1 cm/sec (Rosendahl and Rose 1982). In turn, the slow flow of water served to prolong the annual duration of surface flooding through much of the Everglades and its downstream estuaries until well into the intervening winter and spring dry season, which resulted in multiyear periods of flows in the deeper, freshwater sloughs and into the downstream estuaries.

The word "unique" has been frequently used to describe the pre-drainage Everglades. Writers have referred to the expansiveness of the region, the abundance of such animals as wading birds and alligators, and the occurrence of unusual, rare, or tropical components such as epiphytic orchids, tropical hardwood hammocks, colorful tree snails, crocodiles, and manatees (Beard 1938; Tebeau 1968). These qualitative descriptions did little to reveal the complex ecological relationships in the region. Yet the expansiveness of the Everglades, along with the difficulties of access during the days prior to aircraft, airboats, and motorboats, meant that very little of what characterized the ecology of the pre-drainage wetlands was ever measured or well understood in any quantitative sense. Large-scale planning efforts for drainage, and the actual construction of canals for this purpose, were well underway before any scientist was even aware of the appearance of the interior of the Everglades, much less understood the regional ecology (Stewart 1907). The changes that early drainage projects made on wetlands in South Florida are the reason W. B. Robertson, Jr., recently suggested that the pre-drainage Everglades is "a lost world: one that we will never know" (Davis and Ogden 1994a).

Although Robertson's observation correctly highlights our uncertainty about many details of the pre-drainage Everglades, recently initiated planning efforts designed to attempt large-scale ecological "restoration" of these wetlands have stimulated reasonable agreement among multidisciplinary teams of South Florida ecologists and hydrologists as to which characteristics best define the pre-drainage Everglades. These defining physical and ecological characteristics of the former Everglades, summarized in Weaver and Brown (1993), include: (1) a hydrologic regime that featured dynamic storage and sheet flow, (2) large spatial extent, and (3) heterogeneity in habitat. More recent discussions among Everglades scientists propose broadening the defining ecological characteristics to include: (4) species of vertebrates that have large spatial and multilandscape requirements, and (5) a low-nutrient system, where interannual variability in hydrological patterns served to organize primary and secondary production into pulses at multiyear intervals.

The interplay between the large spatial extent, the strong patterns of seasonal and annual differences in surface water conditions, and the habitat and landscape mosaics worked to increase the richness of the region's fauna and to allow a network of relatively low-nutrient freshwater marshes to support extremely high numbers of many aquatic vertebrate species. For example, habitat specialists such as the Cape Sable seaside sparrow maintained long-term population stability because of the extensiveness of the single landscape feature that was the required habitat for this species (Pimm, Curnutt, and Nott 1994), while snail kites maintained population stability by crossing multiple landscape boudaries as the location of favorable foraging and nesting habitat shifted in response to the dynamic hydrological conditions (Bennetts and Kitchens 1993). The huge nesting colonies of up to 200,000 wading birds that once existed in the southern Everglades may have been possible in such a nutrient-poor system because of pulses of primary and secondary production synchronized by cycles of extensive drought and fire at multiyear intervals. Conversely, low salinities and high rates of primary production in Florida Bay and the mainland estuaries may have been maintained by the relatively heavy flows of freshwater from the interior marshes and swamps during periodic years of above average rainfall.

The key to understanding these relationships and processes in the pre-drainage Everglades was the regional hydrology. The natural patterns of dynamic water storage and sheet flow established the essential ecological quality of the Everglades. It was these hydrological patterns, operating over an extensive area, that made the pre-drainage Everglades a wetter ecosystem than it is today; organized, concentrated, and set the levels of primary and secondary production; determined the timing and location of biological events; established the salinity gradients in the estuaries; and created the substantial network of dry season refugia that were essential habitats for all freshwater animals in the system.

The History of Change

The earliest canals and associated drainage projects in the Everglades basin date back to the period 1881–1894 (Light and Dineen 1994). Drainage efforts increased incrementally through the early decades of the twentieth century, highlighted by the construction of four canals designed to drain water from Lake Okeechobee into small rivers that flowed from the eastern Everglades to the Atlantic Coast. The pace and effectiveness of these drainage efforts sharply increased beginning in 1948, with the creation of the Central and Southern Florida (CSF) Flood Control Project (constructed by the U.S. Army Corps of Engineers and now operated by the South Florida Water Management District). The initial CSF projects included the construction of an eastern perimeter levee to separate the eastern Everglades from urban and agricultural areas along the Atlantic Coast of Florida and a network of canals and levees to drain the northern one-third of the Everglades to create an Everglades agricultural area below the southern shore of Lake Okeechobee.

Following the completion of these structures, the CSF project constructed additional levees in the remaining northern and central Everglades (3,554 sq km), creating five surface-water management impoundments (water conservation areas). These impoundments were designed to achieve major mandates of the CSF project: (1) receive and store agricultural runoff from the Everglades agricultural area; (2) prevent water accumulated in the Everglades from over-flowing into urban and agricultural lands; (3) recharge regional groundwater and prevent saltwater intrusion; (4) store and convey water supply for agricultural irrigation, municipal and industrial use, and natural system requirements in Everglades National Park; (5) enhance fish and wildlife and recreation; (6) receive regulatory releases from Lake Okeechobee; and (7) dampen the effect of hurricane-induced wind tides by maintaining marsh vegetation in the system.

A large portion of the southern Everglades, nearly a million and a half acres, was established as Everglades National Park in 1947. The enabling legislation for the national park, signed into law May 30, 1934, stated, "No project or plan . . . shall be undertaken which will interfere with the preservation intact of the unique flora and fauna and the essential primitive natural conditions now prevailing . . ." A somewhat less rosy picture of the ecological conditions in the portion of the southern Everglades to be included in the national park was presented by Beard (1938), who wrote, "We now have just about all the biological ingredients that were originally present. Wise administration, coupled with the truly amazing fertility of the tropics should begin to show results in about five years. In fifty years, the Everglades National Park is capable of becoming an outstanding place."

Direct responsibilities for the management of the natural resources of the remaining portions of the Everglades north of the park are shared by two state agencies (the South Florida Water Management District and the Florida Game and Fresh Water Fish Commission), one federal agency (U.S. Fish and Wildlife Service), and two Native American tribes (the Seminoles and the Miccosukees). One of three water conservation areas is leased from the South Florida Water Management District by the Fish and Wildlife Service and protected as the Arthur R. Marshall Loxahatchee National Wildlife Refuge. Wildlife resources in two other water conservation areas are managed by the Florida Game and Fresh Water Fish Commission, where much of the ecological research and the more regionally designed ecological and hydrological monitoring is supported by the South Florida Water Management District and the Jacksonville District office of the U.S. Army Corps of Engineers. Other agencies with important regulatory responsibilities in the Everglades include the Environmental Protection Agency and the Florida Department of Environmental Protection.

Contrary to Beard's wishes, the overall management of the greater Everglades system during the fifty years between 1940 and 1990 has resulted in an accelerated rate of ecosystem degradation, rather than a recovery of pre-drainage ecological conditions, in the park and beyond. The competing mandates of the South Florida Water Management District (SFWMD) to manage water to meet urban, agricultural, and environmental requirements over time substantially reduced or altered the water storage and sheet-flow capacities of the Everglades, diverted large volumes of water through lateral canals to the sea, greatly altered seasonal and interannual patterns of hydrological variability, and, for the estuaries located at the southern end of the system, reduced the volumes and altered the timing and distribution of flows entering from the north. An ecologically naive plan, authorized by Congress in 1970, established a water delivery formula that guaranteed that the volume of water provided to Everglades National Park would equal or exceed preset minimum, monthly amounts. Such a water delivery plan based on monthly averages was doomed to fail in an ecosystem dependent on strong patterns of seasonal and annual variability in hydrological conditions as a mechanism for organizing the timing and location of primary and secondary production. During the thirteen years of this "minimum delivery schedule," ecological conditions in the park continued to deteriorate, as evidenced by increasing frequencies of alligator nesting failures and accelerating rates of declines in wading bird populations.

When deteriorating conditions in Everglades National Park reached a crisis level by the early 1980s, the park staff issued a "seven-point plan" designed to begin recovery of more natural hydrological patterns in the southern Everglades. A major part of the seven-point plan was the implementation of a new water delivery formula for the park, one that based the timing and volumes of surface-water flows into the park on actual rainfall amounts over the southern Everglades basin, rather than on less flexible monthly volumes. Unfortunately, in the face of continuing requirements to meet flood-control and water supply needs in the region, the congressionally authorized Experimental Water Delivery Program can not make major improvements in deliveries to the park until large-scale changes in the structural design (canals, levee, pump stations) of the water management system have been completed. These structural changes have been designed to provide much more natural patterns of flows to the park, without increasing the risk of flooding to urban and agricultural areas adjacent to the Everglades. Until these structural changes are made, only small-scale improvements in the water deliveries to the park have been possible. Implementation of two congressionally authorized plans to provide substantial structural improvements in the water delivery system for the park was, in 1996, placed in the hands of a newly created multi-agency team known as the Southern Everglades Restoration Alliance. Ground-breaking ceremonies for the first construction projects occurred in January 1997.

The small-scale improvements in water delivery to the park since 1985 have not translated into any significant recovery of pre-drainage ecological conditions. The numbers of wading birds nesting in the southern Everglades have continued to decline. A new "environmental crisis" has emerged, with the "ecological deterioration" of the large Florida Bay estuary, the downstream recipient of the reduced Everglades flows. Initial expressions of concern by National Park Service scientists following extensive die-offs of seagrass beds in Florida Bay during the late 1980s evoked little response until South Florida newspapers provided major coverage of the problem. More recently, an independent, science review team concluded that, "Florida Bay has undergone changes during the past decade which have been

unprecedented within the period of recorded observation and reflect a degradation of the ecosystem, in terms of its productivity of living resources, biodiversity
and stability. . . . The preponderance of evidence indicates that the ecosystem of
northeastern Florida Bay would benefit by restoring the amount and timing of
freshwater flow . . . (Boesch et al. 1993). Following this science review, the South
Florida resource agencies created a multi-agency, Florida Bay Program Management Team, which has developed what may be the best-integrated, and most
comprehensive, ecological research program in South Florida.

In the northern and central Everglades, storm-water runoff from the agricultural region has carried substantial amounts of phosphorus into the marshes, creating a level of enrichment that has caused the conversion of sawgrass-dominated
communities into dense stands of cattail (Davis et al. 1994). A recent, complex
history of litigation, "settlement agreements," and newly enacted legislation (e.g.,
the 1994 Everglades Forever Act passed by the Florida legislature) involving federal, state, agricultural, and tribal interests, and dealing with the phosphorus
issues, have created substantially increased pressure on science to identify nutrient thresholds in the Everglades. Any future changes in the design and operation
of regional water management projects for the purpose of ecological restoration
will be influenced by concerns for the effects of manipulating water in a regional
system where both the water and the wetland soils contain excessive loadings of
nutrients. More recently, mercury (in the form of methylmercury) has been
detected in freshwater fishes and some higher, aquatic vertebrates, at levels of
considerable concern (Weaver and Brown 1993).

In summary, the Everglades in the 1990s is a much changed ecosystem, compared to what is known of the system fifty or more years ago. Gunderson and
Loftus (1993) calculated that the total area of the true Everglades has been
reduced by almost 50 percent due to the conversion of large portions to agriculture and, subsequently in some areas, to urban land uses. Fennema et al. (1994)
used a "natural system" hydrological model to show that water depth and distribution patterns have been changed, often in major ways, in almost all areas of the
remaining Everglades, compared to the pre-drainage system. Davis et al. (1994)
have shown that of seven major landscape features in the pre-drainage Everglades, three have been entirely eliminated. Bodle, Ferriter, and Thayer (1994)
reported a SFWMD study that showed that about 10,500 hectares of South
Florida wetlands, much of which are in the Everglades, were infested with stands
of 1.0 hectares or larger of the exotic tree *Melaleuca quinquenervia*. Craighead
(1968) suggested that as a result primarily of adverse water management practices, the population of alligators in the southern Everglades may have declined
by 98 percent. The total number of nesting wading birds in the once large nesting colonies in the southern Everglades region declined by over 90 percent
between the 1940s and 1990s (Ogden 1994). Sixteen species or populations of
Everglades vertebrates are now listed as endangered or threatened on federal and
state lists, and an additional eleven Everglades species and populations are listed
by the state as "species of special concern" (Wood 1994).

Two Everglades Assessments

The first comprehensive scientific assessment of the ecological problems in the entire Everglades region was conducted between 1989 and 1994 and culminated in the publication of the book *Everglades: The Ecosystem and Its Restoration,* edited by Steve M. Davis and John C. Ogden (1994b). This assessment was triggered by a concern among scientists and managers for what seemed to be accelerating rates of deterioration of natural systems in South Florida, as evidenced by a growing list of ecological problems (e.g., excessive nutrients in the northern subregions and dying seagrass beds in Florida Bay), coupled with a recognition that the agencies had no adequate means for coordinating and communicating on their increasing commitment to research and monitoring. The book contains a synthesis of existing information, organized around a theme of ecosystem stability, dynamics, and persistence as affected by the spatial and temporal patterns of major, physical driving forces. Contributions from fifty-seven scientists integrate information from a variety of perspectives as a basis for proposing the defining physical and ecological characteristics of the pre-drainage Everglades and explaining how land-use and water management practices have affected these defining characteristics. Identification of the factors responsible for the changes in the Everglades, in turn, offers guidelines for the focus and priorities that should be considered during the planning and implementation of regional restoration projects.

This book was the product of a process that began five years earlier with a cooperative agreement between the National Park Service (Everglades National Park) and the South Florida Water Management District to jointly sponsor a symposium for the purpose of organizing and synthesizing existing information on the physical and ecological characteristics of the Everglades. An important condition of this agreement was the formation of a steering committee that "represents a scientifically independent group which guides the planning of the Symposium and the editing of the Symposium proceedings." A key, early decision of the committee was to apply the principles and practices of adaptive environmental management (Holling 1978; Walters 1986) to the Everglades restoration process. The short-term intent of the symposium was to begin a synthesis and interpretation of existing knowledge of the Everglades system. Much of this information was unpublished at the time of the conference. The long-term goal was to produce a publication that would make this information available, to provide the technical basis and an increased level of momentum for future restoration planning. The series of adaptive environmental management workshops continued during the period of chapter revisions, to assist in the integration of information from the symposium presentations and chapters and to identify causal hypotheses and conceptual strategies for restoration. Summary reports from the workshops became major source documents during the preparation of the restoration chapters in the final section of the book.

Davis and Ogden (1994b) provided the first integrated, regional perspective for system-wide restoration. Prior to the publication of that book, much of the efforts to recover or "enhance" natural conditions in the Everglades basin occurred at smaller spatial scales, in the absence of any regional plan. The most aggressive "restoration" planning during the 1970s and 1980s was focused on Everglades National Park and in upstream portions of the system in the Kissimmee River and Lake Okeechobee basins. Each agency attempted to deal with ecological problems identified in parts of the system that were directly under its purview and that were known to be adversely affecting the ability of that agency to meet its specific mandates. Such an approach, however, failed to address the ecosystem-defining problems of the Everglades and, in many instances, contributed to the further deterioration of those elements of the system that operated across large spatial and temporal scales. Thus the Everglades Symposium and the book, by providing the technical arguments for an integrated effort, were additionally intended to stimulate the development of a single, regional planning process.

A second scientific assessment of the ecological problems of the wetland systems of South Florida was produced late in 1993 (Weaver and Brown 1993). This assessment differed from the first in that: (1) it was prepared by a team of about thirty scientists during an intense three-month process; (2) it addressed restoration issues in all natural systems in South Florida; and (3) it was prepared in response to a specific set of policy questions submitted by a Corps of Engineers planning team responsible for conducting the Central and South Florida Comprehensive Review Study (see below). Although Weaver and Brown drew on the information and conclusions presented in Davis and Ogden, it provided more specific descriptors and recommendations of ecological characteristics, ecological and hydrological restoration objectives, and success criteria for each of nine ecological or geographical subregions in South Florida. Although these two assessments were essentially conceptual in nature, and neither provided plans that would provide a manager with the specific steps to follow in order to achieve and evaluate the objectives of ecological restoration, they were valuable in demonstrating a broad consensus among South Florida ecologists regarding the major problems, and probable solutions, in the Everglades system.

Interestingly, these two assessments were accomplished with only modest funding support from the agencies. The largest direct contribution was $25,000 from the South Florida Water Management District, money that was used primarily to cover costs associated with the five-day, 1989 symposium. Publication costs for the Everglades book were covered by St. Lucie Press. However, the participation by scientists in the assessments, especially during the fast pace of the second assessment, often created conflicts for how participants prioritized their time. Generally, the agencies were slow to recognize the conflicts created when scientists were forced to choose between full research agendas and the expectations of the agencies for their participation on the assessment teams.

Initial Responses

During the period of publication of these two scientific assessments, a major, regional restoration planning effort was initiated by federal and state agencies. The role that the scientific assessments had in stimulating and focusing these planning efforts is not easily defined, in part because the assessments both contributed to, and were an outgrowth of, rapidly growing concern for the fate of the Everglades. For several decades, some agencies and individuals have recognized the pattern of deteriorating ecological conditions in the Everglades and have attempted to correct these problems through initiatives such as the park service's seven-point plan. But efforts to address these concerns greatly accelerated during the 1980s. During that time Bob Graham (then governor of Florida, later U.S. senator) encouraged the creation of the Everglades Coalition, a loose confederation of environmental and agency people, which became an essential catalyst to this accelerated process. Equally important has been the emergence of a much improved ecological understanding of the Everglades system, as the result of the creation by the National Park Service of the South Florida Research Center during the late 1970s and a rapidly growing research and monitoring program at the South Florida Water Management District.

The major, regional ecosystem restoration planning process now underway, the Comprehensive Review Study of the Central and Southern Florida Project, is a direct outgrowth of that accelerated process. Largely in response to activities by the environmental NGO members of the Everglades Coalition, the U.S. Congress in 1992 authorized the U.S. Army Corps of Engineers to "reexamine the Central and Southern Florida Project to determine the feasibility of structural or operational modifications to the project essential to restoration of the Everglades and Florida Bay ecosystems while providing for other water-related needs" (USCOE, 1994). The basic issue to be addressed by this study is to determine whether there are other ways of managing the C&SF Project that will not only eliminate those practices that have been detrimental to the Everglades system but can also contribute to recovering ("restoring") to the extent possible (consistent with water supply and flood protection requirements) the pre-drainage ecological integrity of the region.

The Corps of Engineers initiated the multiphase Comprehensive Review Study (Restudy team) of the C&SF Project in June 1993. To the tremendous credit of the Corps, the Restudy team created to undertake this planning process was both multi-agency and multidisciplinary. Although this team has been viewed as a model for the successful integration of science and management in South Florida, in its initial organization it lacked direct representation from the policy level. And because the ecologists on the team represented several key federal agencies only, the full spectrum of available science was not routinely obtainable. The first eighteen-month phase was a "Reconnaissance Study" designed to: (1) define the ecological problems of the system; (2) formulate an array of conceptual plans for solving these problems; (3) evaluate each of the conceptual plans; and (4) make recommendations for plans or components of plans deserving more detailed study.

Because the two scientific assessments presented a strong level of concurrence regarding the defining characteristics of the natural systems, the nature of the ecological problems in the managed systems, and the conceptual solutions to these problems, they were used as primary guidelines for evaluations and planning during the first phase of the Restudy Project. The assessments contributed to the Reconnaissance Study in two important ways. First, the ecologists and hydrologists on the C&SF Restudy team used the assessments as primary sources for technical information and for system-wide perspectives for meeting objectives 1 and 2 of the study. And, second, in order to meet objectives 3 and 4, the Restudy team developed an ecological evaluation protocol known as the River of Grass Evaluation Methodology (ROGEM), based on the defining physical, hydrological, and landscape components of the pre-drainage Everglades, as presented in the scientific assessments. ROGEM was the primary evaluation tool used by the Restudy team to determine that several of the conceptual plans had the potential to recover important ecological elements of the natural systems.

Problems and Opportunities

Although restoration planners have readily turned to the Everglades assessments provided by Davis and Ogden (1994b) and Weaver and Brown (1993), the initial planning documents produced by the Reconnaissance Study only partially accepted the recommended restoration goals. The assessments identified several broadly defined characteristics of the pre-drainage Everglades that must be recovered if an Everglades-type system is to be restored, including the region's large spatial extent and regional hydropatterns that feature dynamic storage and sheet flow. While several of the conceptual plans proposed by the Restudy team included bold changes in water control structures and operational criteria (extensive removal of internal levees, redirection of flows, etc.) in an attempt to recover sheet flow and improve on the distribution and timing of flows, the plans are largely inattentive to the need for substantial increases both in water storage and in spatial scale. ROGEM verified that, because of these limitations, none of the initial conceptual plans could be expected to recover the full array of defining ecological features of the Everglades systems.

The unevenness in the way that regional assessments have been translated into specific restoration objectives and plans illustrates a lack of understanding, by both scientists and policy makers, of the full role of science in large, complex regional restoration programs. While the managers and scientists in resource agencies have been comfortable with the more traditional role of science for gathering information (i.e., research and monitoring), the other, newer role has been much less well understood and employed. This second role is one of "science application," whereby current understandings of natural systems are synthesized and reorganized into formats that can effectively support the planning and evaluation components of restoration programs.

The scientific community should take the lead in reviewing its role in regional restoration programs and in recommending the steps necessary to fulfill this "science application" role. These steps should correct four "problems" that are created when science is limited to a more traditional role: (1) the lack of regionally scaled, consensus-building processes for organizing and synthesizing technical information and opinion into formats that are compatible with restoration planning and evaluation endeavors; (2) the lack of a clearly articulated set of ecological end points or measures of success; (3) the uncertainties regarding ecosystem responses; and (4) the problems of producing focused scientific guidelines in a timely fashion. The existence of these problems in South Florida has served to illustrate that neither the scientific community nor the management and policy leaders entered the Everglades restoration process with an overall strategy for how science and policy should be linked during the planning and implementation of one of the world's largest ecosystem restoration programs.

Before elaborating on these problem areas, I need to insert with a few general remarks and assumptions. At this relatively early stage in the Everglades restoration program, it is premature to provide a complete evaluation of the issues raised in South Florida, relative to the application of science to management and policy. The participating agencies are increasingly recognizing, in part due to trial and error, that traditional planning and implementation processes are not well formulated to deal with this new world of "ecosystem restoration." While integrated, ecosystem-wide planning efforts have been underway since 1993, and more subregional restoration planning was initiated for Everglades National Park in the mid-1980s, the task of implementing the ecosystem-wide programs at large spatial and temporal scales has yet to proceed much beyond the planning phases. New multi-agency groups and commissions, most notably the South Florida Ecosystem Restoration Task Force and its Management and Coordination Working Group and Subgroups, and the Governor's Commission for a Sustainable South Florida, have been added as ways to better address a multitude of programming tasks and societal issues, respectively. Because the creation of several new policy and management bodies inevitably leads to periods of adjustment in roles and relationships, especially when these bodies must create and manage nontraditional programs, the way that science is used in the South Florida restoration program is evolving. Thus, the ultimate measure of how successfully science influences policy, and the implementation of the Everglades restoration program, remains uncertain.

Given this arrangement, I will assume that much of the initiative for the development of a strategic process for the application of science to the restoration program, including the integration of scientific assessments into planning and policy, must rest with the scientists. I do not assume that managers and policy makers know where and how physical and ecological information should fit into the Everglades program, before deciding the content and pace of ecosystem management and restoration programs. It is clear that planners and policy makers are considering many social, budgetary, and political factors, in addition to science, in developing plans and policy. Much of that information is coming from separate teams and sources. The demands made by large, wetland ecosystems in need of wise management can only complicate the tasks of planners and policy makers when they attempt to integrate complex, and often competing, bodies of information. Given that most planners and policy people lack formal or extensive ecological training, and as a result often do not fully understand the complexities and potential trade-offs associated with ecosystem management decisions, scientists should not expect these people to take the lead in appreciating the strong role that science must play within the policy and management arenas.

The first of the four problem areas exposed by the Everglades restoration program was the lack of a regionally scaled communications and coordination process among scientists, for the purpose of organizing existing understandings of the natural systems in ways that would better support the restoration goals. Two reasons for this problem have been the failure of management and scientists to recognize what is required, in time and personnel, for scientists to fully perform this second role and the lack of an organized, consensus-building process for focusing existing information on the needs of the restoration program. As a result, scientists have lost opportunities to reinforce and refine their primary assessment messages and to learn more about the natural systems. By not being in a position to develop causal hypotheses to explain alterations in the natural systems, and to guide the design and operation of restoration projects in ways that could test these hypotheses, broad policy-level goals and specific restoration proposals have at times been poorly linked.

The state and federal agencies, including the scientists within these agencies, have been slow to recognize that the most effective way to organize existing technical information for application to the restoration process is through multidisciplinary advisory teams of scientists. The task of converting great volumes of research and monitoring data into formats that support the restoration process is large enough that the lead responsibilities must be carried by different scientists from those who are doing most of the field studies, assuming all parties agree that the field research effort should not be reduced. This task is also technically complex. For example, the science advisory teams may need to develop a focused set of restoration-based hypotheses to explain causal relationships and to identify restoration priorities in stressed natural systems. This team must also conduct the technical evaluations required to drive an "adaptive assessment" type of restoration program. These tasks can perhaps best be guided by "senior" scientists, who might be selected using such criteria as number of years of research experience in the regional systems, peer recognition (publications, honors, etc.), and level of advancement within their agency or institution. One important hurdle in organizing multi-organizational science advisory teams is the Federal Advisory Committee Act, which, in the Everglades region, has made it difficult for nonagency scientists to be included as full members of these teams.

The collective aspect of this communication process may be its most important requirement, because of the complexity of the ecological issues, the large number of agency and university scientists, and the importance to the restoration planning process of presenting consensus (or near consensus) views by science. The success of the two Everglades assessments was due to the large amount of technical information presented and to the consensus developed among such a large number of scientists regarding the relevance and organization of this information for the restoration process. In developing the Everglades assessments, each scientist was expected to speak independently of the traditional missions of his or her agency or institution. This nonagency perspective, coupled with the independent refereeing process applied to each chapter in Davis and Ogden (1994b), resulted in a form of collective "peer review" of the ecological issues of the region. It is this perspective and process that has not, for the most part, been successively continued into the detailed planning and implementation phases of the restoration program.

The second problem area has been the lack of a clearly articulated set of ecological end points and success criteria associated with the Everglades restoration program. Davis and Ogden (1994a) described the pre-drainage Everglades and the changes that have occurred in the Everglades system and identified the attributes of the system that should be recovered as a prerequisite to the "restoration" of an Everglades-type ecosystem. But they also pointed out that the system has undergone a substantial and irreversible reduction in total area, as well as the loss of some landscape features, and thus correctly revealed that a healthier Everglades system in the future almost certainly will differ in important features from the pre-drainage system. Neither of the two Everglades assessments attempted to describe what those differences might be and thus left unanswered the question of the important end points that would characterize a restored Everglades. It remains for science to suggest what the ecological attributes in a future Everglades might be, under a range of management options. What are the end points that should be monitored as an indication that the restoration process is on the correct pathway, and how should these end points be measured? The measures of success can be expressed both as directions of change during the restoration process and as the recovery of ultimate targets that are indicative of either a healthier or a "restored" ecosystem.

The establishment of recovery goals and measures of success is essential both to science and to the policy and management areas. The design by science of system-wide monitoring and research programs to determine and understand ecosystem responses to the restoration projects, and to "test" hypotheses about the systems, must be consistent with the features of the system that are considered to be the best measures of success and/or system performance in an "Everglades-type" system. Thus, the priorities in monitoring and research should be set only when a consensus is reached regarding the attributes that best measure the predicted ecosystem responses to management actions. And for policy and management in the Everglades, the current practice of using hydrological attributes as measures of success must be replaced by a practice of using a set of ecological attributes for this purpose. Although most scientists, during the current state of the restoration process, are comfortable with the assumption that the recovery of pre-drainage hydrological patterns will create the pathway that will most likely lead to the recovery of a healthy ecosystem, the uncertainties about the specific ways that a much managed and altered system will respond require that a set of ecological (rather than hydrological) attributes become the ultimate test of success.

The third problem area for science in the Everglades has developed as a result of the message that there are major uncertainties about how the managed system will respond to the restoration programs. Davis and Ogden (1994a) made it clear that, as a result of the substantial reduction in total area of Everglades wetlands, the influences of sea-level rise, the effects of community and landscape alterations, and the unknowns regarding the dynamics of ecological processes, that the ecological responses to the restoration program cannot be readily predicted. It may even be fair to assume that unexpected responses are as likely to occur as are expected responses. For example, I have hypothesized that the pre-drainage Everglades was able to support nesting colonies of several hundred thousand wading birds only because of pulses of primary and secondary production that were organized by multiyear cycles of drought and extensive flooding. While it may be possible to recover these historical hydrological patterns in Everglades National Park and portions of the upstream Everglades, no scientist is certain that the recovery of these patterns will be extensive enough to result in a return of the "super colonies." While scientific honesty requires that these uncertainties be fully acknowledged, this message does not increase confidence among managers and policy makers.

Everglades scientists, as a group, have yet to fully understand, embrace, and advocate a restoration strategy that would effectively evaluate, and deal with, these uncertainties, in a manner which would build confidence among policy makers. Such strategies may well include components of "adaptive assessments" and "adaptive environmental management" (Holling 1978; Walters 1986; Gunderson, Holling, and Light 1995). While it is not my purpose here to describe a preferred strategy, it should be one that includes enough experimental design and operational flexibility to maximize opportunities to learn how the system responds to changes in hydrological patterns. A successful restoration strategy from the perspective of management and policy would be one that could show ecosystem-scale improvements associated with each incremental step of the project. The message from science should be that these improvements could include the acquisition of new information on how the ecosystem operates, based on responses to changes in hydrology, as well as the achievement of some subset of the desired ecosystem responses.

The fourth problem area pertains to the timing of the contributions from science relative to the timing of policy decisions and the implementation of restoration planning and construction projects. For the most part, the temporal scales of these events have not been well matched. In an extreme case, the structural designs for two major restoration projects for Everglades National Park, the Modified Water Deliveries and C-111 canal projects, were planned and approved before scientists were sufficiently organized to recommend the priority ecological targets for these projects or to develop recommendations for an adequate ecological monitoring program. The ensuing catch-up efforts by science, in an attempt to determine if the approved project designs are consistent with the priority ecological objectives, have made it substantially more difficult for science teams to realize their potential as participants in restoration planning and evaluation endeavors.

One contribution to this temporal mismatch has been the argument from scientists that additional "question-driven" field studies, and the development of complex ecological models, must precede any detail design work for restoration projects. Such a view does not realistically recognize the unpredictable pace of politically driven regional planning efforts. Science may never reach the desired level of participation within the policy and management processes until ways are found to create and deliver the relevant science messages consistent with policy and management schedules. In addition to the creation of standing science advisory teams (senior scientists), this level of timeliness undoubtedly will require the bolder application of "best professional opinion" as a means for filling gaps in technical data and for converting broad restoration goals into specific ecological targets.

The Present and Future

The considerable challenge for Everglades science is that it needs to become a more effective driving force and framework for Everglades restoration. The complexity of the Everglades restoration process, in terms of the number and differing missions of the participating agencies, and the dynamics and uncertainties pertaining to the Everglades ecosystem, requires that the planning and implementation of Everglades restoration become guided by a single, technically approved element of the process. Much of that element is provided by science. It may be unlikely that a successful, regional restoration program for the Everglades can be designed except by a process that includes full participation by a single, multidisciplinary team of scientists, guided by a single strategic plan regarding the application of science to the restoration process. Science is in the best position to guide the development of a restoration program that: (1) is based on an understanding of the characteristic functions and components of an Everglades-type system, and is focused on the recovery of those defining characteristics, (2) contains the experimental designs and adaptive measures necessary to deal with the uncertainties of ecological responses, and (3) incorporates a comprehensive, regional monitoring program.

Between 1995 and 1997, scientists and policy makers in South Florida took major steps to improve the effectiveness of science in the Everglades restoration program. An initial set of restoration success measures was developed by a multi-agency "science subgroup" (Science Subgroup 1997). This initial science advisory team was created in 1993 by the multi-agency South Florida Ecosystem Restoration Working Group. The subgroup, much too large, uncertain of its mission, and organizationally isolated from the full working group, was disbanded in 1997. In its place the working group created a smaller, better constituted Science Coordination Team (SCT) made up of a mix of senior managers and senior scientists. The SCT was chartered to deal not only with traditional issues of research priorities and design but also with the newer issues associated with the organization and application of technical information in planning and evaluating restoration projects.

Prior to the establishment of the SCT, the South Florida Water Management District, which, with the U.S. Corps of Engineers, has lead responsibility for designing and implementing the structural and operational components of the Everglades restoration program, created two senior ecologist policy advisor positions in its executive office. These senior scientists, working jointly with scientists from the University of Miami Rosenstiel School of Marine and Atmospheric Science, organized a series of multidisciplinary workshops for the purpose of developing a set of conceptual ecological models for each of the major natural landscape features in South Florida. The development of these conceptual models became an extremely effective means for creating scientific consensus regarding the important ecological components and links in stressed natural systems, for defining a set of causal hypotheses to explain the altered systems, and for identifying the major ecological end points and success measures for the restoration program (Gentile 1996; Ogden et al. 1997). The conceptual models, for the first time, established clear links between the broad restoration goals established by the policy makers and a set of measurable, ecological restoration targets at a range of spatial, temporal, and hierarchical scales.

The new SCT has adopted the conceptual ecological models as a basis for refining the ecological success measures for the restoration projects, and for developing a single, integrated, ecological monitoring program for the entire Everglades basin. The SCT is also developing, for the first time, a regional science plan for South Florida. The science plan is intended to identify ways of collecting and integrating information produced by empirical studies and ecological models for more effective application to restoration projects and to describe a strategy for organizing the implementation and evaluation of restoration projects into tests of hypotheses on causal linkages in Everglades systems. And it should show feedback processes for adjusting restoration plans and the supporting conceptual ecological models, based on responses by the natural systems during and following the implementation of each restoration project (adaptive assessment).

Concurrently with the actions of the new SCT, the C&SF Restudy Project (discussed earlier) has moved into a more advanced planning phase designed to present a Comprehensive Restoration Plan to the U.S. Congress by July 1999. The

Restudy Project has, responding to lessons learned during earlier planning phases, created a multidisciplinary, plan evaluation team to develop a set of performance measures (i.e., success criteria: based on ecological and water supply targets) and to use these measures to evaluate a range of alternative plans. Most of the ecologically based performance measures used by the Restudy evaluation team are derived from the major ecological linkages and end points identified in the conceptual models. The organization of this plan evaluation team, and the set of ecological performance measures developed by the team, is the first important large-scale test of how successfully these multidisciplinary teams (about thirty members) will perform in guiding the planning and evaluation processes of the restoration program.

In conclusion, policy makers, managers, and scientists must develop a strongly integrated, horizontally organized process for linking science into the Everglades issues. The "restoration" of large, complex ecosystems, imbedded in even larger social systems, requires the adoption of a new planning strategy, one that is characterized by a greater level of responsibility for scientists, operating as an equal partner with policy makers and managers at the upper levels of the planning, implementation, and evaluation processes.

Literature Cited

Beard, D. B. 1938. *Wildlife reconnaissance.* Everglades National Park Project. U.S. Department of Interior, National Park Service. Unpublished Report. 106 pp.

Bennetts, R. E., and W. M. Kitchens. 1993. *Estimation and environmental correlates of survival and dispersal of snail kites in Florida: 1993 progress report.* Gainesville: Florida Cooperative Fish and Wildlife Research Unit, University of Florida.

Bodle, M. J., A. P. Ferriter, and D. D. Thayer. 1994. The biology, distribution, and ecological consequences of Melaleuca quinquenervia in the Everglades. In *Everglades: The ecosystem and its restoration,* edited by S. M. Davis and J. C. Ogden. Delray Beach, FL: St. Lucie Press.

Boesch, D. F., N. E. Armstrong, C. F. D'Elia, N. G. Maynard, H. W. Paerl, and S. L. Williams. 1993. *Deterioration of the Florida Bay ecosystem: An evaluation of the scientific evidence.* Washington, DC: National Fish and Wildlife Foundation, National Park Service, and South Florida Water Management District.

Craighead, F. C., Sr. 1968. The role of the alligator in shaping plant communities and maintaining wildlife in the southern Everglades. *Florida Naturalist* 41:2–7, 69–74, 94.

Davis, S. M., L. H. Gunderson, W. A. Park, J. R. Richardson, and J. E. Mattson. 1994. Landscape dimension, composition, and function in a changing Everglades ecosystem. In *Everglades: The ecosystem and its restoration,* edited by S. M. Davis and J. C. Ogden, eds. Delray Beach, FL: St. Lucie Press.

Davis, S. M., and J. C. Ogden. 1994a. Towards ecosystem restoration. In *Everglades: The ecosystem and its restoration,* edited by S. M. Davis and J. C. Ogden. Delray Beach, FL: St. Lucie Press.

Davis, S. M., and J. C. Ogden, eds. 1994b. *Everglades: The ecosystem and its restoration.* Delray Beach, FL: St. Lucie Press.

Fennema, R. J., C. J. Neidrauer, R. A. Johnson, T. K. MacVicar, and W. A. Perkins. 1994. A computer model to simulate natural Everglades hydrology. In *Everglades: The ecosystem and its restoration,* edited by S. M. Davis and J. C. Ogden. Delray Beach, FL: St. Lucie Press.

Gentile, J. H., ed. 1996. *Workshop on "South Florida ecological sustainability criteria". Final*

Report. Coral Gables, FL: Center for Marine and Environmental Analyses, Rosenstiel School of Marine and Atmospheric Science, University of Miami.

Gunderson, L. H., C. S. Holling, and S. S. Light. 1995. *Barriers and bridges to renewal of ecosystems and institutions.* New York: Columbia University Press.

Gunderson, L. H., and W. F. Loftus. 1993. The Everglades. In *Biodiversity of the southeastern United States: Lowland terrestrial communities,* edited by W. H. Martin, S. G. Boyce, and A. C. Echternacht. New York: John Wiley and Sons.

Holling, C. S. 1978. *Adaptive environmental assessment and management.* London: John Wiley and Sons.

Light, S. S. and J. W. Dineen. 1994. Water control in the Everglades: A historical perspective. In *Everglades: The ecosystem and its restoration,* edited by S. M. Davis and J. C. Ogden. Delray Beach, FL: St. Lucie Press.

Ogden, J. C. 1994. A comparison of wading bird nesting colony dynamics (1931–1946 and 1974–1989) as an indication of ecosystem conditions in the southern Everglades. In *Everglades: The ecosystem and its restoration,* edited by S. M. Davis and J. C. Ogden. Delray Beach, FL: St. Lucie Press.

Ogden, J., S. Davis, D. Rudnick, and L. Gulick. 1997. *Natural systems team report to the Southern Everglades Restoration Alliance.* West Palm Beach, FL: South Florida Water Management District.

Parker, G. G., G. E. Ferguson, and S. K. Love. 1955. *Water resources of southeastern Florida.* Water-supply paper 1255. Washington, DC: U.S. Geological Survey.

Pimm, S. L., J. L. Curnutt, and M. P. Nott. 1994. *Population ecology of the Cape Sable seaside sparrow (Ammodramus maritima mirabilis). Annual report 1994.* Knoxville: University of Tennessee, Department of Zoology.

Rosendahl, P. C., and P. W. Rose. 1982. Freshwater flow rates and distribution within the Everglades marsh. In *Proceedings of the national symposium on freshwater inflow to estuaries,* edited by R. D. Cross and D. L. Williams. Washington, DC: U.S. Fish and Wildlife Service, Coastal Ecosystems Project.

Science Subgroup. 1997. *Ecologic and precursor success criteria for South Florida ecosystem restoration.* Report to the Working Group of the South Florida Ecosystem Restoration Task Force. Jacksonville, FL: U.S. Army Corps of Engineers.

Stewart, J. T. 1907. *Everglades field survey report.* Unpublished ms. Washington, DC: U.S. Department Agriculture.

Tebeau, C. W. 1968. *Man in the Everglades: 2000 years of human history in the Everglades National Park.* Coral Gables, FL: University of Miami Press.

USCOE (U.S. Army Corps of Engineers). 1994. *Comprehensive review study. Reconnaissance report: Central and Southern Florida Project.* Jacksonville, FL: Jacksonville District.

Walters, C. J. 1986. *Adaptive management of renewable resources.* New York: McGraw-Hill.

Weaver, J., and B. Brown, eds. 1993. *Federal objectives for the South Florida restoration.* Miami, FL: Science Sub Group of the South Florida Management and Coordination Working Group. November 5.

Wood, D. A. 1994. *Official lists of endangered & potentially endangered fauna and flora in Florida.* Tallahassee, FL: Florida Game and Fresh Water Fish Commission.

Science Review

Frank J. Mazzotti

Recent bioregional assessments of the Greater Everglades–South Florida Ecosystem have been a culmination of a series of efforts involving many agencies and institutions in the region. The first began as a symposium on the Everglades organized by the U.S. National Park Service and South Florida Water Management District in 1989. This "Everglades Symposium" had the immediate purpose of compiling, synthesizing, and interpreting existing knowledge on the Everglades system. Five years later the book *Everglades: The Ecosystem and Its Restoration* by Davis and Ogden was published, completing the long-term objective of making this information available to provide momentum and a technical basis for ecosystem restoration efforts. Criticized more for what was not included in the Everglades book than for its contents, Davis and Ogden succeeded admirably at their daunting task. Inspired by the Everglades book and increasing evidence of ecosystem collapse, the U.S. Army Corps of Engineers initiated its Comprehensive Review Study of the Central and Southern Florida Project in 1993 and as part of this planning effort completed its assessment "Reconnaissance Report" in 1994. The Science Subgroup of the Federal Interagency Task Force on the South Florida Ecosystem also completed assessments, and together with the Engineers' report, greatly expanded the boundaries of their assessments, defining a Greater Everglades–South Florida System. These bioregional assessments form the basis for this review.

These assessments, conducted to provide direction and impetus to restoration, have not yet had the desired effect on policies being made in the region. To focus on "ways to make information from scientific analysis more usable for policy makers and managers," this review will discuss obstacles affecting our ability to apply the scientific knowledge that we have struggled to gain.

Summary of Results of Bioregional Assessments

Themes that linked these assessments included describing the pre-drainage ecosystem, enumerating changes that have occurred to the Everglades, characterizing current conditions, and providing guidance and initiative for ecosystem restoration activities. These assessments did not address social or economic issues.

The natural ecosystem of South Florida was a mosaic of wetlands, uplands, and shallow water marine areas. The ecological richness and productivity of the pre-drainage system were dependent on a large (about 28,000 sq. km.) spatial scale, a hydrologic regime characterized by dynamic storage and sheet flow and heterogeneity in habitat. The natural ecosystem of South Florida has been altered by anthropogenic changes in the hydrology of the watershed. Initially the changes were focused on drainage, flood control, and water supply. These hydrological alterations of the Kissimmee–Lake Okeechobee–Everglades watershed enabled massive land-use changes and population growth that further altered the South Florida ecosystem.

Agriculture and residential developments have increased the demand for water while diminishing the capability of the system to supply it. Consequences of these human activities have become increasingly evident. An engineered system of canals, levees, and developed areas has reduced and fragmented habitats, uncoupled uplands from wetlands, and freshwater from marine areas. The combination of impounding and sending freshwater to tide prevents sheet flow and dynamic storage, alters hydroperiods, and reduces groundwater recharge. Marine and fishery resources have been affected. Populations of landscape-dependent species such as panthers and wading birds have declined, while the spread of generalists and non-native species is burgeoning.

Summed together, the effects of these changes threaten not only the rich biological heritage of a naturally functioning ecosystem but also a human population whose water supply, economy, and quality of life are dependent on it.

The assessments underscored the interrelationship of these effects and went beyond an analysis of water quality and endangered species to describe the disruption of ecosystem processes that threatened both the human and the natural communities of the region. It is important to recognize the consensus among the assessments that these changes continue to threaten and degrade the ecological integrity of South Florida. Even with the increased understanding these comprehensive assessments have provided, it must be understood that management and policy decisions will inevitably be made based on incomplete data and uncertain outcomes. Yet clearly we cannot wait until all the data have been collected, for instead of gaining the knowledge needed to save the Everglades, we will have documented its death. Although specific objectives and plans for ecosystem restoration remain unarticulated, it has been possible to synthesize the above-mentioned observations, to provide guidance for policy makers to keep the South Florida ecosystem from further degradation, and to prevent the loss of restoration options.

Immediate recommendations to environmental decision makers for reducing or preventing further degradation of natural resources can be lumped into the following categories:

- *Scientific integration*: scientific knowledge and peer review should be incorporated at all stages of research, planning, and environmental decision making.
- *Management integration*: the hierarchy of agencies with policy and decision-making powers should work together in an integrated setting with a shared vision and goals for a regional plan for South Florida ecosystem restoration.

- *Policy integration*: land- and water-use planning should be better integrated and should not compromise restoration alternatives.
- *Immediate action*: Improve water quality, reduce loss of spatial extent of habitats, and develop management plans for non-native species.

The assessments have proven useful to those planning ecosystem restoration, but, unfortunately, their recommendations circulate in rather narrow confines and are not directly delivered to those making land- and water-use decisions.

The Scorecard

Current events in South Florida ecosystem restoration efforts provide a scorecard to see how well science has influenced policy in South Florida.

It should not surprise anybody who has been involved in conservation planning and adaptive management efforts that the path to following the previous recommendations has been rocky and not always followed. This raises the question that if relatively simple prescriptions to avoid further damage from well-defined ecosystem threats cannot be followed, what hope do we have of implementing complicated, time-consuming, and expensive restoration alternatives?

Until recently, litigation and threat of litigation have driven research, planning, and management efforts in South Florida. This has resulted in an unbalanced focus on water-quality issues. Water quality is certainly important, but not to the exclusion of other issues, which has been the case. There is a distinct impression, held by the general public and reinforced by the media, that if agricultural runoff is cleaned up, then the Everglades will have been restored. If it were only that easy. As important as water quality is (and water of appropriate quality is essential), cleaning farm runoff does nothing to address the two defining conditions (spatial extent and hydrological patterns) of pre-drainage Everglades identified by Ogden (this chapter) that "must be recovered if an Everglades-type ecosystem is to be restored."

As Ogden has pointed out, restoration goals and success criteria have not been defined. No unified vision for restoration is evident, and the various entities charged with managing existing natural areas have not come to agreement on restoring the whole ecosystem, but rather continue to focus on restoring their own parts. Instead of a single, coordinated, cooperative, integrated effort at planning ecosystem restoration, we have separate but overlapping federal and state programs and committees. There are literally so many committee meetings, workshops, and planning sessions that if individuals devote their time solely to attending these meetings (and many do), they cannot attend them all. Again, it should be no surprise that redundant and sometimes even conflicting activities result from such disconnected efforts.

Contrary to repeated exhortations over the last five years of the importance of a regionwide, comprehensive plan linking land- and water-use planning and regulatory decisions, no systematic, landscape-level effort is forthcoming.

Further, no systematic effort has been made to transfer the scientific information collected in the bioregional assessments to land- and water-use policy and decision makers. Ogden (this chapter) is correct in assuming that policy and decision makers are not waiting for scientific input. We should not expect decision makers to wade through lengthy tomes in search of relevant knowledge. It is the responsibility of scientists and planners to directly and clearly convey information to policy and decision makers. The absence of restoration leaders at county commission meetings is in sharp contrast to the substantial number of lawyers and lobbyists representing development interests who are perpetually present. Another problem is that political decision makers like to listen to lawyers and lobbyists but are not necessarily interested in listening to scientists.

An example is provided in melaleuca, one of the most notorious and detrimental non-native species in the region. Although years of research and practical experience in melaleuca biology provide a realistic strategy for its management in Florida, recommendations for restoration and management of melaleuca-infested wetlands are being ignored by key decision makers. It is not the accumulated scientific knowledge of melaleuca that is driving legislation, but rather the development community's attempts to eliminate mitigation requirements for wetlands with melaleuca. It is remarkable that five years of recommendations by scientists and consultants to develop a regional, comprehensive, wetland conservation plan have resulted in little action; yet a legislative request by developers, based on the science fiction that wetlands with melaleuca have no biological value, has sent a bevy of agency staff scurrying in response. Clearly, scientific knowledge lacks the influence on policy making that political and development interests enjoy.

Obstacles and Remedies

The ultimate measure of how successfully science influences policy in the Everglades remains unanswered. The following obstacles must be removed for science to have an effective voice in environmental policy and decision making:

- lack of well-defined goals and objectives and measurable success criteria;
- lack of an overall integrated science team with a regional focus;
- lack of peer review for the planning process;
- lack of a comprehensive, regional, land- and water-use plan;
- lack of a systematic effort to educate policy and decision makers about scientific knowledge gained from bioregional assessments.
- institutional limitations.

It is tempting and true to say that the remedy is simply to turn the obstacles around. That is, Why don't we simply set goals, integrate and peer-review our science and policy, combine land- and water-use planning, and educate decision makers? The optimist's answer is that it will be done next week; the pessimist's is that no one really wants to. It is well recognized that a major cause of failure for ecosystem restoration and conservation planning is the lack of clearly defined goals and measurable objectives. The only reason not to set goals and objectives is so that no participant can be held accountable for failure to reach them. As for integrating and peer-reviewing science and planning; this means that different stakeholders must give up the ability to control both science and planning. Rather, each stakeholder is seeking to control information to benefit his or her own agenda, not restoration as a whole. Combined land- and water-use planning is not likely to happen because that would mean managing growth, rather than managing to continue to grow, which is what current policies promote. Finally, if no wants to be accountable and if all parties want to do what they want to do, why educate anyone? The inability of institutions participating in Everglades restoration to escape the limitations of their own agendas and philosophies may be a formidable barrier to ecological improvement.

In spite of this, there is a strong, sincere commitment among all of the agencies and institutions involved to restore the Greater Everglades–South Florida ecosystem and to provide for a sustainable South Florida. What is needed now is a clear vision of the future of South Florida, along with bold leadership required to integrate disparate efforts.

The Future

Will efforts to restore the Everglades ecosystem be successful? Will South Florida be able to chart a path to a sustainable future? Uncertainties outweigh guarantees. Although there is political commitment in the form of regional, state, and federal funding to restore the Everglades, the political will to change the way land use is determined is still lacking. This incongruity is exemplified by two articles appearing side by side in the June 25, 1996, *Miami Herald*, which stated that while the federal government was announcing a "land buying spree" in western Broward County to prevent continued sprawl and to protect an "imperiled water supply" (and the importance of leadership from local government to accomplish this ambitious scheme), local government was busy approving land-use changes for additional development in the same area. Clearly, this contradiction cannot continue.

More than restoration of the Everglades system is at stake here. This is a test of our national if not our global ability to create a sustainable future. Following on the heels of lessons hard learned from controversies in the Pacific Northwest, Everglades restoration has become the flagship effort of the federal government to reconcile conflicts between economic development and environmental protection. The science (understanding of the ecosystem) is for the most part well articulated, noncontentious, and credible. The Greater Everglades–South Florida case is a true test of our ability to make good, science-based policy decisions.

Management Review

Susan D. Jewell

The Everglades means different things to different people. If it could be neatly bounded and categorized, it might be easier to manage. But we must abide by quasi-scientific, quasi-political boundaries established for the purpose of managing under state laws. The South Florida ecosystem, geographically defined as the Kissimmee River–Lake Okeechobee–Everglades–Florida Bay watershed, has been politically designated by the state as the South Florida Water Management District. It encompasses lakes, rivers, springs, freshwater marshes, swamps, tree islands, pinelands, upland ridges, salt marshes, mangroves, and seagrass beds, all integrally tied together. More than six million residents are squeezed onto upland and coastal areas of the region, causing intense political and resource pressure. Land managers have been making decisions here for more than a hundred years, but their souls, roles, and goals have changed.

In the late 1800s, local politicians and developers appointed themselves as managers of the land. They considered the Everglades a worthless swamp unless it was drained. Their souls told them that the right thing to do was to turn the Everglades into a place that produced short-term, readily tangible benefits to people. Thus, their roles became draining of millions of acres of marshes to plow fields, create ranches, and build houses (with little, if any, dissention). Their goals, most likely, were to attract potential constituents and customers. What is different today?

For one thing, science has led to a new understanding of the ecosystem and its potential seventh-generation benefits. For another, several crises, such as floods, droughts, and loss of biological diversity, have erupted from the extensively engineered system and caused a loss of resilience. A third difference is that the legal system has increasingly become the fulcrum on which to solve land-use disputes.

Who Manages the Everglades?

The current managers of the Everglades are largely federal natural resource agencies, state natural resource and water management agencies, large and small agricultural businesses, private developers, and Native American tribes. Their roles are in conflict. Some managers (primarily those from agriculture and development) still desire short-term, readily tangible results. Federal and state agencies now strive to carry out the goals of scientists (attenuated in some cases by politicians), which vary from long-term ecosystem restoration to ecosystem sustainability to regional (primarily social) sustainability. The tribes (Miccosukee and

Seminole) have maintained their long-standing, long-term commitments to preserve their cultures, but they have only recently acquired some management rights to their reservations—their souls have not changed, but their roles have.

The burgeoning human population is pressing against the visible and invisible walls of the Everglades. The narrow upland corridor is close to build-out limits in Palm Beach, Broward, and Miami-Dade counties. A bird's-eye view reveals that the construction adjacent to the perimeter levee has no buffer zone for most of the eastern Everglades; in fact, many developments are built directly on former Everglades. These same developments require additional supplies of water which are not available from the Everglades and additional flood protection. Land-use planners in these three counties are still issuing permits for massive housing complexes, despite being unable to ensure water supplies or flood control.

A fundamental problem in managing the southern Florida region is that there is a proportion of recent immigrants compared to long-time residents. Whether newcomers have immigrated from another state or another country, their ties to this land are rarely as strong as those of the people whose families have lived here for generations. They have not seen the original condition and may not feel a need to protect what is left of it. Some of these people are managers and land-use planners; most are voters. Enlisting the concern of the newcomers is imperative, although it is only part of the solution—the souls of all must want to protect the Everglades.

Many agencies share responsibilities for an individual land area in the region, which frequently causes conflicts in management directives. For example, USFWS manages the Arthur R. Marshall Loxahatchee National Wildlife Refuge (NWR) in the northern Everglades, most of which is owned by the SFWMD, while the U.S. Army Corps of Engineers authorizes the refuge's water regulation schedules. As a result, water passing through the refuge is regulated by three agencies for flood control, water supply, and conservation of habitats. It took approximately five years to get approval once the refuge requested a change in the water regulation schedule. Another example is Everglades National Park, which has an agreement with the Miccosukees that acknowledges the tribe's claim to a small portion of the park to live on. Differences in management opinions have led to several lawsuits filed by the tribe against the park, although both groups want to protect the Everglades.

Land managers at various points along the approximately 200-km. length of the flowing Everglades perceive the problem from differing viewpoints. Directly downstream of the agricultural areas, such as in Loxahatchee NWR, water quality is the major concern. But farther downstream, such as in Everglades National Park, the northern and central Everglades have filtered out the pollutants and water has been diverted for water supply, so the managers face water-quantity and timing worries.

New problems continually arise, and managers must be able to prioritize and keep pace with them. A slow, insidious problem is easily waylaid by a fast-acting and highly visible crisis. For example, while state and federal managers have been confronting water-quality degradation, new exotic plants with explosive growth patterns have appeared and have had to be dealt with swiftly, often with substantial investments of staff time and financial resources. Managing crises allows little time or funds to integrate solutions, so actions proposed as immediate repair may be in conflict with restoring system-wide processes. Money may be wasted on short-term cures as the resource base continues to erode.

How Bioregional Assessments Affect Management

Water-quality degradation, primarily due to agricultural practices, is one of the many serious hydrologic issues facing the Everglades. An early bioregional assessment, although never dubbed as such, was a collection of scientific data that provided the basis for the federal lawsuit filed against the State of Florida in 1988. Because the state had not been enforcing its own water-quality standards for several decades, the federal government sued the state on behalf of Loxahatchee National Wildlife Refuge and Everglades National Park. The assessment was in the process of being formalized for court when the state formally conceded by signing the federal Settlement Agreement in 1991.

Part of the settlement is currently underway, although delays have occurred due to the legal volleys. Legal battles have been waged since 1991 by groups who think the settlement is too strong or not strong enough. However, the water-quality lawsuit has provided more restoration activity than any other assessment to date. Over $800 million will be spent on the water-quality and hydropattern restoration. It is the largest ecological restoration project ever undertaken in the world. One filtration marsh has been completed, and five others (eventually totaling 40,000 acres) are being constructed. The water-quality lawsuit provided the impetus for most of the ongoing Everglades restoration plans.

Despite the lawsuit settlement and the Everglades Forever Act in 1994, which was the state's effort to expand and solidify the settlement, much of the pollution is allowed to continue. Farmers have little economic incentive to cooperate in reducing pollution once the water leaves their land. Coordination of economic and ecological goals would be strengthened by demonstrating their shared dependence on a common resource base.

The Everglades ecosystem has been the subject of three bioregional assessments during the 1990s. Davis and Ogden (1994) described the results from one, a compilation of historic and scientific understanding. It did not include social or economic factors. A subsequent assessment was produced by the Science Subgroup of the South Florida Ecosystem Restoration Task Force (Weaver and Brown 1993). With its multidisciplinary membership, economic and social components, specific regional focus, and a high level of authority, the task force has all the factors that are critical to implementing a successful restoration plan in South Florida. Several tiers of subcommittees bring the issues to the proper levels of expertise. The task force promises to be the best tool to integrate science, management, and policy in South Florida.

A third assessment, the Central and Southern Florida Restudy of the Army Corps of Engineers (U.S. Army Corps of Engineers 1994) was initiated by Congress's Water Resources Development Act of 1992, before the Ogden and Davis and the Weaver and Brown assessments were published. While it is a planning and action tool for improving the "plumbing system" in South Florida, it must assess the problems before it can prioritize, gain public acceptance of, and implement actions. The Corps is utilizing input from environmental and economic sectors. Congress's mandate to complete the project ensures that action will be taken to reduce, if not reverse, current hydrologic problems.

What the Everglades Needs Now

These assessments have all contributed greatly to the working knowledge of the scientists, managers, and policy makers who can make educated decisions based on the collective information. The urgent need now is *to make those decisions*. However, fear of legal retribution may be preventing this. Deadline exceedances of the ongoing filtration marsh construction are keeping federal and state lawyers busy. Every move from conservationists seems to invite the threat of legal action from opposing groups, until it seems the lawyers are making the decisions for the Everglades. It is time for a legal assessment.

The *souls* of land managers in South Florida want to work toward a sustainable ecosystem restoration, yet their *roles* must allow them the necessary flexibility within their jurisdictions in order to achieve coordinated *goals*.

ACKNOWLEDGMENTS

I would like to thank Burkett S. Neely, Jr., former manager of Arthur R. Marshall Loxahatchee National Wildlife Refuge, for providing me with insight into the management problems of the Everglades.

Policy Review

Maggy Hurchalla

Florida is a young state. The attempts to manage the vast wetlands of South Florida are almost as old as it is. The recent assessments of the Everglades Ecosystem culminate over ninety years of trying to manage the big swamp by consensus. In 1905 the consensus was to drain the swamp. Those efforts failed miserably, although the engineered plumbing system has permanently altered the hydrology of the system. The efforts failed because they did not consider nature's extremes. In South Florida it rains fifty inches in an average year, but it can rain half that much or twice that much in any given year.

More recently, we have reached a consensus that the natural system should be restored, to regain what we lost when we altered the natural hydrology of the Everglades. The first attempts to implement restoration failed, again because of extremes. Returning average water flow back to the River of Grass will not restore the Everglades. The sawgrass marsh needs its extremes of wet and dry. The complexity and timing of such extremes cannot be easily engineered.

For twenty-five years there has been a consensus building among scientists and environmentalists that the Everglades system was being drained and diverted and engineered to the point that it would soon be dead. But a community of cattails and blue-green algae is not really dead. Just as a corpse has changed from a living organism to a community of decomposers, so the Everglades was changing from a sea of grass to a roadside ditch. It was not dead, but soon it would not be the Everglades anymore.

Political Embrace

John Gordon (chapter 3) describes a gulf between science and policy that is sometimes bridged only by a rising current of hot air. The rising air of scientific concerns from the last two decades finally gathered into a political initiative by a Florida governor, dramatized by scripted photo opportunities of politicians mucking about in the wetland wilderness. The script worked. The governor became a U.S. senator, and since then every Florida governor has felt it necessary to wade into the Everglades and embrace its salvation. The script worked for the Everglades, too. The education and commitment that came out of campaigning on an environmental theme put a knowledgeable and committed friend of the Everglades in the U.S. Senate.

Yet the political embrace did not lead to a consensus on a plan for restoration. We knew it was broken, and we thought we wanted to fix it. The rising hot air that convected into a rumble boom, when science truly connected to politics, came not from the scientists and the policy makers but from the ecosystem itself. In 1986, Lake Okeechobee sustained a massive blue-green algae bloom. It looked like a corpse; it smelled like a corpse; the floating fish were definitely corpses. The lake had failed visibly and dramatically, because of the extremes of human use.

After the drama of Lake Okeechobee, the Everglades didn't have to play dead to get attention. The Everglades are internationally recognized as a place worth saving. They are a World Heritage site; they are in Kipling's *Just So Stories*. With a federal wildlife refuge and the Everglades National Park downstream from the ailing lake, federal officials argued that conservation of the Everglades system was mandated by law. The screaming algae blooms in the lake made clear that the ecosystem was in trouble and that the ecosystem was interconnected.

Sustained political action might not have been possible without the legal standing of the federal parts of the system, the international reputation of the Everglades, and the tangible rumble-booms from the ecosystem where toxic algae translated scientific principles for the masses. Throughout the interconnected system of the Kissimee River, Lake Okeechobee, the Everglades, Florida Bay, and surrounding coastal estuaries, the ecosystem was complaining. When Lake Okeechobee calmed down, Florida Bay started pitching fits with algae blooms and seagrass destruction that was too gross to ignore.

Understanding Water

Twenty years of scientific hot air convected in understandable events and consequences. We were throwing away too much water; making what was left too dirty for man or beast; and overallocating, on a first-come, first-serve basis, to all who asked for yet more water. The ecosystem was complaining about scarcity of water while the social system was continuing to mess up and waste what was there. The rumble-booms of the ecosystem made it clearer than scientists had ever been able to that the social system and the ecosystem were connected and that to solve the problems of Everglades water flow, society must connect to the environment through better water management.

Florida controls its water through a blend of eastern and western water law that gives substantial power to the state. Yet that power is made meaningless by legal requirements that regional water management agencies must provide sufficient water for all reasonable beneficial uses. There is not enough water to supply all present and projected uses. New commitments of water continue to be made, and new drainage systems continue to waste water to tide. Just as the Northwest has overallocated its old-growth forest and faced scarcity at the edge of the Pacific, Florida has overallocated its water while dumping what was left in the Atlantic and the Gulf.

At the headwaters of the South Florida hydrological system sits a wealthy and powerful dairy industry. The cow manure was obvious, the politics were tortured, but the scientific evidence of cause and effect, though slow in coming, forced a change in water quality at the top of the wetlands system. A strong Dairy Rule for water quality went into effect in 1989. A federal buyout program removed a significant portion of the herd from the Okeechobee watershed. Further downstream stands the industry of Big Sugar with hundreds of acres in sugar cane production. Its impacts are just as obvious on the Everglades as the cow manure had been on the lake; the politics are even more tortured. But because of scientific information from bioregional assessments, Big Sugar is now a nervous but positive player in Everglades restoration.

The process to save and restore an ecosystem has been politically painful and extremely expensive. Although the ecosystem has been singularly successful at pitching fits and winning attention, the whole restoration might have foundered on costs if it had not been for the next big rumble-boom, this time from the federal government. In 1988, the Republican U.S. Attorney for South Florida filed suit against state water management agencies answerable to a Republican governor. Again, the issue was water quality. The lawsuit was an unsophisticated answer to Everglades' problems, serving as a two-by-four for the donkeys and elephants of politics who preferred to do nothing and hope the problems of the Everglades would go away.

While overallocation of Old Forests and Old Water has striking similarities, there is a notable difference in the lawsuits that played such a significant role in both assessments. The Northwest lawsuit stopped allocation of the public resource until critical elements of that resource were protected; there would be no timber leases in the federal forests if there was no credible species protection plan. In Florida, by contrast, nothing stopped. The lawsuit in the Northwest produced an industry shutdown, a ninety-day assessment, and a presidential decision. The lawsuit in Florida produced a bloom of task forces and bureaucratic committee meetings that seem as though they will go on forever. Wetlands continue to be drained; water continues to be overallocated.

Speaking the Same Language

You might remember the tower of Babel. God realized that if people could speak the same language, they could accomplish *anything*. So he set them to babbling so that no one could understand each other, and the tower to heaven was never built. We want to save the Everglades with a tower to be built by the Corps of Engineers, but we are still babbling at each other. Public understanding of the task and public commitment to it are not yet in place. For that we need to speak the same language and reach consensus. We need to translate the technical language into languages we can all understand, and we need to translate the public consensus back to the technical builders of the tower.

In spite of the fact that civility and common language are necessary to build, for those in a hurry, filing lawsuits seems quicker and more efficient. Public involvement seems like a lot of trouble, fraught with the suspicion that disagreement is based on malicious intent as well as downright ignorance. Yet public involvement is central to successful assessment and restoration. Scientists need to understand that in dealing with people who disagree, it is better to presume their ignorance than to presume their malicious intellectual dishonesty. Patiently teaching, preaching, and talking together works well to dispel ignorance, *and* it helps moderate viciousness. With the ecosystem bellowing about its pain and the lawsuit acting as a two-by-four, patience, trust, and civility have allowed the bad cops of the disagreement to fight while the good cops negotiate solutions.

The first breakthrough was in 1991, when a new Democratic governor surrendered his sword to the Republican U.S. Attorney. With millions of dollars going to lawyers for both sides and nothing happening to improve water quality, the state agreed to settle the federal lawsuit on terms that would save the Everglades. In 1991, the legislature passed the Everglades Protection Act. In 1994, the settlement of the lawsuit was further implemented in the Everglades Forever Act.

In 1994, the governor appointed forty-five people to the Governor's Commission for a Sustainable South Florida. It was a diverse group that included doctors, lawyers, Indian chiefs, environmentalists, and farmers. At first it was reminiscent of the tower of Babel and seemed to personify the endless multiplicity of studies by committees with exhausting names. Within a year, after repeated facilitated discussions ("Death by Consensus," it was labeled by participants), the commission concluded unanimously that South Florida was not sustainable on its present course.

The Water Resources Development Act of 1996 enshrined the Governor's Commission into federal law and gave it the task of acting as sounding board and guide to the Corps of Engineers' Restudy of the Everglades' water system. The commission has been an experiment in translating science to civilians and civility to Florida's bitterest political battle. Optimists hope that this forum will provide enthusiasm, values, and politics to the federal public works process and carry it to broader public forums.

Yet the Everglades is not saved, and all is not sweetness and light. In 1996, politics exploded equivalently to the ecosystem's roars. Environmentalists took a ballot initiative to the voters, asking that a penny-a-pound sugar tax, an Everglades Trust Fund, and a polluters-must-pay principle be added to the state constitution. Big Sugar reacted furiously and effectively with a wildly expensive advertising campaign against the tax. The tax failed. The other two initiatives passed. Both sides declared victory. Innocent bystanders wondered if civility could be restored in our lifetime.

In 1997, there was more indication that the scientific message about ecosystem sustainability was not getting through. State legislation passed that reaffirmed that the water management districts must identify water sources for all reasonable uses, now and in the future. The idea that all the water we want might not be an option was not discussed.

Science and Politics

Although science has not solved all the social problems of the Everglades, it has set a hopeful new direction. Science can continue to help dispel conflict by demonstrating rigorous intellectual honesty. It can help win lawsuits by presenting evidence with a proven factual base. It can help build credibility for the enthusiasm, values, and politics that will ultimately be needed to make the political and social equation of ecosystem restoration work.

Scientists who remain detached from the political debate risk the detachment of their supporters. Scientific leadership, not just research, is necessary to achieve long-term goals in ecosystem restoration. Scientists should treat politicians like substance abusers. The first treatment frequently does not work. Don't give up. Repetitive visual demonstrations of environmental cause and effect are important and convincing. Politicians are neither as stupid nor as clever as they seem. They grab for simple solutions and will need to be taught that the environmental decisions are not yes and no, but an ongoing acknowledgment of mosaics, diversity, patterns, and pulses.

Remember the tower of Babel. Translation is essential if scientific information is to build into public understanding. Those impatient with the process need to acknowledge that in a democracy, lasting solutions are found only through public understanding and support. Remember, too, Winston Churchill's words: "Democracy is the worst form of government imaginable, except for the others."

CHAPTER 9

Northern Forest Lands Assessments

Case Study
Perry R. Hagenstein

Science Review
David E. Capen

Management Review
Henry L. Whittemore

Policy Review
Laura Falk McCarthy

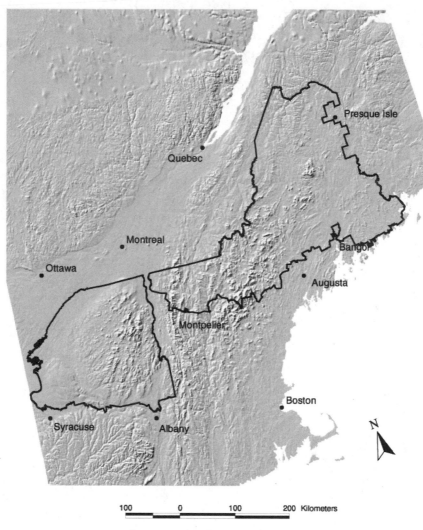

100   0   100   200 Kilometers

Northern Forest Lands Bioregional Assessment

Case Study

Perry R. Hagenstein

In New England we remember 1988 with some fondness. The economy was booming and so was the real estate market. The governor of Massachusetts was running for the presidency based, as he said, on his economic success with the "Massachusetts Miracle." But in the region's northern forests, there was a growing sense that long-standing uses of the forest were being threatened by changes. Development values exceeded timber production values on much of the forests in the region, and simple economics was forcing changes in forest ownership and uses.

Northern New England is one of the few places in the country with extensive and contiguous private forest holdings in a relatively few ownerships. While this ownership pattern in itself brings no assurance that forests will be treated with some attention to overall landscape values, it helps make this possible. Thus, when a forest products firm, Diamond International, was dissolved, and its forests in northern New England and New York were put on the market to capture the value of the land for development, the conflict between economics and some environmental concerns was brought to a head. The response was to study the situation.

In the first phase, the Northern Forest Lands Study was conducted by the U.S. Forest Service with guidance from a task force appointed by the governors of the three northern New England states and New York. As stated in the report in the spring of 1990, the study laid the foundation for interstate communication and cooperation "to explore a common solution" to the threats facing the northern forest.

The exploration was done in a second phase, also supported by federal funds, and directed by the Northern Forest Lands Council, a regional group with four members from each state appointed by the four governors and one representative of the U.S. Forest Service. The report was published in September 1994 (Northern Forest Lands Council 1994). Together, the two reports assessed the forest situation in a region of some twenty-six million acres that stretches from the Atlantic shore in downeast Maine across much of Maine and northwestern New Hampshire and Vermont through the Adirondacks in northern New York almost to Lake Ontario.

The first study was conducted by a small study team of Forest Service employees. The second was the responsibility of a small staff hired by the Northern Forest Lands Council to direct the study. Both reports relied heavily on public meetings and contractors to help in determining the scope and focus of the studies, to define problems and hypotheses, and to assemble information. In a sense, the first of the two studies was responsible for problem definition and data collection and the second for analysis and recommendations for action.

Without the seeming immediacy of the development pressures on the northern forest, the resources for the two studies would not have been made available. But other issues were also simmering. The increased intensity of timber management practices, especially on the half of the total area that was owned by forest industry firms or in other large holdings, was causing concern among environmental groups. Symbolic issues, such as banning log exports and protection of old-growth forests, both issues of little real importance in the Northeast, were imported from regions where they may have been cause for real concern. Closing large tracts of private forest to free use by recreationists (free use was a New England, not entirely an Adirondacks, tradition) was seen as a growing threat. There was also the matter of federalism. The role of the federal government in resource ownership and management as in dealing with forest resource issues had a somewhat checkered history in New England and New York. The federal government's prospective role in the northern forest was viewed with concern in some circles.

Development pressure in the northern forest region comes largely from the "sports," the evocative term Bill Burch has resurrected to describe the recreationists from outside the region (Burch 1992). Over the years, the number of sports has grown from a relative handful of the rich with lodges and extensive holdings, plus a few dedicated outdoorsmen who visited the isolated hunting and fishing camps, to a veritable horde of recreationists who can now drive their cars and campers to most parts of the region. With the change in the makeup of the sports, the recreation constituency for the northern forest has changed from one that was generally satisfied with the ownership patterns and uses to one that includes many who hardly notice, and perhaps welcome, change.

Getting Underway

It was a presidential election year when the Diamond International sale took place in 1988. This always bodes well for getting federal dollars into New England. Senators Patrick Leahy of Vermont and Warren Rudman of New Hampshire got Congress to ask the Forest Service to study and report on the likely impacts of changes in land and resource ownership in the northern forest and to define alternative strategies to protect the long-term integrity and traditional uses of the land. At the time, forestland in northern New England was selling for about the value of the standing timber if the expected use was growing timber. The land itself was seen as having little value for growing timber. But forestland with any potential for recreation, especially that near lakes, was often selling at much higher prices. In the overheated market for recreation homesites, even some forestland with relatively low recreation potential was being priced well above its market value for growing timber.

At the same time, financial managers were becoming more shrewd in handling forest industry assets to meet Wall Street demands for higher profits. Corporate raiders were eyeing some big forest industry firm and, when they had the chance, splitting forestland from processing facilities so that each part could be judged on its separate ability to generate profits. For the forestland segments, this meant further splitting off and selling those lands for which market values were higher than timber production values. The threat was acute in the northern forest region with its many lakes and generally low timber-growing values.

Prompted by those who were concerned by this situation, Senators Leahy and Rudman wrote to the Forest Service chief stating, "We are seeking reinforcement rather than replacement of the patterns of ownership and use that have characterized these lands." The question they wanted answered was, as put by David Dobbs and Richard Ober in their book, *The Northern Forest*: "Is it possible to protect a great forest without destroying the best parts of the resource-based economy and culture that both arise from and contribute to the land?" (Dobbs and Ober 1995).

The Forest Service team that conducted the first study was aided by a task force of representatives of the four governors. The task force made its own separate report and specifically recommended that the effort be continued by a council of governors' representatives from the four states. The Congress adopted this recommendation, provided funding for four years, and charged the Northern Forest Lands Council with making recommendations for action to the Congress, the governors, and state and local elected officials. The initial study served as the fact-finding stage, while the council's efforts in the second stage were aimed at analysis and building a constituency to support its recommendations.

The Bioregion

The northern forest region was defined for these studies on the basis of political rather than ecological criteria. The basic idea was to include the core of the more or less contiguous private industrial forest area of the northern part of the four-state region. As a bioregion, the northern forest includes parts of several major forest vegetation types as well as portions of the St. Croix, St. Lawrence, Hudson, Connecticut, and Merrimac river drainage, and of several smaller rivers that flow through Maine directly into the ocean. It is separated from similar areas in Canada by the international border, even though Canada provides a market for logs from the northern forest, is the home of several forest products firms with forestland on the U.S. side of the border, and is the source of large numbers of recreationists who visit the northern forest.

The northern forest region as defined for these efforts left out the two national forests that are just south of the northern forest—the Green Mountain National Forest in Vermont and the White Mountain National Forest in New Hampshire and Maine. But it included the five-million-acre Adirondack Park in New York, which includes about 1.5 million acres of commercially available timberland in relatively large private holdings and some 2.4 million acres in the

Adirondack Forest Preserve. This preserve is state-owned forest within the park, and disruptions, including logging, are banned in the preserve by the New York State Constitution.

Including the northern part of New York in the northern forest region gained some congressional support for funding the two studies but also brought some contentious issues to the fore. These were prohibiting logging in the Forest Preserve, a wild zone marked with a blue line, and the moderately restrictive land-use zoning practices of the Adirondack Park Agency on private land within the park. As the Northern Forest Lands Study process progressed, one idea that emerged was identifying in some way, perhaps with a "green line," some core areas as a focus for some policy initiatives. This, however, drew fire, especially from timber interests, who saw it as being similar to the Adirondacks blue line or as foreshadowing other park or ecological reserve proposals. The Adirondacks Preserve had long been a whipping boy of these interests, and they did not want the idea to be spread to other parts of the region.

Despite the lack of clearly defined boundaries based on ecological criteria, the northern forest of New England and New York is a coherent region in other respects. The region is heavily dependent on wood-products harvesting and processing and on the aforementioned sports for economic sustenance. Some of the major paper companies had operations spread across the region. New Englanders are comfortable with the idea of the "north woods," with reference to the northern reaches of the three northern New England states, as are New Yorkers with reference to the Adirondacks region. Combining the northern reaches of the four states into a single northern forest region to address common policy issues made good sense to many of the various interests.

At the same time, differences in the place of the northern forest in each of the states created some problems in reaching a regional consensus on policy recommendations. As defined for these studies, the region occupies some three-quarters of Maine, but only about one-fifth of New Hampshire, one-third of Vermont, and one-quarter of New York. This disparity in the place of the northern forest in each state's affairs was raised often throughout the process, especially by Maine's representatives. Some of those from Maine asked, "Why should Maine have its cloth cut to fit the desires of states with less at stake?" The forest industry people from Maine, in particular, were concerned that recommendations for the whole region not be crafted to favor the notably greener interests in the other states.

Wildly disparate views on the potential role of the federal government in the future of the northern forest also posed problems in reaching consensus. On the one side were the proponents of large parks and reserves in what was largely a privately owned forest region. Some of the proposals had been discussed for years, but the northern forest studies provided a good opportunity to introduce more grandiose ideas. The federal treasury was the obvious place to get funds for such proposals. On the other side were the property rights forces that were adamantly opposed to any extension of the federal presence in the region. Anomalously, in 1997 these two forces together succeeded for quite different reasons in defeating a referendum in Maine that would have put some limits on timber harvesting practices on private forests. But for the northern forest study process, they created a deep division that was difficult to bridge in the effort to find consensus. Their efforts attracted a good bit of attention in the media, but their goals received scant attention during the study process.

The Policy Issues

The big issue for the two studies was stated in the letter from Senators Leahy and Rudman—how to reinforce the existing patterns of forest ownership and use in the northern forest region. The first study—the Northern Forest Lands Study— identified several strategies that might be considered, either singly or in some combination, for addressing the big issue. These were:

- *Land-use controls.* How can state and local land-use zoning and controls on development be used to maintain current patterns of forestland use?
- *Public and quasi-public land acquisition.* What is an appropriate role for public and quasi-public (e.g., nonprofit organizations and conservation easements) acquisition of forestland and rights in forestland, and how would this be funded?
- *Incentives to encourage keeping private forests in current uses.* What is the role for federal, state, and local tax policies as incentives for forest management or as disincentives for development?

Twenty-eight specific strategies were identified and described in some detail. Together, they covered a wide range of possible government actions and some quasi-government or nongovernment actions. Few of the suggested strategies were wholly new, and they were presented more as ideas for further investigation and definition than as a distinct package ready for adoption. The report also contained estimates of dollar costs per acre for some of the strategies and a brief description of the kinds of benefits that could be expected for each strategy. Because of the broad variation in specific situations that could be expected for any of the strategies, even the cost estimates for the most definitive strategies gave only a very general indication of what could be expected. The general lack of substantial previous research that could have supported these estimates limited their usefulness in judging potential program costs.

The second study report—that of the Northern Forest Lands Council— addressed much the same issues in thirty-seven major recommendations aimed at forestalling the same problem covered in the first report—the breakup and conversion of privately owned forests in the northern forest region. The recommendations were put into four groups: (1) fostering stewardship of private land; (2) protecting exceptional resources; (3) strengthening economies of rural communities; and (4) promoting more informed decisions. In contrast to the suggested strategies in the first study report, the recommendations in the second report are quite specific. They also deal with a series of issues that received scant attention in the first report—a reflection of the shift from a Forest Service study to one conducted by the representatives of the governors of the four states in the region.

The sponsorship of the study by the governors heightened the role of the various interest groups in lobbying for their favorite positions. The recommendations reflected this, especially insofar as they tended toward a balancing of interests. Thus, those wanting major new parks and reserves were given a recommendation for protecting some exceptionally important ecological and recreation resources, while the forest landowners were given recommendations to ameliorate property and timber income taxes. More direct involvement of the governors and their appointees to the Northern Forest Lands Council shifted the political dynamics from the relatively stand-back posture of the first study to one that clearly involved cutting some deals by the members of the council.

None of the recommendations came close to addressing the issue posed by the kind of market and financial pressures that led to the breakup of Diamond International, the event that gave impetus to the two studies in the first place. But in true political fashion, these recommendations kept the interests at the table by giving each of them something and withholding something else from the others.

The Science

The two northern forest studies were not, nor were they intended to be, scientific in the sense that they developed new knowledge or in the sense that a team of scientists conducted the studies. Both studies relied to a degree on available information that was based on research. But the research base reflected John Gordon's conclusion at the 1995 Society of American Foresters convention that there is a "lack of richness" in the existing science for making such assessments. Gordon's conclusion is probably more applicable to the northern forest region than to the forest regions of the Northwest and South, in part because of the limited extent of federal forests, which have been the subject of many studies in other parts of the country, and the relatively limited role, except in Maine, of the timber industry in the states' economies. There have been, for example, no regional analyses of the timber situation in the northern forest region that are as detailed and as "numbers rich" as those that have been done in the West and South.

Both study reports relied heavily for information on help from outside the immediate study teams, on issue-specific forums, on regular public meetings throughout the region, and, to a lesser extent, on contract studies. To examine in this chapter all of the science-based work of the two projects would be trying. The alternative approach used here is to examine the character of the science used in three of the major issues addressed by the Northern Forest Lands Council— maintaining and enhancing biological diversity, enhancing forest-based economic development opportunities, and modifying state and federal taxation policies to provide incentives for forest landowners to hold and manage forestland.

Two of these issues are social science related. The first of these, enhancing forest-based economic development opportunities, focused on both wood products and forest-based recreation and tourism. The proposals in the report were wide ranging, but there was little that was new or that was not reminiscent of economic development boosterism generally. The report showed little evidence that science had played much of a role in developing the report's conclusions.

While research on matters related to local economic development issues has never been robust, a forum on the topic that was sponsored by the council appeared to focus on the nostrums of economic development boosters and to neglect what has been learned from the research that has been done on the topic in other regions. For example, the notion of capitalizing on value-added manufacturing was raised, but, as with most other assessments, little attention was paid to explaining why value-added manufacturing plants locate where they do, a researchable topic. Understanding this is a prerequisite to developing sound policies to attract such plants to regions such as the Northeast. Without such understanding, proposals for attracting value-added manufacturing are devoid of practical policy content.

The second social-science-related issue is modifying state and federal taxation to provide incentives to landowners. Tax policies and their possible effects on land conversion and on forest management practices were clearly a topic of central concern to the Northern Forest Lands Council. Federal tax policies in general are arcane and are aimed at a broad range of situations rather than just at forests and timber specifically. Perhaps as a result, the contract reports that dealt with federal taxes concentrated on explaining these policies and how they apply to forests and especially to timber. They did this well, but they contained little quantitative information on the extent to which landowners would be affected or on the overall costs and benefits of using tax policies to pursue the goals of the Council. In contrast, the reports on property taxes provided at least some analyses of the impacts of property taxes on the economics of owning and managing forestland for some assumed situations.

The important research question for tax policy aimed at forests is, "How do changes in the amount and timing of taxes change the behavior of forest landowners?" With respect to the federal income tax, substantial research on this question is simply lacking. The work that has been done has mainly addressed the matter in terms of theory—how landowners would behave if they were to maximize short-term income. But there has been remarkably little effort directed at the actual actions of real landowners in various circumstances. Forest landowners would presumably benefit from lower tax rates. But corporate and individual forest landowners now have different federal tax rates on income from harvesting timber and could be expected to benefit to different degrees from changes in tax rates. Although this may have been evident to specialists examining the reports, there were no comparisons of expected costs and benefits from such changes.

This lack of analysis is not particularly surprising in that research on how forest landowners actually respond to incentives provided by the federal income tax code has been very limited in general. As noted in one of the contract study reports, there is "no conclusive theoretical or empirical evidence that suggests that favorable income tax treatment will increase rotation ages or intensify management on any given forest stand." These are the kind of results of changes in taxes that would perhaps indicate some benefits to the public. The lack of such analyses does not provide a firm foundation for recommendations to change federal tax policies. The Northern Forest Lands Council simply accepted the conventional wisdom that reducing taxes would be beneficial.

The third issue, maintaining and enhancing biological diversity, is an issue on which the Northern Forest Lands Council had difficulty in reaching even the modest recommendation that "the states should develop a process to conserve and enhance biodiversity across the landscape." The emphasis in this recommendation was on assessing the status of biodiversity and biodiversity protection, on starting with public and quasi-public lands first, and then, if needed, assisting private landowners with information and then with financial incentives to encourage them to "conserve" biodiversity on their lands.

In reaching this recommendation, the council relied on the results of a forum on biological diversity and several papers on the character and status of biological diversity classification and protection programs. Numerous specific biodiversity issues were raised in the forum and the papers, along with a set of wide-ranging recommendations.

The conclusion reached by those who presented information to the council on conserving biodiversity achieved a modicum of agreement that there are serious information gaps. No one appeared able to explain what private forest owners should be expected to do to integrate biodiversity concerns into their forest management decisions, how this would affect meeting their goals for their properties, or how this would improve societal well-being. Each of these are researchable questions. Perhaps because public concerns with biodiversity are a relatively recent phenomenon (although scientists and managers have long been concerned with them under different guises), there simply has not been time to develop the necessary research protocols and to conduct the necessary research. And perhaps because of this, those who provided information to the council fell back on suggesting approaches for dealing with biodiversity concerns, such as protecting natural areas (there are relatively few truly natural areas in New England), that seem stale. Whatever the reason, the science foundation of the biodiversity recommendation was weak.

The voluntary approaches to encourage landowners to conserve biodiversity were to add protecting biodiversity as an objective in a variety of programs that are also used, or have been proposed, for use in meeting other objectives. The programs include conservation easements, registration of special natural areas, property tax abatement to recognize keeping land from being developed, estate tax abatement, the Stewardship Incentive Program (which pays landowners who manage their lands for long-term environmental and other benefits), the Forest

Legacy Program (which authorizes the federal government to pay for conservation easements from willing landowners in selected states, including those in the northern forest), public acquisition in fee simple, and transferable development rights (TDRs). The regulatory approaches were state regulation of forest practices that would include the goal of conserving biological diversity and sensitive-area protection, such as that requiring setbacks along water bodies or other core areas. Again, this is the more or less standard list of approaches offered by advocates of conserving biological diversity.

In general, the state of science about forest resources is weaker than what is needed to answer key questions about protecting biodiversity, as well as for other policy issues posed by the northern forest studies. For example, an important question with respect to maintaining some desirable level of biodiversity in the northern forest is: "Who bears what costs and for what ends?" Especially in a forested region that is largely in private ownership, this question cannot be brushed aside. Although maintaining biodiversity may be important, it clearly must be weighed against other important social goals. In addition, the goals of the private landowner must be considered if efforts to maintain biodiversity are to be effective. While maintaining biodiversity may be very important to society overall and of dominant importance in some cases, there are undoubtedly cases where the marginal costs of maintaining an additional increment of biodiversity may be judged to be intolerable.

Conclusions

The Northern Forest Lands Council chose to build a consensus around its recommendations rather than to build a strong foundation of science for its report. In doing this, it reflected the pragmatism of downeast Maine. As Bill Burch noted, the interest was in "the nature of the problem and how to get it solved" (Burch 1992). Scientists are, of course, also interested in getting problems solved, and some of the matters facing the council have yet to be resolved by science. But the issues faced by the council were largely policy based. While solid, science-based information is usually a prerequisite for sound policy decisions, the council, in effect, went ahead with the already available science. Had the council pursued a substantially different course, one that questioned the adequacy of the existing information and used public input to weed out the unacceptable policy choices, or had it tried to bolster its decisions with new science-based information, consensus would probably have been elusive.

As it was, the council had difficulty reaching agreement on its final recommendations. David Dobbs and Richard Ober point out that the council has been "chastised" by many environmentalists because its recommendations were not bold enough (Dobbs and Ober 1995). But, echoing Bill Burch, they go on to note the pragmatic character of the council and its assigned task. Senators Leahy and Rudman were not looking for more confrontation when they first asked the Forest Service to conduct the Northern Forest Lands Study. They bounded the study carefully, and this carried over into the Northern Forest Lands Council and its report.

Epilogue

The goal of the initial Northern Forest Lands Study was "to explore a common solution" to the threats facing the northern forest. This was achieved by the Northern Forest Lands Council to the extent that common solutions that could apply across the region were explored. But once the council was disbanded after making its report, taking action was left largely to the individual states. The federal government's presence, mainly in the form of the U.S. Forest Service, has not led to any strong multistate regional initiatives.

During the course of the two studies, the most vociferous of the interests in the study results formed near the polar ends of the spectrum of interests. Several of the mainstream and fringe environmental groups formed the Northern Forest Coalition to represent an amalgam of their views. There was no parallel organization at the other end of the spectrum, which included a range of property-rights, small landowner, and generally libertarian views. The more conservative state-level conservation groups and the major forest industry interests stayed largely in the middle.

These splits continue to be evident as issues arise that involve the northern forest. The most obvious such split was in the referendum vote in the fall of 1997 on a proposal to limit clear-cutting and restrict some other forest practices in private forests in Maine. This was a long-simmering issue in Maine, one that had been exacerbated by heavy logging in forests decimated by the spruce budworm in the 1970s. The Northern Forest Lands Council avoided the issue in large part by focusing on changes in forest use. The proposed compact, which was supported by the forest products industry, two of the major state conservation groups, and the governor, was defeated narrowly by an anomalous convergence of lobbying by environmentalists and property-rights interests. The kind of consensus in support of a middle-of-the-road approach that had been the goal of the Northern Forest Lands Council did not carry over in this particular vote. Whether the consensus will have staying power as other issues come to a head remains to be seen.

One force for regional action has been Senator Leahy, a Democrat, who, along with Senator Rudman, a Republican, provided the initial support for the Northern Forest Lands Study. It was Senator Leahy's initiative that created the Forest Legacy Program as part of the 1990 Farm Bill. This program, which provides federal funding in selected states for public acquisition of conservation easements that prevent development of forestland, addresses the kind of concerns over development of forestland that are at the heart of the northern forest issue. But limited funding for the program has kept it from being a major force in limiting forest land-use changes.

In August 1995, Senator Leahy, joined by Republican Senators William Cohen and Olympia Snowe of Maine, Judd Gregg of New Hampshire, and James Jeffords of Vermont, introduced a bill, the Northern Forest Stewardship Act, to "implement the recommendations of the Northern Forest Lands Council." In his statement accompanying the bill, Senator Leahy said that "the Council's study was driven by a desire to achieve something." He offered the effort of the council as "a model for meeting the conservation challenges of the country" and said that his bill, along with the proposed Family Forestland Preservation Act, "goes no further than, nor falls short of, the Council's proposals for the Northern Forest lands."

The bill starts with a declaration of principles taken directly from the council's report. Another section defines "principles of sustainability" taken from the council's report that are to guide the Forest Service in working with the states. The other eight sections of the bill provide authority related to formation of marketing cooperatives, formation of a Northern Forest Research Cooperative, establishment of an interstate coordination mechanism, providing assistance in labor safety and training, providing assistance in land conservation, encouraging the states to protect landowners from liability for users of their lands, encouraging Congress to create a funding mechanism for conservation of nongame wildlife, and authorizing appropriations to carry out the bill's provisions.

Senator Leahy's bill now has numerous added congressional sponsors from the New England and New York delegations in both the Senate and the House of Representatives. But the bill also has strong opposition from private-property–rights advocates and has failed so far to be reported out of committee. In recent years, Congress has tended to pass more legislation specific to a region—the congressional action that established the Northern Forest Lands Council is an example. Nevertheless, the odds that Senator Leahy's bill will pass in anything close to its present form appear low. That the bill closely parallels the recommendations of the council and that it has been introduced at all speak well, however, of the efforts by the council to build a consensus for addressing the issues of the northern forest.

LITERATURE CITED

Burch, William R., Jr. 1992. *The Maine woods—property, primary production, play and profit in a global, post-industrial environment.* Keynote address: First Munsugan Conference, College of Forest Resources, University of Maine, Orono. October 13.

Dobbs, David, & Richard Ober. 1995. *The northern forest.* White River Junction, VT: Chelsea Green Publishing Co.

Northern Forest Lands Council. 1994. *Finding common ground: Conserving the Northern Forest.* Northern Forest Lands Council, Concord, NH.

Science Review

David E. Capen

This assessment began with unusual circumstances: there was no crisis—no conflict with habitat of an endangered species, no immediate concern about a shortage of forest-based resources, and no conspicuous clash between developers and special-interest groups opposed to development. Instead, there was just a wake-up call—a warning that a remote, relatively undeveloped part of this country might be threatened with a different pattern of land ownership and therefore a different way of life. The northern forestlands situation was clearly distinct from other bioregional assessments because the initial objective was not to restore what had been lost but to plan a strategy to keep things the way they were.

Northern Forest Lands Study

Senators Patrick Leahy and Warren Rudman must have intended the Northern Forest Lands Study to result in a product of policy that was not built on science, because the initial study was a modest effort. A small staff with minimal funding led the first phase of the assessment. There was no ambitious scheme for assessing biological resources or even the more tangible physical resources of the region. There still is not a consistent set of GIS maps for the region. It is significant, however, that on-the-ground data collection and mapping activities were strongly opposed by many of the private landowners in the region, who perceived the initial study as work that was designed to support a public "land grab." In hindsight, the assessment's humble beginning probably was the key to the longevity of the planning process and its acceptance by a diverse constituency.

Instead of identifying problems, designing studies, collecting data, conducting analyses, and drawing conclusions, the study team identified the problems, then focused directly on solutions. The report of the Northern Forest Lands Study identified twenty-eight strategies for land conservation. Each was presented as an hypothesis, then evaluated in some detail by summarizing the pros and cons of past experience in the four states of the northern forest region. However, the report did not draw a conclusion or present a solution to the problem of maintaining the status quo in the region. The study team did recommend a more formal assessment and inventory of northern forest resources, however.

Northern Forest Lands Council

The more comprehensive study of the land conservation strategies recommended by the Northern Forest Lands Study was coordinated by the Northern Forest

Lands Council, a product of the governors' task force that paralleled the work of the initial study team. The council sought to *solve* the problem, not just *identify* solutions. But here the issues broadened. Advocacy groups and individuals—including some well-known scientists—became involved. They were not just suggesting different solutions; they were raising more issues. Sustainability became a common theme, as did forest health, biological diversity, and ecosystem management. Members of the council soon found themselves debating issues that many had never even considered.

This study was somewhat better funded, with $4.5 million appropriated by Congress for the four-year period from 1991 to 1994. About half of this sum was allocated for resource inventory, but these monies were granted directly to the four states involved in the assessment, inhibiting a coordinated effort to collect data. There was no guidance for consistent collection of data and no central clearinghouse for dispersal of information (the council's office closed upon completion of its report).

Numerous small research projects were initiated by the Northern Forest Lands Council, but few were of a truly scientific nature. Although reports from these projects and transcripts of public discussions about the reports were published in a hefty technical appendix, the thirty-seven recommendations presented in the council's final report in 1994 were products of exhaustive public meetings where social and political traditions in four states demanded compromise policy. Science was reflected in the process mostly because a number of scientists chose to be involved, not because of a science-based design for the assessment.

Recent Assessments

Assessment of the northern forestlands did not end with publication of the Council's report; in fact, this was only the beginning. Among the council's recommendations were statements calling for establishment of benchmarks of sustainable forestry and plans for conserving biodiversity. Details of these recommendations called for scientific assessment, but responsibilities for these studies fell back to state governments and to an active community of nongovernmental conservation organizations. The result has been an impressive flurry of assessment activity and conservation planning.

In Maine, an Ecological Reserves Steering Committee has worked independently of the federally supported northern forest studies to assess representative ecosystems in the state and to design a system of ecological reserves in Maine. The initial task was to delineate 15 biophysical regions and 102 different ecosystems, and the goal is to design a system of reserves that assures the protection of representative ecosystems in each of the 15 regions. At present, a detailed inventory is being conducted of public and private conservation lands to determine the extent of current protection for the 102 ecosystems. Following this inventory, the committee will almost certainly propose additional areas to be protected as reserves, because only 5 percent of Maine's landscape is conserved by public ownership.

Maine's effort to identify a system of ecological reserves—and scientists associated with this effort—contributed significantly to a recommendation in the report of the Northern Forest Lands Council that endorsed the establishment of ecological reserves. Similar reserve-design efforts in New Hampshire and Vermont did not develop until after the council's report in 1995. In New Hampshire, a self-appointed committee, led by scientists at the University of New Hampshire, compiled and published an assessment of the state's biodiversity. Concurrently, the director of the state's Division of Forests and Lands appointed two committees to address needs for ecological reserves. The charge was explicit: "To determine the need for reserves and to assess the current system of conservation lands in satisfying this need." Typical of New Hampshire's frugal state government, however, there were no funds to support this effort, so the assessment has been coordinated by the New Hampshire Chapter of The Nature Conservancy, and funds have been successfully raised from private foundations. The Scientific Committee will soon submit a thorough analysis of physical and biological resources within the state and identify important resources that are not protected by current conservation programs.

In Vermont, the commissioner of the Department of Forests, Parks, and Recreation responded to recommendations of the Northern Forest Lands Council by asking the state legislature to convene the Forest Resources Advisory Council (FRAC) and charge it with the task of recommending measures to assure sustainable forest practices in the northern region of the state and establishing benchmarks to gauge sustainability. One of its most significant recommendations resulted in a 1997 law that restricted clear-cuts of greater than forty acres, angering private landowners and stimulating extensive posting against trespass on private land. This is a setback to the cooperative attitude that existed when the Northern Forest Lands Council completed its work.

Meanwhile, another group gathered in Vermont to assess the state's biological diversity and to design a system of priority conservation areas, similar to the efforts in Maine and New Hampshire. The Vermont Biodiversity Project (VBP) convened in 1996 but did so without directives or funding from state government. The VBP is compiling the most detailed assessment of the state's physical and biological resources to date and is using it to design a network of suggested conservation lands in each of the state's biophysical regions.

In New York, the long-standing Adirondack Park Agency continues with its mission, which includes assessment, being influenced much less by recommendations of the NFLC than the other three states. Scientists in these four states exchange information frequently and, on one occasion, met for a day-long workshop simply to compare notes on scientific assessments. Despite the different origins of and somewhat different charges to the committees responsible for biodiversity assessments in the northern New England states, their work is aligned with recommendations of the Northern Forest Lands Council, and committee members remain aware of the importance of the entire bioregion.

A look back at the assessment of the Northern Forest Lands results in an image of a traditional New England town meeting. Everyone had an opportunity to participate. Although progress was slow, a foundation of consensus was the final product. It will be instructive to see if a consortium of state agencies, special committees, and nongovernmental organizations can work toward a common goal of a scientific assessment of the bioregion without the involvement of a federally funded study team to provide resources and coordination.

Management Review

Henry L. Whittemore

The vast, largely privately owned forests of northern New England have provided a long tradition of openness and use-for-all as if the land were publicly owned. This tradition was perceived to be threatened when nearly one million acres of timberland appeared on the real estate market in 1988 and was purchased by a group of developers at a price that exceeded the value of its timber inventory. The Northern Forest Lands Study, and later the council, were convened to prepare the region for future significant shifts in land ownership that might challenge the underpinnings of the region's natural, economic, and cultural environment.

The council began its work against a complex backdrop that blended social issues (private land ownership, protection of "public" values, and continuation of traditional use patterns) with biological issues (timber growth and yield and wildlife population dynamics) and with economic issues (fiber and solid wood supply and demand in the global forest-products marketplace). Economics, biology and societal values: the stage was set for some contentious debate, and the region became tuned into forestry issues as it had never been before.

The mission of the Northern Forest Lands Council (NFLC) was "to reinforce the traditional patterns of land ownership and uses of large forest areas in the Northern Forest of Maine, New Hampshire, New York and Vermont." This was to be accomplished by "enhancing the quality of life for local residents through the promotion of economic stability for the people and communities of the area and through the maintenance of large forest areas; encouraging the production of a sustainable yield of forest products; and protecting recreational, wildlife, scenic and wildland resources."

The following is an assessment of the Northern Forest Lands Council—a critique from a management perspective that can shed some light on both the NFLC and regional efforts in other parts of the country. The NFLC suffered from some notable blind spots, but the process of engaging in a difficult, controversial, yet constructive debate in the northern forests established an environment for discussion that will serve the northeast region well for years to come. There have been many creative, collaborative problem-solving efforts in the Northeast since the NFLC report was published. While they may not all be a direct result of the content of the report, successful resolution of forestry issues in the Northeast will be predicated upon the understandings built during the NFLC process.

Results

The findings and recommendations contained in the report of the NFLC, *Finding Common Ground: Conserving the Northern Forest*, present a conceptual strategy for achieving the mission. What is missing in the report—by design, not by accident—is a tactical outline of action steps for achieving the recommendations in each of the four states. It suggests a destination but offers no map to get us there. In spite of its lack of an implementation plan, the NFLC report is a valuable document for bringing northeastern forestry into the twenty-first century by embracing tenets of sustainability and promoting strong working relationships among a diverse group of related interests.

Forest Practices

The most obvious omission of the NFLC report, from a forest management perspective, is the near total lack of discussion of forest practices. This was a purposeful omission when the council was established through federal legislation. Some argue that this is a fatal omission, a failure to seize an important opportunity to discuss critical management practices and policies. Others claim that the choice not to focus on forest practices is what salvaged the process of the NFLC and allowed council members to find any common ground at all. Certainly, the level of detail in discussions on forest practices would have been excruciatingly difficult, perhaps impossible when considered across all four states.

As the Northeast continues to consider implementation of the recommendations of the NFLC, flexibility in addressing forest practices is often raised as a key factor. Just as the NFLC resisted taking a prescriptive approach to forest policy by not recommending (or even debating) regulatory changes, so do many of those involved resist onerous policy changes that are based on government regulations. Rigid regulations will yield strict compliance with only the letter of the law. Land managers who are involved in building voluntary agreements and measures of accountability will better understand and be better prepared to adhere to the spirit of such agreements. This is not to suggest that forest practice regulations are outmoded or should be stricken from existing law. It is to suggest the existing laws should be well enforced before new laws are implemented. A notable example is the existing regulation regarding liquidation, or "cut-and-run" harvesting. Debate over this enormously complex issue has begun in all four of the NFLC states. One measure of the success of the NFLC will be the extent to which its report has created an environment in which participants will be able to agree on an appropriate approach to liquidation harvesting.

Business Context in Public Policy

Another omission in the report is its failure to address issues of the northern forest in an economic and business context. The report does a fine job of discussing local economies and stating their importance. But a deeper discussion of the economics—the business realities of owning and managing timberland, a mill facility, or a contracting business—is nearly absent. This is of concern because the policies that might stem from this report will have far-reaching business implications. The report acknowledges the importance of these issues (especially as they influence local economies) but does not set up a framework to guide business, policy leaders, and the citizenry to understand and weigh the trade-offs required in balancing the management of natural, economic, and social resources. And this is exactly what forest managers are required to do in nearly all decisions. If the NFLC report has changed anything significantly in the northeast region, it is the huge extent to which it has served to heighten public awareness of forest management. No longer are forests managed solely to produce wood. Forest managers are responsible for a far more complex suite of interactions among biological, economic, and social systems.

Balancing Interests

To remain credible, bioregional assessments need to convince participants that business and environmental outlooks have been fully considered. This balance can be very difficult to maintain. Environmental organizations can be quick to assert that the assessment process is too cozily held "in the hands of industry." Business entities can be equally concerned that these assessments mask an underlying motive of environmental preservation no matter what the business costs.

The council was composed of four people from each state, each representing one of four constituencies: forest landowners, environmental interests, state conservation agencies, and local communities. Substantial differences and disagreement existed among representatives of similar constituencies from each state. The issues they faced were a unique mix of science and emotion that pervaded the debate of how best to manage the northern forest. The quantitative parameters of the forestry debate—things like harvest-to-growth ratios; shifts in forest inventory; dynamic supply-and-demand shifts in a global forest-products marketplace—are enormously difficult for participants in the debate to grasp. Equally difficult is the qualitative, emotional context that frames the debate, characterized by the tension between the property-rights advocates on one hand and society's demand on the other that the "public values" of the forest be safeguarded (preserved, or at least well managed).

Offsetting this complexity, however, is the fact that New England's town meeting style of government and longer history than other regions in the nation provide an intuitive basis for airing widely divergent views in a constructive manner. Many of the players know each other and have strong professional relationships that carry them beyond the disagreements. This is important to forestland management, as it defines the cultural environment in which we do business. It is this environment in the Northeast that allows the delicate balance among and across stakeholder groups to be maintained—no matter how tenuously at times.

Outcomes

Nonregulatory Change

New Hampshire's High-Elevation Memorandum of Understanding (MOU) is a good example of a successful, nonregulatory outcome of the NFLC. This MOU, signed by seven landowners and by the New Hampshire Department of Fish and Game, contains specific provisions that address road building, regeneration requirements, residual stand densities, erosion control, and harvesting methods for more than thirty-three thousand acres of high-elevation forestland. The voluntary document will likely be more meaningful than prescriptive legislation as a tool to protect habitat and forest community types while still allowing forest management. A regulatory approach would have been opposed by some who would manage high-elevation stands only grudgingly according to the detail of the law.

Maine's Forest Compact

An example of an unsuccessful regulatory approach to forest policy is the compact for Maine's forests that was defeated in a statewide referendum vote in 1997. Briefly, the compact was proposed by Maine's governor, the forest products industry, and mainstream environmental groups as a creative alternative to an earlier Green Party initiative to ban clear-cutting and impose other regulatory constraints on forest practices. The compact failed to gather enough votes to become law. The results of the voting suggest that the public wants change in how Maine's forests are managed but disagrees over what kind of change would be beneficial. The process was overtaken by the more strident environmental groups on one side and by property-rights advocates on the other. Although the Northern Forest Lands Council established a strong track record of reconciling disagreement through public discourse and involvement, its example of inclusiveness was heeded too late in this process.

The section "Fostering Stewardship of Private Land" of the NFLC report addresses some issues of daily management of forests. The nine principles of sustainability are well articulated and, if followed to the full extent of the spirit in which they are written, could do much to alter current forestry practices. However, there is no implementation plan to achieve these principles, no analysis of economic and environmental costs and benefits of following them. To some extent, it was the lack of this kind of tangible evidence that doomed the forest compact in Maine to defeat: in the absence of sound facts, the emotions of the debate from both sides swept the public in a wave of confusion to simply vote no.

Conservation Easements

The Northeast has developed a useful tool to achieve many of the principles of sustainability: conservation easements—they protect important public values on private lands and provide another example of how business and environmental perspectives converge to benefit many interests. Conservation easements—from willing landowners—are a tried-and-true tool for land conservation in the Northeast and are an excellent means of maintaining the working forest while balancing business goals with environmental protection objectives. Historically, the terms of conservation easements restrict development rights and mining rights and provide an affirmative right to the public for responsible recreation access to privately owned lands. Increasingly, conservation easements are restricting forest practices by limiting clear-cut size and basal-area removals in riparian areas, on steep slopes, and above certain elevations.

An Ongoing Forum

The final section in the NFLC report, "Post Council Action," has likely had the most direct impact on forest management. This section calls for state (not regional) roundtables, continued dialogue, leadership and support to implement the recommendations of the council, increased research, and proactive federal legislation to implement the recommendations. Each of the four states has taken positive steps to enact the roundtable suggestion. New Hampshire has several active working groups discussing biodiversity, bioreserves, sustainability, forest practices, and a recent recodification of the forestry laws in the state. New York is addressing the issues through existing state and nongovernmental organizations such as the Adirondack Council, Adirondack Park Agency, Department of Environmental Conservation, and local government groups. Vermont has established a Forest Resources Advisory Committee, and Maine has a similar council on Sustainable Forest Management, as well as a Biological Diversity Forum, to follow up on NFLC recommendations. While Maine's sustainability council was interrupted during the process of the compact debate, many observers believe it will again become an important foundation for continuing the difficult discussion. All of these efforts are positive and important for the future, as they help to build a scientific basis for mutual understanding of the principles of sustainability.

The questions remain, however: How will the NFLC's abstract conceptual principles be implemented in the field? Will forestry practices in fact be changed as a result of all this dialogue? Perhaps the most pressing question to come from the NFLC process is whether policies, practices, and the biological, economic, and social realities of timberland management in the northeast can ensure that the forests remain viable as timberland and are not converted to nonforest uses.

Conclusions

A central theme in the debate over the northern forests is a desire for future stability of forest ownership and management. The forest provides jobs and wildlife habitat and contributes to the region's high quality of life. Yet the natural processes that build and sustain the forest are dynamic, and society's demands for stability may be unrealistic. The path to stability is not at all clear, but several approaches seem critical: address issues on a case-by-case basis; avoid rigid pre-scriptive approaches where possible; exercise common sense; empower local deci-sion makers; be inclusive of a broad range of interests; offer guidelines instead of rigid policies; focus on business and environmental and social issues; promote mutually beneficial partnerships; and—in all of this—recognize the paradox of the demand for stability in inherently dynamic systems. Unrealistic demands for predictability from any side in the debate will simply be unattainable. Society's desire for stability in the forest and in the economic, biological, and social systems it supports is often at odds with the very dynamic nature of the forest, the econ-omy, and our society.

As currently drafted, the NFLC report will probably not have much influence on daily forest management issues, but the report has set up a dialogue, height-ened a regional and state-by-state awareness, created an environment that is con-ducive to cooperation, and fostered relationships that might help to ensure the future stability of our forests, forest-dependent economies, and high quality of life in northern New England.

Policy Review

Laura Falk McCarthy

The best lessons from the Northern Forest Lands Study and the Northern Forest Lands Council are in their process, not in their technical or scientific work. The effort was initially an opportunity to address changes in land ownership and to study the possibility of additional public land acquisition. Many people in the region, including forest landowners, local officials, and forest industry, labeled the study a federal land grab. Major environmental groups tried to elevate the land-use changes to an issue of national importance. The prospect of national attention further alarmed local interests, who believed they might lose control over the future of their region. These political dynamics, stemming from the region's cultural traditions and land ownership patterns, shaped the process by which the study and the council conducted their work.

The Context

The environment for finding common ground changed substantially as the process of determining the future of the northern forest region progressed. Early on, when conservationists and environmentalists joined forces with Senators Patrick Leahy and Warren Rudman to authorize the Northern Forest Lands Study, public land acquisition was assumed to be the likely outcome. The International Diamond land sale was viewed as a tremendous opportunity to preserve the remote and largely unspoiled northern forest, the greatest conservation opportunity since the Weeks Act was passed in 1911. While this view was prevalent among environmental interests, residents of northern forest communities were largely unaware of its potential implications.

The Governors' Task Force on Northern Forest Lands, which was formed to assist the Northern Forest Lands Study and ensure the involvement of diverse interests, included a few individuals who were opposed to a public acquisition agenda. The twelve-member body was made up of representatives from the timber industry, state government, and environmentalists from the four states. The Maine timber industry made it clear at the first meeting that it would never support public acquisition of timberland in Maine. In contrast, New York's representatives described the Adirondack Park as an example of public support for land acquisition and a model of working forest protection that could be applied to the entire northern forest region. The New York task force members, environmental representatives from the other states, and conservation-minded landowners from New Hampshire and Vermont created a majority of support on the task force for exploring public acquisition options and creative conservation solutions, such as easements allowing timber harvesting.

The political environment changed after the Draft Northern Forest Lands Study was released two years into the process. The draft study provided the public its first opportunity for meaningful input into the options being studied. Environmentalists were outraged that the study was not taking a serious look at the region's biodiversity and was not considering the opportunity to protect large, intact, and interconnected forested landscapes. On the other hand, property-rights advocates in New York were alarmed at the prospect of more "blue lining" and set out to warn northern forest residents in the New England states about the real-life hardships of living in the Adirondack Park. From this point onward, the political environment was highly contentious.

The Process

News of polarized communities from forestry debates in other regions of the country influenced the task force and study staff to involve people in the process. For example, the Northern Forest Lands Study coordinator countered the federal land grab perception by meeting face to face with the people who expressed the most concern. The task force organized public involvement sessions in each of the states to ask people about their vision of the northern forest in fifty to one hundred years.

Two years into the first phase, the study completed a problem statement that framed the issue as perceived by people who contributed to the process:

> Change threatens traditional land uses and lifestyles in the Northern Forest of New England and New York. Economic pressures on landowners, changing land values, and rising tourism have led to increasing development and are threatening the traditional uses of the land. In some areas the pattern of forestry, farming and recreation that evolved over three centuries is giving way to other uses. The important values enjoyed by people across the Northeast for generations are at risk.

The problem statement accurately captured the sentiments of people in the northern forest and beyond, and it spoke equally to property-rights advocates and environmentalists. Framing the problem in this way helped carve out some initial common ground, but it also indicated that the region was a long way from agreement on solutions.

In the second phase of the effort, people who feared that public interest in the northern forest would infringe on their private property rights became especially vocal. The Northern Forest Lands Council found considerable opposition to land acquisition as a solution. It also found people expressing the view that a coordinated, four-state effort was unnecessary government interference in an issue that should be resolved locally. The council responded with innovative methods of public involvement. It provided people with opportunities to voice their concerns and reduced tensions enough to allow the council to move forward with recommendations for change.

The council's approach was to combine research from acknowledged expert sources with the beliefs and personal expertise of people interested in the future of the northern forest. It established seven subcommittees to conduct assessments on a variety of topics and invited experts and key stakeholders to participate in the work. The subcommittees contracted new studies, reviewed existing information, and held issue forums with invited participants. Concurrently, the council organized a Citizen Advisory Committee in each state. Members were selected to include the full range of perspectives found in the northern forest region. The Citizen Advisory Committees served as a sounding board for the council through every stage of its work.

A wider public audience was invited to review and evaluate the research conducted by the council subcommittees. The council published research findings and a preliminary set of policy options and asked the public for extensive feedback. In large type on the cover of the publication was the message, "Your prompt feedback on these Findings and Options is critical to our development of draft recommendations." Inside the publication the council wrote, "Seeking your input at this stage is unconventional for advisory bodies of this kind, which often postpone public participation until draft recommendations have been developed. The Council felt uncomfortable taking such an approach." In effect, it provided a forum for thousands of people to help evaluate the facts and to express their own perceptions about changes taking place in their region. Many of the thousand responses were emotional, but, nonetheless, they allowed the council to integrate scientific information about the region with people's belief systems and values.

The council used the feedback on the findings and preliminary options to develop consensus recommendations. These were published as a draft report. Again, the public was invited to share their thoughts about the council's assessment of the situation and what to do about it. Even though the council's proposals were already a blend of expert analysis and local values, by providing another opportunity to review the material before it was finalized, the council reinforced the message that individual participation in shaping the region's future was important.

The Outcomes

The council's final report included thirty-seven recommendations. Several were adopted in proposed federal legislation, and many of the recommendations for state actions are being implemented. The council's process led it to adopt recommendations that were justified by scientific evidence and that already had public support. For example, among the council's most popular recommendations were its calls for strengthened state current-use taxation programs, changes to federal estate tax policies, and increased frequency of U.S. Forest Service data collection about forest conditions. The majority of recommendations advocated state initiatives that would affect the entire state, not just the northern forest region, or the continuation of existing federal assistance programs.

Interestingly, the two recommendations that were the most difficult for the council have received considerable attention since the council disbanded. The Maine timber industry, assisted by the property-rights movement, successfully prevented any large-scale public land acquisition. Together with a few other council members, the Maine timber industry was equally determined to prevent any meaningful discussion of forest sustainability or conservation of biological diversity. However, the public pressured the council to examine these issues.

Recommendations to address biodiversity and forest sustainability have resulted in significant policy activity. Maine, New Hampshire, and Vermont all initiated statewide biodiversity assessments. In Maine, the biodiversity assessment led to design of ecological reserves on state-owned lands to preserve a range of natural communities and ecosystems. New Hampshire and Vermont are also planning systems of ecological reserves. In 1997, the Vermont legislature placed restrictions on clear-cutting and a permanent moratorium on herbicide use in forestry.

Implementation of the council's recommendations for changes in federal policies has been difficult and disappointing. Changes in the federal estate tax on forested properties were included in the 1997 Taxpayer Reform Act and are the only changes in federal policy to date to result from the council's work. The Northern Forest Stewardship Act, introduced by Senator Leahy in 1994, was intended to draw visibility to the council's recommendations and initiate the implementation phase. The bill still has not been passed. The council hoped its joint efforts would lead to greater influence over federal actions; however, many of its recommendations, such as increasing funding of the Forest Legacy program, have gone unheeded.

Conclusions

In addition to implementation of some of the council's recommendations, the assessments compiled during the northern forest process continue to form the basis for state and local resource policy making. Investigations by the study and the council, and the experts they enlisted, frequently concluded that information was insufficient. The lists of information needs have been heeded. For example, in addition to the statewide biodiversity assessments, New Hampshire launched a study of its remaining spruce-fir habitat, while Vermont looked at the impacts of its property tax structure on forestland ownership. The information collected is too general to be used in resource management on what is still mostly private land, yet the assessments have paved the way for coordinated landscape management by willing landowners.

The legacy of the northern forest assessments is their process for involving people in determining the region's future. Three lessons are clear. First, assessments that blend analysis of scientific information with a process to involve the public in reviewing information may build broader credibility and support for their conclusions. The interpretation of scientific information is colored by people's values. It was important for residents of the northern forest region to have a role in evaluating research findings. Second, regional assessments need to consider the land ownership pattern, the regional culture, and the local political climate, both as topics to include in the assessment and in designing the process for conducting the assessment. In the northern forest, the predominance of private land created a need to involve people at state and local levels. Involvement was encouraged at every stage of the process—from determining what topics should be assessed to conducting the assessment, interpreting the information, and using the findings to develop policy recommendations. Third, the six-year northern forest process resulted in a series of incremental changes because it did not attempt to accomplish sweeping reforms. The Northern Forest Lands Council's recommendations were pragmatic, and achievable, and generated widespread support. As a result, many are being implemented and are helping the region adapt to changing circumstances.

CHAPTER 10

Southern California Natural Community Conservation Planning

Case Study

Dennis D. Murphy

Science Review

Peter A. Stine

Management Review

James E. Whalen

Policy Review

Ron Rempel, Andrew H. McLeod, and Marc Luesebrink

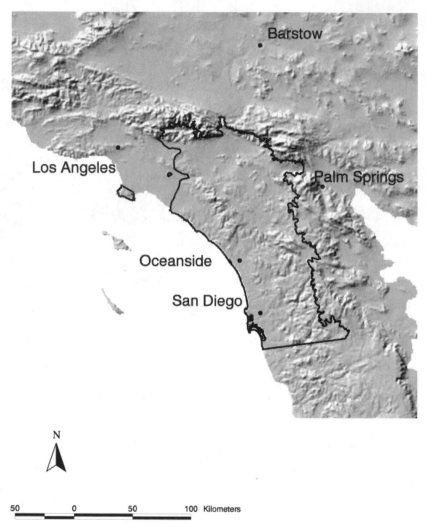

Southern California Natural Community Conservation Planning (NCCP)

Case Study

Dennis D. Murphy

Followers of emerging bioregional assessments 'may not initially recognize the relationship between such assessments and the scientific effort supporting Southern California's Natural Community Conservation Planning (NCCP) process and other efforts described in this volume. In contrast to assessments that attempt to characterize regional biodiversity on largely public holdings, NCCP is being carried out on private land, much of which is slated for residential and commercial development. While regional assessments of public lands generally have substantial baseline data, most of the biological information for private landscapes is found in locally focused environmental impact documents that assess lands on the urban fringe. These reports tend to be spare of detail and void of regional context, and, perversely, they assess only lands that will soon be lost to development. Add a usual lack of pertinent local scientific expertise, limited funding for ongoing studies, and daunting time lines for land-use decision making, and private lands planning efforts must operate on an especially limited base of ecological information.

Whether current attempts to engage both species conservation and economic growth on the same private landscape will succeed at a meaningful geographic scale—on the order of hundreds to thousands of square kilometers—is a wide open and oft-debated question. But the challenge looms inevitably in the face of impending major landscape changes and has the regulated community, environmental organizations, and policy makers in government fully occupied. Recent Endangered Species Act reauthorization bills propose dramatic changes in its implementation on private land, with landowner-friendly provisions that sport catchy labels like "no surprises" and "safe harbors" promising to facilitate multiple-species planning, and with inducements for property-owner compliance, including small land parcel waivers and prelisting agreements. Importantly, such provisions must be developed with both regional species conservation needs and local socioeconomic considerations in mind.

Since many of these progressive proposals find their roots, and their first field tests, in Natural Community Conservation Planning, application of biological information in this path-breaking regional effort warrants our careful consideration. Considering that nearly half of all federally protected species exist solely on private land and perhaps four-fifths occur at least in part there, biologists who have contributed to public lands planning exercises will find it hard to duck the challenge on private lands for long. As I hope the following discussion conveys, science in support of conservation on the private landscape demands a real creativity and, honestly, a scientific approach that can make the most of limited empirical and theoretical tools.

NCCP and Its History

Early in 1991, California Governor Pete Wilson proposed Natural Community Conservation Planning as part of his omnibus Resourceful California program, an effort created on the one hand to resolve the tension surrounding federal mandates to protect habitat for imperiled species and on the other to meet an increased demand for new housing and infrastructure (Governor's Office 1991). As a direct result of that initiative, the California legislature passed the Natural Community Conservation Planning Act of 1991. The statute authorizes individuals and government agencies to prepare Natural Community Conservation Plans based on agreements with and guided by the California Department of Fish and Game. Each plan should promote "protection and perpetuation of natural wildlife diversity while allowing compatible and appropriate development and growth" (California Fish and Game Code).

The NCCP program, paralleling regional Habitat Conservation Planning (HCP) efforts elsewhere across the country, represents a paradigm shift away from existing models of strictly reactive, project-based permitting that emphasize a species-by-species approach to the protection of biodiversity. Typically, this previous approach has done little to stem the tide of rapidly diminishing biotic diversity on private land; a myopic focus on individual development projects has led to further species attrition and fragmentation of remaining habitat. It also has sparked expensive, protracted administrative and legal battles, as individual landowners fighting the broad charge of regulations focus on what they view as most egregiously unfair—the Endangered Species Act of 1973.

With NCCP as its pilot effort, initiatives on private land have shifted toward front-loaded, ecosystem-based planning to identify and protect important habitat areas and their resident native species in advance of land development. The explicit program goal is to avoid species listing and economic displacement. Eventually, specific areas identified through this planning process are expected to become part of a scientifically validated system of reserves, including corridors and linkages with other natural lands, that will be managed for long-term protection of multiple species and other ecological values.

NCCP promises something for everyone involved in endangered species conflicts. It reduces the likelihood of additional species listings on the same landscape by offering protection to organisms before they become threatened or endangered. This provides both economic and environmental returns. Identifying important habitat in advance of development introduces more certainty into the regulatory permitting process, an issue of great concern to landowners and developers (Thornton 1991). Many scientists and environmentalists support the concept; by protecting biotic diversity at the level of a natural ecological community, a more thorough and effective conservation strategy emerges (Brown 1991). Regulators are served as well. Exhausted by the trench warfare of project-by-project permitting, they desire to shift much of the responsibility for day-to-day implementation of endangered species mandates to local jurisdictions (Dwyer, Murphy, Johnson, and O'Connell 1995b).

Pilot Program in Southern California

Formal implementation of the NCCP pilot program started in September 1991, with a Memorandum of Understanding between the California Department of Fish and Game and the United States Fish and Wildlife Service (California Department of Fish and Game 1991). The agencies selected the coastal sage scrub of Southern California as the pilot natural community for NCCP. Coastal sage scrub is a vegetational community that supports a diverse assemblage of plants and animals native to coastal and subcoastal southwestern North America. A majority of its historical extent has been lost to urban sprawl and agriculture. Residing in the remaining habitat patches are more than twenty vertebrate species that have been candidates for federal protection, several invertebrate candidates, and nearly one hundred plant species that are of special conservation concern (Natural Diversity Data Base 1992). Because coastal sage scrub exists within a natural mosaic of habitats that includes grasslands, chaparral, oak woodlands, and riparian habitat, a regional sage scrub NCCP will by extension conserve a yet broader range of plant communities and the animals that inhabit them (figure 10.1). As originally conceived, the geographic area of the NCCP pilot program was to cover approximately six thousand square miles of coastal sage scrub and other natural communities within and surrounding an urban matrix in five California counties—Los Angeles, Orange, Riverside, San Diego, and San Bernardino—and was to provide conservation through statutory authority for the broadest possible scope of biotic diversity.

In 1993, the Department of the Interior established additional federal authority and interest in promoting conservation planning of this type when it listed the California gnatcatcher, a bird endemic to this region and to coastal sage scrub, as threatened. It employed a seldom used section 4(d) special rule provision of the federal Endangered Species Act to commit formally to share lead agency status for conserving the species with the state. Special rules are applied only when a species is listed as threatened (versus endangered) and give the secretary of the interior and the state greater managerial discretion when determining a strategy to protect a species. At that time, Secretary Bruce Babbitt provided a now oft-quoted endorsement of the fledgling state program when he blessed it, saying: "The only effective way to protect endangered species is to plan ahead to conserve the ecosystem upon which they depend. . . . This [NCCP] may become an example of what must be done across the country if we are to avoid the environmental and economic trainwrecks we have seen in the last decade" (USFWS 1993b).

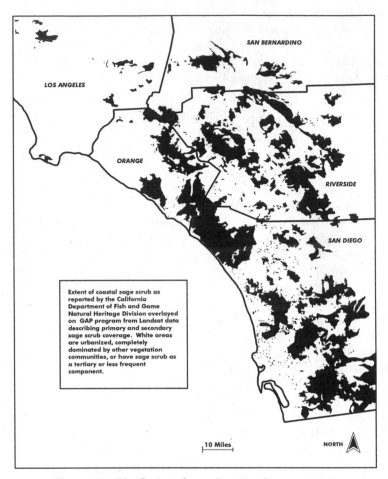

Figure 10.1. *Distribution of coastal sage scrub across coastal and subcoastal California.*

The New Approach

Certainly the stage was set for Babbitt's trainwrecks. Section 9 of the federal Endangered Species Act prohibits "take" of listed species wherever they occur and broadly defines take to include acts that "harass, harm, pursue, hunt, shoot, wound, kill, trap, capture, or collect" imperiled species. Importantly, legal and regulatory interpretation of this "harm" provision of the statute is understood to include acts that modify habitat in ways sufficiently severe as to make likely the death or destruction of a species (Meltz 1994). Following a listing, there can be substantial limits placed on private land uses, including development, farming, and resource extraction. Only when a habitat conservation plan or other forms of management agreements have been completed and approved by the Fish and Wildlife Service can a landowner receive a section 10(a) permit to take an imperiled species "incidental" to otherwise legal activities. But the process to obtain a section 10(a) permit is arduous, expensive, and time consuming and has fueled much private-sector anger and opposition to the act (Brown 1994).

The 4(d) rule accompanying the California gnatcatcher listing creates an "interim take" mechanism, releasing some of the economic tension while regional planning grinds forward. Advised by consulting scientists, regulators placed a 5 percent cap on loss or conversion of existing coastal sage scrub habitat during an interim land planning period (see Reid and Murphy 1995). The cap was intended to function as a pressure valve, releasing smaller, degraded, or isolated patches to development but inhibiting further subdivision of the limited remaining high-quality habitat. While such losses were expected to result in the incidental take of a limited number of gnatcatchers, the figure represented a conservative estimate of the amount of habitat that could be lost without foreclosure of future conservation options. The cap ensured maintenance of important wildlife corridors and overall contiguity of the remaining coastal sage scrub habitat while implementing agreements were being prepared. Nominal mitigation measures and eventually effective management of the remaining coastal sage scrub to better sustain species of conservation concern would compensate for the 5 percent loss (CDFG 1993b; USFWS 1993a).

The end of the interim period would be signaled by the acceptance by the agencies of long-term NCCP implementing agreements for subregional planning areas. Comparable to HCPs, which traditionally have focused on individual species on one or a few landholdings, these agreements were to incorporate scientific studies, maps of the resources to identify habitat areas essential to the persistence of the gnatcatcher and other targeted species, and provisions for long-term management of conserved habitat areas (CDFG 1993b; Meade 1995). In return, the state and federal wildlife agencies would explicitly acknowledge that the agreements meet the requirements of a state management agreement and a federal habitat conservation plan, under the state and federal Endangered Species Acts, respectively.

This new approach comes none too soon. With rapid urbanization of natural lands severely threatening biotic diversity, Southern California has been identified as one of four national "hot spots" for imperiled species (Dobson et al. 1997). San Diego County alone is home to twenty-four plant and animal species that are formally protected or are experiencing significant declines (Fairbanks and Toma 1994). Orange County has thirty-five plant and animal species that are listed or are severely imperiled but not yet protected (Meade 1995). Upward of 90 percent of native grasslands, wetlands, and other natural habitats have been lost as a result of urban and agricultural development in the region (Noss 1994).

That biological reality runs smack into a human demographic one. More than 50 percent of California's population live in the subcoastal south, more than half of that in the NCCP area. San Diego and Orange counties are respectively the second and third most populated counties in California (Fay 1995). By the year 2040, the NCCP counties are expected to double their 1990 population (USFWS 1993a). Rapid development of the remaining land base stemming from these forecast population increases will obviously and quickly exacerbate the problem of imperiled species. A 1992 California Department of Fish and Game report not surprisingly described the development that attends population growth, be it residential, industrial, or commercial, as the primary cause of adverse impacts on federal- and state-listed plants and animals (CDFG 1992; Bowman 1990). Also not surprising is that a substantial portion of the private land ownerships on Southern California's urban fringe have investment-backed expectations of future development.

A thorough discussion of the potential legal trainwreck where the tracks of species protection and constitutional private-property guarantees may soon intersect is beyond this presentation. To date, no such case has found its way to the Supreme Court. Nonetheless, the proper mix of conditions exists in Southern California. In 1979, the Ninth Circuit Court in *Palila v. The Hawaii Department of Land and Natural Resources* formally expanded the scope of the Endangered Species Act to private land. In 1995, the highest court affirmed *Palila* in the case *Babbitt v. Sweet Home Chapter of Communities for a Greater Oregon* and in so doing validated the U.S. Fish and Wildlife Service's prohibition of acts that significantly modify or degrade habitat. But the Supreme Court has equally staunchly come down on the side of landowners in a passel of property rights cases, including several that test those rights in the context of other prohibitive policies designed to protect the environment (Dwyer and Murphy 1995; Dwyer, Murphy, and Ehrlich 1995a). Moreover, the court has affirmed the right of economically aggrieved parties to challenge agency decisions under the act in *Bennett v. Spear*, which seems to make a showdown inevitable. But for the habitat conservation planning process, the requirement for habitat protection on private land runs headlong into Constitutional prohibitions against federal condemnation of property without just compensation.

No means existed to remedy conflicts between species conservation and economic development on private land until 1982, when Congress revised the act to allow for incidental take of protected species under section 10(a) of the Endangered Species Act, after a landowner submits a satisfactory HCP. While habitat conservation planning does allow some take of individuals of a species, it was conceived to provide for more comprehensive biological management in return (see generally Beatley 1994). In an early institutional embrace of the oxymoronic concept of "sustainable development," Congress clearly hoped that such plans would find compatibility between development and conservation (Mann and Plummer, 1995).

Unfortunately, habitat conservation plans seem not to have provided the hoped for remedy. A review of the scholarly literature finds it replete with pointed criticisms of HCPs. Houck (1993) believes that habitat conservation planning is "taking place in a largely ad hoc . . . fashion," with the result being "a gradual attrition of habitat to a baseline of [species] survival." Bosselman (1992) notes that the HCP process does not begin to operate until a species is listed as threatened or endangered. Even then the plans tend to protect only listed species, doing little to protect other species that may inhabit a particular declining habitat and might soon themselves need special protection. One major complaint about HCPs, past and present, is that they have provided minimal regulatory certainty to landowners. Thornton (1991) portrays the process as "severely troubled," criticizing "the inability of the HCP to remove legal risks associated with the subsequent listing of species not addressed in the [original] HCP." A pervasive complaint among proponents of species protection is that habitat conservation plans have been poorly informed scientifically and require inordinate amounts of time for completion.

Because the Supreme Court has shown a propensity for protecting landowner rights against regulatory "takings" of property of certain types and extents, the U.S. Fish and Wildlife Service engineers HCPs to avoid legal challenges. To speed processing, and to avoid court, FWS tends to use available scientific information, eschewing demands for extensive additional studies that might reduce uncertainties about species' fates under alternative planning scenarios. In other words, the sort of assistance that empirical data and sophisticated analyses have provided regional planning efforts on public-lands tends to fall outside of the private-lands planning framework. Scientific support to private-lands planning is not necessarily unreliable; it is, however, often limited.

Science and NCCP

With that background, suffice it to say, attempts to bring reliable scientific knowledge to the first Natural Community Conservation Planning effort have not satisfied everyone. Indeed, a number of scientists have been vocal critics of the scope of biological input to the NCCP's effort to date. Certainly, the lack of empirical data on species of concern has forced decision makers to exercise their responsibilities with the spare "best available information." What was developed by scientists to support the program vision was a hybrid—part multiple-species, multiple-habitat HCP on a very large spatial scale and part innovative, locally based conservation planning (Reid and Murphy 1995). And, criticism notwithstanding, the program moves forward absent of substantive legal challenge, suggesting that the scientific contribution to date has met at least minimal standards.

The state of California's Resources Agency (the administrative umbrella over the Department of Fish and Game) initiated the program by engaging a five-member multidisciplinary team of conservation biologists and biogeographers drawn from academic institutions. It dubbed the team the Scientific Review Panel (SRP) and engaged its members to create conservation guidelines to provide an ecological framework to the process (Murphy 1992b). By hiring a team of "independent" scientists to drive the development of the program, California hoped to ensure that its claims that NCCP would be "scientifically justified" and "scientifically based" (CDFG 1991). Five early goals of the panel were to: (1) define the coastal sage scrub planning area, (2) recommend appropriate techniques for future data gathering and analysis, (3) define subregional areas that consider landscape features and biological parameters, (4) create a conservation planning framework and management guidelines for developing a viable system of regional reserves, and (5) produce subregional conservation planning guidelines (Murphy 1992b). The first three of these goals were relatively readily met, but due to a shortfall of spatially specific information, an alternative approach had to be employed with the last two goals.

Setting about its task, the Scientific Review Panel identified three "target" species for study—the California gnatcatcher, the coastal cactus wren, and the orange-throated whiptail lizard. The panel studiously avoided labeling the vertebrates "umbrella" species. The three species were selected because each was thought to be particularly habitat specific to coastal sage scrub and all were thought to have resided in the portions of the scrub that had previously suffered the greatest losses. As such, the panel thought that they could serve as surrogates, offering some level of protection for vertebrates and invertebrates that were candidates for federal protection and for scores of rare plants dependent on the scrub ecosystem (Scientific Review Panel 1993; CDFG 1993b; Murphy 1992a).

The target species approach provided an essential focus for the NCCP program because it simplified the biodiversity challenge, helping landowners and local government understand that broader conservation goals were achievable and that provision for regulatory coverage of multiple species was possible. But the strategy has some obvious limitations. The SRP recognized that no individual plant or animal species could serve as a fully effective umbrella, conferring protection to all co-occurring species. Additionally, no single species is an entirely reliable indicator of coastal sage scrub conditions across the regional planning area (CDFG, 1993). Also needing attention are species whose distributions do not coincide snugly with the sage scrub community and whose continued existence cannot be assured through the protection of representative examples of that community alone.

The necessary use of surrogate species did little to remedy the greater problem of a pervasive data debt—for even these targets rudimentary demographic information was lacking. The panel established rigorous survey guidelines for those species by the panel, but little immediate progress in data collection was made. With a housing recession in full swing, landowners were carrying out few environmental studies, and with no clear mandate for government intervention to gather data from private property, the mainstays of modern biological input to planning—population viability analyses, spatially explicit population modeling efforts, and such—could not be carried out.

Nevertheless, an early successful application of existing data allowed the science to guide an effort to "subregionalize" the NCCP program area. Ideally, the entire planning area would have been treated as a single conservation planning unit, but the five-county area included more than fifty cities and thus was administratively unmanageable as a whole. Not legally authorized to draw de facto political boundaries on maps, the panel (using geographic information system products from the California Gap Analysis Program) instead directed placement of those boundaries by identifying thirteen habitat "patches" of the highest quality to serve as "subregional focus areas"—areas that when treated as integrated units would assist in meeting planning goals (figure 10.2). These habitats have the most extensive areas of primary and secondary coastal sage scrub cover, adjusted for "edginess" and percentage of perimeter bordering natural vegetation. The panel suggested that boundaries be drawn in intervening areas of lower quality or noncoastal sage scrub, thereby reducing further fragmentation of the best intact, remaining habitat. In essence, the panel directed the determination of program boundaries without itself drawing the lines.

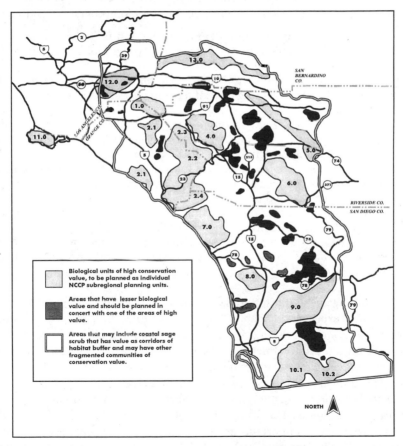

Figure 10.2. *Subregional planning areas. Shaded numbered areas are core habitat planning foci. Areas without numbers are satellite planning areas to be planned for with core areas. Nonhabitat matrix is white.*

Bridging the Data Gap

Recognizing that the systematic data shortfall limited the possibility of completing a scientifically defensible reserve design and management plan in the immediate future, the Scientific Review Panel chose instead to recommend an ambitious program of research with six interactive tasks (see Stine, this chapter). By nearly any measure the research agenda was ambitious. Limited by funding, now five years after it was proposed, results are mixed, although information important to planning has accrued, including better understanding of wildlife and habitat relationships, distributions and relative abundances of some of the rarest species, and the environmental factors that regulate others. Additionally, substantial attention was to be paid to issues central to the ongoing management of coastal sage scrub, a vegetation community shaped by cycles of frequent fire and regrowth. The Scientific Review Panel encouraged the creation of an oversight committee to coordinate a management and restoration program to consider:

- the role of fire in natural ecosystem dynamics and processes, including the application of managed burns and the control of ignitions of accidental and vandal origin;
- identification of restoration unit sizes, including identification of maximum areas that are restorable using current techniques and focus on patch enlargement techniques;
- identification of coastal sage scrub responses to soil conditions in restoration efforts, with focus on soil structure, soil nutrient levels, organic matter content, water-holding capacity, and soil compaction;
- identification of appropriate seeding, outplanting, and irrigation techniques, with focuses on proper mixes of species, seeding techniques, and timing of applications of seed and irrigation;
- identification of techniques to encourage native herbaceous species and to discourage the establishment of exotic species;
- establishment of realistic success criteria to evaluate management and restoration efforts considering sage species diversity and cover and use by target species.

The management and restoration committee was expected to design multifactorial field experiments at appropriate spatial scales using explicit and repeatable scientific methods to aid in differentiating among alternative techniques. Since the outcome of treatments will vary with physical circumstances, local vegetation composition and structure, and other unique conditions, each subregional planning unit was expected to contribute to the regional management and restoration research effort. A centralized effort has not yet been initiated.

Despite the limited data from portions of the overall planning area, a number of considerations of importance to conservation planning emerged from available empirical information and natural history observations. The Scientific Review Panel found that the coastal sage scrub community is inherently transitory over substantial parts of its range; where natural successional processes are disrupted, it must be managed to retain its ability to support species of concern over the long term. Correspondingly, the species that depend on coastal sage scrub show transitory residence. Many characteristically have short lifetimes, high potential rates of reproduction, small home ranges, and dramatic population fluctuations and are highly susceptible to local extirpation. The panel found no evidence to show that continuous coastal sage scrub areas constitute optimal habitat for sage-dependent species. Rather, scrub areas with open canopies in matrices with other habitat types may best provide for target species and other species of concern.

The panel also found that small habitat patches surrounded by development, even if appropriately managed, are usually destined to lose significant fractions of their species diversity in coming decades. It found that certain subhabitats, such as those at low elevations, those close to the coast, and those with reduced slope, were already in precarious situations and likely have been subject to multiple recent local extirpation events.

Furthermore, the NCCP planning area is embedded in a matrix of lands that has been greatly impacted by past and ongoing human activities. Natural and anthropogenically generated disturbances will continue; many of those disturbances serve to reduce the capacity of coastal sage scrub habitats to support many species of concern. Areas designated as reserves are thus unlikely to be self-sustaining (that is, provide for natural, dynamic ecosystem processes) or to be capable of maintaining viable populations of target species without active management. The ability of individual patches of coastal sage scrub habitat to be managed effectively over the long term is viewed as a critical factor in the prioritization of conservation efforts.

To allow for reasonable land development while avoiding foreclosure of future planning options, three key objectives necessarily had to be met during the interim data acquisition period. First, as described previously, development was limited to 5 percent or less of the landscape that is currently occupied by coastal sage scrub. Development was to strive not to disproportionately impact any "environmental subunit" (defined by vegetational subcommunity, elevation, slope, aspect, latitude, distance from coast, and substrate) within each subregional NCCP planning area. Second, development should, to the extent feasible, avoid likely hot spots of biotic diversity on the basis of habitat size and isolation criteria (as described in the "decision tree" in the Draft Conservation Planning Guidelines, figure 10.3). Third, development was not to sever existing open-space landscape linkages between subregional focus areas.

To meet these objectives, habitat not occupied by target species and even some landscape that was not coastal sage scrub had to be conserved. The coastal sage scrub plant community is scattered throughout the planning region in patches that are interdigitated with other natural habitats. Although each subregional plan was to focus on coastal sage scrub, other contiguous natural communities are necessary to provide for healthy functioning reserve areas and to facilitate gene flow, migration, and recolonization between biodiversity hot spots and

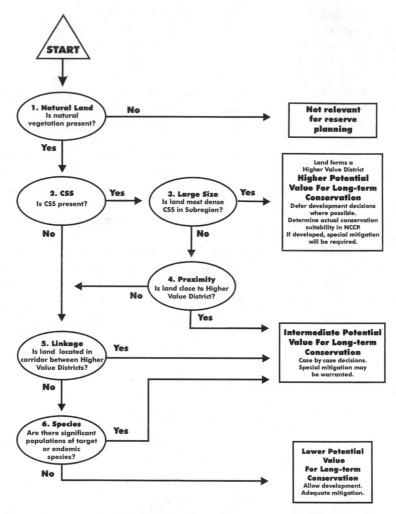

Figure 10.3. *Evaluation logic flow chart for determining habitat value for planning purposes.*

other conserved patches of coastal sage scrub. In other words, the best plan for the coastal sage scrub ecosystem is one that integrates effective protection for other natural ecological communities in Southern California.

In sum, the short-term goal of the NCCP during the interim planning period of data acquisition was to avoid foreclosure of future conservation planning options. Again, without direct empirical justification for subregional reserve designs of explicit size and shape, the Scientific Review Panel lifted from the planning effort for the northern spotted owl in the Pacific Northwest a set of conservation planning "tenets" (see Murphy and Noon 1992; Wilcove and Murphy 1991) and adapted them to assist coastal sage scrub planning at all spatial scales in regional and subregional planning. The tenets are:

- *Species that are well distributed across their native ranges are less susceptible to extinction than are species confined to small portions of their ranges.*

The regional planning area encompasses all of the known remaining coastal sage scrub habitat in California south of the northern extent of metropolitan Los Angeles. Ultimately, the panel suggested that NCCPs for coastal sage scrub habitats in areas from Ventura County northward should also be carried out and that international efforts should focus on preserving this community and the species it supports in Baja California as well. Such a strategy is the only one that will insure maintenance of the full range of genetic, morphological, behavioral, and ecological variation within coastal sage scrub species. This geographic variation is necessary for those species inhabiting sage scrub to adapt to future environmental changes.

- *Large blocks of habitat containing large populations of the target species are superior to small blocks of habitat containing small populations.*

While the persistences of all populations of coastal sage scrub species are subject to the effects of normal random environmental events (environmental stochasticity) and catastrophes such as wildfires and severe drought, small populations are disproportionately threatened by random variations in birth or death rates (demographic stochasticity) and random changes in genetic composition (genetic stochasticity). Subregional conservation planning designed to mitigate the combined effects of these factors was to focus on large blocks of high-quality coastal sage scrub habitat.

- *Blocks of habitat that are close together are better than blocks far apart.*

An arrangement of habitat patches that facilitates dispersal of individuals among patches is necessary to encourage the natural dispersal that contributes to demographic "rescue effects" (whereby populations are supported by migrants) and continued genetic interchange. All else being equal, patches of coastal sage scrub that are closer to other patches of coastal sage scrub are more likely to support habitat-specific species for longer time periods than will more isolated patches.

- *Habitat that occurs in blocks that are less fragmented internally is preferable to habitat that is internally fragmented.*

Viable populations of many coastal sage scrub species are likely to require large blocks of undisturbed habitat where the presence of disruptive edge-dwelling species, such as cowbirds and house cats, is minimized. Habitat highly fragmented by disturbed or developed lands has relatively little conservation value for species that exhibit high habitat specificity. Species that are susceptible to the deleterious consequences of edge are more likely to retain populations in habitat patches that tend to be rounded or squared than in patches that are more elliptical or rectangular, where those patches are surrounded by disturbed or developed land. In such circumstances, small, linear strips of habitat that maximize the ratio of edge to area are least desirable.

• *Interconnected blocks of habitat serve conservation purposes better than isolated blocks, and habitat corridors or linkages function better when the habitat within them resembles habitat that is preferred by target species.*

Interpopulation dispersal, as mentioned earlier, is important for regional species persistence. Thus, the incorporation of historic and other open-space corridors that provide habitat for plants and animals and that can serve as conduits for interpatch movement is central to conservation planning. Corridors were to be designed to serve coastal sage scrub–specific species and other species that contribute to the maintenance of ecosystem function.

• *Blocks of habitat that are roadless or otherwise inaccessible to humans better serve to conserve target species than do roaded and accessible habitat blocks.*

Human contact is thought to be a major cause of decline in certain coastal sage scrub–specific species, thus populations in habitats that are inaccessible to motorized recreation or similar activities are more likely to persist than those in habitats where human access is less restricted.

The panel intended that the recommended field research studies and analyses be carried forward before subregional plans were finalized, but the press of development and the undesirability of a blanket moratorium on construction meant that science and land planning would move forward simultaneously. For that reason, the application of the planning tenets has been the single dominant "scientific" exercise in setting more permanent reserve boundaries. This application has been combined with GIS-based maps of vegetation types and the physical landscape, survey data that identify areas of high target-species density and maps of the habitats of extremely rare or narrowly distributed species. An arguably elegant set of six conservation planning rules is thus being used to reduce the likelihood that the current data deficit will result in future species losses.

The simplistic NCCP approach, however, makes adaptive management an obligatory program component. Plan amenability is essential; new information will have to be integrated into "finalized" subregional plans well into the future. And, since the NCCP program is operating under the Fish and Wildlife Service's "no surprises" policy, which promises landowners who meet statutory requirements for the protection of targeted species relief from the economic consequences that may result from new information or changed circumstances, sources of governmental funding to pay for plan amendments will be critical to the program for its duration.

With adaptive management promising to provide the necessary safety net for the NCCP, as well as other regional conservation strategies, one as yet unrealized, scientific challenge looms. That is, the development of a monitoring program that brings to planners the information necessary to adjust management, or even amend reserve boundaries, if necessary. Scientists will have to engineer a data acquisition and analysis program at appropriate (as yet undetermined) spatial and temporal scales with requisite statistical rigor. Furthermore, managers and landowners must realize the vital importance of monitoring in closing the loop of adaptive management. A meaningful monitoring strategy has to address important information gaps, risks and uncertainties, and hypotheses inherent in management plans and reserve designs. Without monitoring, or without a clearly focused monitoring strategy, it will be nearly impossible to ascertain the appropriateness of actions in achieving conservation goals or to adapt and refine plans accordingly.

Conclusion

A universal application of bioregional assessments in providing the foundation for ecosystem management and conservation planning requires reconsideration where landscapes are dominated by private property. New administrative initiatives are encouraging multiple-species, ecosystem-based regional planning efforts under the Endangered Species Act, but the necessary scientific information to inform these efforts lags far behind. As a result, private-lands planning rarely enjoys the benefits of guidance gleaned from previous research and synthesis.

The application of simple rules drawn from biogeography and population ecology, combined with natural history observations and limited studies on target species, can provide a rudimentary framework for planning. Nonetheless, many reserve design activities will move forward absent guidance, which means that conservation plans will need to incorporate research into management activities so that future decision making can be scientifically informed. The application of adaptive management tools in the service of amendable plans will be the means of providing the best possible technical information to conservation efforts on private lands.

LITERATURE CITED

Beatley, T. 1994. *Habitat conservation planning, endangered species, and urban growth.* Austin: University of Texas Press.

Bosselman, F. P. 1992. Planning to prevent species endangerment. *Land Use Law* (March), 3–8.

Bowman, R. I. 1990. Evolution and biodiversity in California. In *California's wild heritage: Threatened and endangered animals in the golden state*, edited by Peter Steinhart. Sacramento: California Department of Fish and Game.

Brown, B. 1991. Landscape protection and The Nature Conservancy. In *Landscape linkages and biodiversity*, edited by Wendy Hudson. Washington, DC: Island Press.

Brown, C. 1994. Landowners rally for laws protecting rights. *Dallas Morning News*, p. 43-A, August 28.

California Department of Fish and Game (CDFG). 1991. Memorandum of understanding by and between the California Department of Fish and Game and the United States Fish and Wildlife Service regarding coastal sage scrub Natural Community Conservation Planning in Southern California. Sacramento.

CDFG. 1992. Annual report on the status of California state listed threatened and endangered animals and plants. Sacramento.

————. 1993a. Memorandum to jurisdictions within the NCCP area. Implementation of the NCCP interim coastal sage scrub habitat loss process and the federal Endangered Species Act 4(d) special rule for incidental take of the California gnatcatcher. Sacramento.

————. 1993b. Southern California coastal sage scrub Natural Community Conservation Planning process guidelines, amended. Sacramento.

————. 1995. Natural Community Conservation Planning: Innovation in multi-species protection in the coastal sage scrub habitat of southern California, report to the legislature. Sacramento, August.

Dobson, A. P., J. D. Rodriguez, W. M. Roberts, and D. S. Wilcove. 1997. Geographic distribution of endangered species in the United States. *Science* 275:550–553.

Dwyer, L. E., and D. D. Murphy. 1995. Fulfilling the promise: Reconsidering and reforming the California Endangered Species Act. *Natural Resources Journal* 35:735–770.

Dwyer, L. E., D. D. Murphy, and P. R. Ehrlich. 1995. Property rights case law and the challenge to the Endangered Species Act. *Conservation Biology* 9:725–741.

Dwyer, L. E., D. D. Murphy, S. P. Johnson, and M. O'Connell. 1995b. Avoiding the trainwrecks: Observations from the frontlines of Natural Community Conservation Planning in Southern California. *Endangered Species Update* 12(12):5–7.

Fairbanks, J., and L. X. Toma. 1994. Room to roam. *Planning* 10:24–26.

Fay, J., ed. 1995. *California almanac.* Santa Barbara, CA: Pacific Data Resources.

Governor's Office. 1991. Wilson proposes resourceful California—Major environmental plan to invest in our children's natural inheritance. Press release, Sacramento.

Houck, O. A. 1993. The Endangered Species Act and its implementation by the U.S. Departments of Interior and Commerce. *University of Colorado Law Review* 64:277–370.

Mann, C. C., and M. L. Plummer. 1995. *Noah's choice: The future of endangered species.* New York: Alfred A. Knopf.

Meade, R. J., Consulting. 1995. Draft Natural Community Conservation Plan & Habitat Conservation Plan, County of Orange, central and coastal subregion, parts I & II: NCCP/HCP. Santa Ana, CA: County of Orange, Environmental Agency.

Meltz, R. 1994. Where the wild things are: The Endangered Species Act and private property. *Environmental Law* 24:369–417.

Murphy, D. D. 1992a. *NCCP special report no. 2: The California coastal sage scrub community conservation planning region.* Sacramento: California Department of Fish and Game.

————. 1992b. *NCCP special report no. 1: The California coastal sage scrub Scientific Review Panel: its purpose and approach.* Sacramento: California Department of Fish and Game.

Murphy, D. D., and B. R. Noon. 1992. Integrating scientific methods with habitat conservation planning: Reserve design for northern spotted owl. *Ecological Applications* 2(1):3–17.

Natural Diversity Data Base. 1992. Descriptions of habitat types for species associated with coastal sage scrub in the Natural Community Conservation Planning regions of Southern California; sensitive species associated with Southern California coastal sage scrub identified by the Scientific Review Panel "survey guidelines." Sacramento: California Department of Fish and Game, May 3.

Noss, R. F. 1994. California's ecosystem decline. *Defenders* 69(4):34–35.

O'Leary, J. F., S. A. Sesimone, D. D. Murphy, P. F. Brussard, M. S. Gilpin, and R. F Noss. 1994. Bibliographies on coastal sage scrub and related malacophyllus shrublands of other Medieterranean-type climates. *California Wildlife Conservation Bulletin* No. 10. 51 pp.

Reid, T. S., and D. D. Murphy. 1995. Providing a regional context for local conservation action: A Natural Community Conservation Plan for the Southern California coastal sage scrub. *BioScience* 45:584–590.

Scientific Review Panel (SRP). 1993. A description of scientific elements and conservation prescriptions of the Natural Community Conservation Planning program for the coastal sage scrub community of Southern California. On file with authors.

Thornton, R. D. 1991. Searching for consensus and predictability: Habitat conservation planning under the Endangered Species Act of 1973. *Environmental Law* 21:605–666.

USFWS. 1993a. Final environmental assessment of the proposed 4(d) rule to define the conditions under which incidental take of the coastal California gnatcatcher would not be a violation of section 9 of the Endangered Species Act. U.S. Department of the Interior, Washington, DC.

———. 1993b. Gnatcatcher to be listed as "threatened"; Interior's Babbitt promotes regional conservation efforts. News release, U.S. Department of the Interior, Washington, DC.

Wilcove, D. S., and D. D. Murphy. 1985. Conservation strategy: The effects of fragmentation on extinction. *American Naturalist* 125:879–887.

Wilcove, D. S., and D. D. Murphy. 1991. The spotted owl controversy and conservation biology. *Conservation Biology* 5:261–262.

Science Review

Peter A. Stine

The goal of the Natural Community Conservation Planning Program is far reaching: to provide a mechanism for linking a regional conservation perspective with local land-use authorities, addressing the conservation needs of species parcel by parcel (i.e., local jurisdictions) while keeping the entire range and status of the species in mind. The initial program for NCCP, focused on the coastal sage scrub community of coastal Southern California, has proven to be a significant challenge for a fledgling program as ambitious and new as NCCP. My treatment of this subject will be confined to the scientific aspects of data gathering, development, and synthesis necessary to support the program.

Scientific Review Panel

At the outset of this program, the Resources Agency of the state of California, in cooperation with the U.S. Fish and Wildlife Service, Department of Interior, established a panel of scientists to assess the body of existing knowledge and synthesize this information into recommendations for the respective wildlife resource management agencies of these state and federal government entities. The five-member panel included Dennis Murphy from Stanford University, Michael Gilpin from the University of California at San Diego, Peter Brussard from the University of Nevada at Reno, Reed Noss from Corvallis, Oregon, and John O'Leary from San Diego State University. The recommendations of this group were expected to assist the wildlife resource agencies with their responsibilities for carrying out the NCCP program, including the array of decisions dealing with selection of lands to incorporate into a reserve system. These are enormously difficult decisions under any circumstances, particularly when land values are so high and development pressures strong. Expectations of the esteemed group of five scientists were high.

Major Findings of the SRP

Initial efforts of the Scientific Review Panel focused on setting a biogeographical context for the conservation strategy and assembling available information. Members of the SRP produced "Special Reports" in 1992, summarizing our knowledge of the status of coastal sage scrub and defining subregions of extant coastal sage scrub. The status report (O'Leary et al. 1992) depended largely on a handful of studies conducted in the 1970s and 1980s by academic scientists as well as some data collected by consultants on behalf of landowners and local jurisdictions. This collection of studies, composed of work from a variety of sources, constituted the body of available information. The SRP discovered that comprehensive data sets were difficult to find, and detailed autecological information on even high-profile species such as coastal California gnatcatchers (*Polioptila californica californica*) was virtually absent.

As it assembled the data, mostly in the form of "gray" literature from incomplete data sets and reports, the SRP concluded that there were few well-developed data sets regarding biological elements of coastal sage scrub. Of the existing data sets known, some were difficult to access because the data were collected on private lands. It became clear to the SRP that additional data should be developed to provide the scientific foundation needed to support land-use decisions that would result in wholesale changes to landscape and the remaining coastal sage scrub habitat on that landscape. Lacking all of the desired biological information, the SRP was faced with providing the best scientific guidance possible with the available information. The political and economic interests involved in this program would dictate the pace of land-use planning.

At this point, the task of the SRP shifted toward summarizing what information was available, providing conservation recommendations, and developing a research agenda for development of the additional information deemed most important for the program. In 1993, the SRP published its findings, which included: (1) delineation of a list of twenty-five vertebrate species, seven invertebrate species, and sixty-one plant species considered to be sensitive and associated with coastal sage scrub habitat; (2) development of standard survey guidelines; (3) description of some basic research and monitoring needs for the purposes of developing data for NCCP planning; (4) a detailed summary, in outline form, of the existing data; (5) a bibliography of the published and unpublished data; and (6), most notably, the Conservation Guidelines. The SRP recognized that the existing information was incomplete, even for some of the better understood elements of coastal sage scrub, and that it would need to take a different approach to providing scientific guidance for the NCCP program.

Conservation Guidelines

Although the SRP did not intend to provide specific prescriptions for delineating of a reserve system, the Conservation Guidelines became the focal point for the recommendations it could offer. The lack of detailed data limited what the SRP could conclude, and the Conservation Guidelines relied heavily on more general conservation biology tenets. The SRP concluded that such generally accepted theoretical tenets would provide robust guidance and prove more useful than any guidance that could be immediately derived from available biological information. These guidelines provided the standards by which specific, local land-use plans and conservation strategies were formulated and judged with respect to compliance with the NCCP program.

The Conservation Guidelines contained several sections intended to provide a scientific foundation for conservation planning in the NCCP planning area. These sections included: (1) an introductory explanation of the basic features of coastal sage scrub ecology; (2) a discussion of management and restoration practices that should be addressed as part of the overall conservation program; (3) conservation planning guidance, including a description of the major research needs; (4) recommendations for an interim strategy while additional information is being assembled; and (5) an evaluation process for ranking remaining lands in terms of their long-term conservation potential. There was no peer review of these guidelines. (6) The Conservation Guidelines were issued by the Department of Fish and Game and immediately became the official guidance to all the participating local governments and other prospective applicants in this program. The resource management agencies (California Department of Fish and Game and U.S. Fish and Wildlife Service) were the responsible parties for ensuring that these guidelines were adequately followed. The scientific community, including the SRP, remained peripheral to the day-to-day activities of implementing the program.

The interim strategy was a necessarily important feature of the guidelines, due to the interest in proceeding toward resolutions of land-use disputes. The interim strategy concept was based on the conclusion of the SRP that 5 percent of the remaining coastal sage scrub landscape could be removed while additional information was being gathered without appreciably reducing the ability of the ecosystem to sustain itself. The interim strategy contained several assumptions, notably that the evaluation process would identify the potential reserve lands of high and intermediate value, and that decisions to develop such lands would be deferred until either the long-term plan for the subregion (i.e., local planning area) was completed or the specific lands were judged by the resource management agencies to be noncritical. The 5 percent was intended to come from what was deemed the least valuable habitat, and the need for an interim strategy should only last a short time.

The interim strategy took on greater significance than first anticipated. The SRP made it clear initially that it might be three to six years before enough information would be available for it to conclude objectively that larger portions of the landscape were expendable. For all subregions, this delay in firm land-use decisions was unacceptable, and the resource management agencies made increasing use of the interim strategy and the evaluation process. The interim strategy was expected to be needed only until subregional plans (i.e., long-term plans addressing the entirety of a geographic region) were completed.

Research Agenda

Additional information needs were described in the research agenda of the Conservation Guidelines. This agenda formed the basis for identification and prioritization of future research, of what the major holes were in our body of scientific knowledge. They were summarized into six major, interrelated items that when completed should resolve unanswered questions that bear on the conservation of the coastal sage scrub (CSS) target species. It was the expectation of the SRP that upon completion of these research requirements adequate guidance for long-term planning should be available. If the interim guidelines were followed, there should not be any crucial losses of habitat. These research items are described below.

Biogeography and Inventory of Coastal Sage Scrub

So that the geographic extent and distribution of coastal sage scrub vegetation and the individual species can be understood, the entire region should be mapped and data incorporated into a GIS at a scale of 1:100,000; the individual subregions should have detail of 1:24,000 scale. Additional data layers of importance include land use, land ownership, topography, and climate.

The majority of these data layers were developed by local governments in cooperation with the state and federal governments over a period of time from 1990 to 1994. Regional data were available through the California Gap Analysis program. These data sets represent some of the most extensive and detailed GIS data ever assembled for a regional conservation planning program.

Trends in Biodiversity

Planning decisions require better information on the effect of reserve size and adjoining land uses on biodiversity. The relationships of species richness or composition and habitat patch area and the effects of isolation should be investigated. Indicator taxa (such as CSS-dependent bird, small mammal, and butterfly species) should be employed due to time and funding constraints.

Researchers from the University of California at San Diego and Riverside, supported by the Biological Resources Division of the U.S. Geological Survey, have assembled a team to address the major taxonomic groups of vertebrates, including reptiles, amphibians, small mammals, and birds. Relatively little is known about the size and configuration of habitat patches that will sustain populations of these species. Researchers have established data collection stations at twenty-four locations within San Diego, Riverside, and Orange counties. This project represents an unprecedented effort, in terms of the geographic scope and sampling intensity, in the coastal sage scrub ecosystem and perhaps anywhere in the United States. The sampling design intends to capture variation within and between sites inherent in coastal sage scrub throughout the region.

Dispersal Characteristics and Landscape Corridor Use

More information about dispersal limitations of CSS species is needed for planning efforts to identify adequate linkages between reserves and to reveal trade-offs between increasing reserve size and improving corridors. Data from several locations within the planning region during both breeding and nonbreeding seasons should be gathered on target species, mountain lions, coyotes, and representative small mammals and invertebrates. The research program described in the item immediately above is addressing this issue peripherally. One additional research project, involving researchers from the University of California at Santa Cruz, is exploring this issue.

Demography and Population Viability Analysis

One test of the potential effectiveness of reserve systems is population viability analysis. Time-series data on the population biology of the two target bird species should be gathered in at least half of the subregions and from representative physical circumstances that span those found across the regional distributions of the species. Data should include territory size, time budgets, reproductive success, survivorship, emigration, and immigration, with separate data obtained for males and females where possible.

Intensive monitoring of coastal California gnatcatchers has been going on in various locations since the late 1980s. Unfortunately, with few exceptions, these efforts have not been continuous, and data normally represent a maximum of two or three years. The total effort on this task is incomplete; additional funding will be necessary if the full objectives of the task are to be met.

Surveys and Autecological Studies of Sensitive Animals and Plants

Basic information on the location, abundance, distribution, and natural history of vertebrate and invertebrate candidate species for federal protection and CSS-associated plant species of special concern should be gathered from select sites throughout the planning region. We need a better understanding of the specific habitat affinities of the full array of less well-known vertebrates that depend largely on coastal sage scrub habitat.

The first twenty-seven locations referenced earlier have been selected to capture as much of the variability throughout the region as possible. Each site has ten to twenty individual trapping arrays to attempt to capture the within-site variation inherent in local stands of coastal sage scrub and adjoining chaparral and grasslands. The combined study was conducted between 1995 and 1998, collecting data during all seasons. Detailed, site-specific habitat variables have been collected to help describe the subtle landscape features, at different levels of spatial resolution, that have an influence on distribution and abundance of each taxon. Extensive data analysis is in progress.

Genetic Studies

The maintenance of genetic variation is critical to the long-term viability of species inhabiting CSS and will be an important aspect of monitoring populations under an NCCP program. Declining genetic variation will be one symptom of inadequate linkages between reserves and may signal a need for changes in reserve management. Baseline data for comparison with future conditions should be gathered at the earliest possible opportunity. The reptile and amphibian component of the Biological Resources Division/U.S. Geological Survey research work is collecting tissues for genetic analysis. This, however, is being done only opportunistically. No specific research on this issue is currently underway.

Some additional research, not specifically identified in the Conservation Guidelines but nevertheless relevant to the NCCP program, is also underway. Notably, there is research ongoing at several universities, and the U.S. Forest Service is investigating the role of fire in the function of the coastal sage scrub ecosystem. Fire is a naturally occurring component of the CSS system, and researchers are examining its particular effects on regeneration of vegetation and the persistence of vertebrate and invertebrate components of the system. This research will be valuable in assisting fire managers and resource management agencies with their responsibilities for providing adequate protection to the human environment from wildfires.

Conclusions

What is the role of scientific information in the support of a regional conservation planning effort like the NCCP program? Everyone agrees that a program like NCCP must be based on sound scientific information. Resource management agencies require scientific data and assistance in interpretation of those data so they can have confidence in their decisions. In the case of the coastal sage scrub community, there will be little opportunity for adjustment of many of the land-use decisions that are authorized under the auspices of the NCCP program. It is a fairly simple question to pose: How much is enough? Resource management agencies are required to make informed judgments on a wide array of questions dealing with reserve design. These include such questions as: How much habitat is required to conserve a given species? What configuration of protected lands maintains the integrity of the reserve? What are the compatible adjoining land uses? Should isolated patches be linked? What are the implications of individual subregional plans to the persistence of a species throughout its range? These are some of the crucial issues upon which the scientific information is asked to shed light. Ultimately, the decisions based on this scientific information will be responsible, both legally and in reality, for the persistence or extirpation of countless numbers of species.

Such questions ask a lot of the scientific community. The natural world is an infinitely complex system. Scientific investigation is designed to distinguish important influencing factors from those that have little bearing on any given independent factor under investigation. Yet there are thousands of influential variables in a natural system, and the ability to control other variables in an experimental, laboratory-like fashion is enormously difficult. The questions posed above are among the most difficult to investigate in a scientifically sound manner, and they may require many years of field work before the salient results are not obscured by natural environmental variation.

When a conservation issue of the magnitude of the NCCP program arises, we often find that the available information is sparse, unconnected, partially or largely untested (i.e., anecdotal as opposed to empirical), and the sampling scheme (both spatially and temporally) is often not adequate for drawing sound conclusions regarding the long-term conservation of a species. However, it does constitute the "best available information" and is usually used as such.

Some would argue that this dilemma is inescapable, and we should simply learn to make the best of it. Indeed, it is impossible to answer the question "How much land is enough (to sustain the natural community)?" in a strictly scientific sense. This is more a policy decision than a biological one. Indeed, a process like the NCCP program contains several decision-making filters besides scientific guidance. There are many subjective, nonscientific components to deciding what land must be conserved and what lands can be developed. The problem is too complex for science to provide a strictly objective answer. There are literally thousands, if not hundreds of thousands, of ecologically relevant variables that may influence the outcome of such a decision. Ultimately, it is also a value judgment. For example, what level of certainty are we willing to accept to assure us that we will have California gnatcatchers in the United States fifty years from now? Science can and should provide insights to guide, but these insights will be mixed with other considerations through established decision-making mechanisms of regulatory agencies.

I argue, however, that both scientists and land-use and resource managers have a responsibility to minimize the errors of the present and the risks of the future. Scientists need to be more active in communicating the uncertainty that is inherent in the information they provide and what the appropriate uses of that information may be. Land-use and resource managers should refine their vision of the information needs of the future and give scientists the time needed to develop reliable information. Policy makers and the public, through efforts of researchers and land-use and resource managers, need to make the effort to become better informed regarding what is learned from scientific research.

Management Review

James E. Whalen

Over the last seven years, the San Diego region has been engaged in a precedent-setting effort to implement its version of the state's pioneering Natural Community Conservation Planning process, through the subregional Multiple Species Conservation Program (MSCP). My firm has represented businesses, large landowners, and utilities in the job of integrating the goals and objectives of business interests into the region's new conservation agenda. It has been challenging.

The NCCP is designed to protect dozens of threatened species, among them the redoubtable California gnatcatcher, in the first ever landscape-level urban conservation planning effort in the five-county coastal scrub ecosystem of Southern California. It is precedent setting in the nation, as well, for the ambitious scope of its planning and attempts to satisfy overwhelming objectives, such as buying land with prices in the hundreds of thousands of dollars per acre and higher. The region has been grappling for some time with the question of whether this comprehensive approach could resolve the flaws of the past, primarily biologically focused planning failures. Though it looks as though the NCCP may, at last, be the right answer, the details have been bedeviling. Some critics have suggested that NCCP is basically a political decision rather than a biological solution. This begs the question of whether the rules of science have been either overplayed or underemphasized in their application to NCCP preserve planning. Has there been too much science or not enough? An answer satisfactory to everyone is elusive.

The Issue

California has been experiencing noticeably shrinking numbers of native plants and animals, despite increasing laws to protect them. To stem this problem, either a new system of protecting species is needed that is more comprehensive and efficient than the previous methods or the California and federal Endangered Species Acts need to be repealed. Clearly, repeal is unlikely, but neither can we continue with business as usual, with over forty-three plants and animals poised for endangered listing in the San Diego area.

The NCCP's proposed solution is to protect some very large blocks (some bigger than forty thousand acres) of interconnected urbanizing habitat in three counties (San Diego, Orange, and Riverside) that would be forever sequestered from development. The challenge is to safeguard nearly one hundred species spread over wide areas of the region while also addressing political realities and accommodating needed economic activity.

Background

Consider the following factors shaping the NCCP in Southern California:

- *Species diversity.* The diversity of plant and animal species in Southern California is among the highest in the continental United States, due to an exceptionally large range of habitat types. Where else can a person ski in powdery snow and two hours later swim in the ocean after driving through meadows, forests, scrublands, river valleys, and desert? Unfortunately, without a detailed long-term plan of conservation, Southern California's biological diversity will disappear into memory. Rapid urbanization in the region over the last thirty years has led to a long-term decline in Southern California habitat viability and, consequently, to a high number of species being considered for protection by the wildlife agencies.

- *Economic meltdown and restructuring.* Major economic changes since 1991 have led to great uncertainty among the working people of Southern California. While these changes were caused largely by the consolidation and shrinkage of the defense industry, particularly in aerospace, employers of all stripes have been "right-sizing" and redefining business in the face of severe global competition. To be sure, the kinds of changes taking place in Southern California in the early 1990s have been occurring elsewhere in the United States, but they have been much more pronounced in our region, similar to those in the industrial Midwest during the 1970s when the Rust Belt was tightened. As in the Midwest, our changes are fundamental and permanent. Difficult as it has been, one beneficial outcome of this post–Cold War shift has been to force community leaders, particularly elected ones, into new ways of looking at old problems such as growth management.

- *Land values.* Land values in Southern California are among the highest in the nation, even after years of recession. Since early 1997, raw land prices have surged over 20 percent in some areas. Coastal land sells for $400,000 per acre or more, and in some parts of coastal Orange County, individual lots go for as much as $2 million. By no fault of nature, some of the rarest habitats, such as vernal pools and southern maritime chaparral are found in coastal areas, leading to the observation that the finest development land is invariably the best habitat as well.

- *Political conservatism.* Despite California's reputation of being avant-garde, rock-solid political conservatism has always set the tone in San Diego. This political conservatism, once analogously characterized as "Iowa with palm trees," most likely stems from the agricultural and military roots of the county. To this day, the San Diego County Farm Bureau exerts influence commensurate with its status as one of the top five economic engines in the region.

This conservative sensibility has fostered a vocal, ideological private-property–rights movement. Further, a well-informed, Libertarian-driven citizen faction has recently created controversy over the public funding of several proposed new facilities, casting doubt over the odds for developing any significant public places in the foreseeable future. Among the eighteen cities and the unincorporated area in San Diego County that are constantly jockeying for advantage, only the City of San Diego has ever been able to exert a hegemony, and that has sometimes come at the price of animosity from the others. Consequently, very little ever gets done at the regional level.

As local policy makers are being forced to recognize the need to address the limits inherent in a growth-based economy in a finite county, the impetus for change has become inevitable. The San Diego region has avoided the crowding and congestion of Los Angeles, yet its problems are the same, just scaled down. Unlike Los Angeles, enough habitat remains in San Diego County to be considered viable. Accordingly, the U.S. Fish and Wildlife Service and the California Department of Fish and Game, operating under the mandates of their respective Endangered Species Acts, have joined the fray. In exchange for the cooperation of local jurisdictions with the wildlife agencies, the agencies have promised to return land-use regulation to local jurisdictions, once they successfully play by the new rules of the NCCP. This delegation of authority is explicit and central to the attraction of NCCP to the landowners. Environmental groups are chary of what they see as this "dereliction of duty" on the part of the agencies, and they feel let down since they have traditionally believed they have more influence with their kindred souls in the wildlife agencies than they do with local governments. Lawsuits by radical environmental groups have been filed to eliminate this key benefit of the NCCP.

Despite over $7 million spent on vegetation and species mapping, the amount and quality of scientific data have been a subject of debate as long as the planning process has been underway. As a practical matter, any legal action challenging the adequacy of the decision making will run squarely into the assertion of "best professional judgment" of the wildlife agencies, probably a good thing, due to the impossibility of developing "enough" empirical information to make decisions. All the while the wildlife agencies have been assiduously pressing their advantage in individual landowner negotiations and in the face of signed agreements insisting on parity between landowners and regulatory bodies.

Status

On the local, regional, and state levels, the timing has been right for resolution of the conflict between open space and development, but keeping the consensus has been exquisitely precarious. Mainstream environmental organizations, such as The Nature Conservancy and, to a lesser extent, the Sierra Club, are displaying conflicted but increasing interest in making the process work. A rare confluence of political forces may have created a setting in which intuitively derived solutions actually have a chance at succeeding.

To hash out the major policy issues before sending a product to the elected officials, a deliberative body was appointed by Mayor Susan Golding of San

Diego. This strategy of assembling major stakeholders into a working group has been a success. However, conflicting ideologies periodically set up roadblocks that are difficult to avoid, and the process has yet to fully resolve these ideological conflicts. Most of the issues we have confronted with the private-property–rights movement and the Earth First activists come from a blend of unrequited ideology and ignorance. And though the ideologies are ostensibly irreconcilable, the ignorance can be reduced through public education.

The public education process is in its earliest stages, with the Zoological Society of San Diego expending significant amounts of grant money to start a mass-market baseline information program on the benefits of wildlife conservation. When the grant funds are exhausted, the Zoo's MSCP outreach essentially stops, and public approval for increased funding for education efforts has been slow in coming.

The process was not without planning errors either. Notwithstanding several notifications sent to all of the households in San Diego, public information levels are low and changing only slowly. The cozy insularity of the working group, even though it was composed of a broad cross section of interest groups, initially led to a false sense of consensus with the larger world outside the conference room of the Multiple Species Conservation Program. Agreement on difficult balancing acts such as mitigation ratios falls on indifferent ears outside of the MSCP meeting rooms. The paucity of empirical evidence on how best to manage species numbers has led to environmentalist fears that the wildlife agencies made deals at the expense of the resource. But there has been no evidence that the information used by the agencies to make preserve decisions was not the best science available or that it was insufficient for good decisions.

Many landowners are not happy about the MSCP because they feel it is a regulatory taking; other, more informed property owners appear resigned to it. Ultimately, the changes to land-use planning in the area lead to more land for open space, and that open space is connected biologically. Whether landowners develop, hold, or trade mitigation or development credits or sell their land has not changed much. It comes down to immediately higher development costs, which translate into higher housing costs in San Diego and throughout Southern California. Nationally, NCCP may be seen as the biggest land grab ever, yet national conservation groups have worked to eliminate the assurances package, which is the basis for local support of NCCP.

All of these changes are managed by local jurisdictions, which are contractually obligated upon signing implementing agreements to ensure that development activities occur in rough step with the assembly of open space into preserve areas. For most jurisdictions participating in the MSCP, the policy decisions reached in the working group and by elected officials have been (or will be) embodied in ordinances, which will bind the parties and force involvement whether a property owner wants to be involved or not. Costs are supposed to be spread fairly, but the public has been given three years to come up with its share.

Science has been employed most effectively in the areas of population biology, preserve management, and reserve design. But much remains to fulfill the crucial need for long-term, successful adaptive management. There are those who believe it is too late to salvage the program; others fervently hope it is too late. But these voices resonate only from the sidelines. By most accounts, NCCP planning is moving ahead, paid for in large part by grants from charitable foundations interested in efforts of this sort. Nevertheless, many people in Southern California do not comprehend the magnitude of the compromise being forged. A recent poll conducted by The Nature Conservancy indicated that 81 percent of those asked believe that the unbuilt hills and mesas outside of town are public land and are, therefore, protected for conservation. But in fact, most of the land with conservation value in the San Diego area is privately held. Many of those landowners have other plans for their hills and mesas that do not include conservation of open space. This disconnection between the public's predisposition to pay and the working group's desire to move ahead needs to be resolved as soon as possible.

At the regulatory level, most wildlife agency field personnel and local planning department staffs still act as if landowners are the enemy. Many opposed an interim process—the 4(d) rule—that allowed a 5 percent loss of coastal sage scrub, although only a small fraction of the habitat available for take was actually utilized before the adoption of the subregional plans eliminated the need for the interim process. Some property owners felt they were treated shabbily by agency staff, and to them it presaged the way the MSCP will be run, once it is fully approved.

This created serious and still unsolved problems, particularly since almost every fair-minded person involved recognizes the central role property owners play in any solution. This also makes it hard to collaborate in reaching negotiated solutions when the landowners have to anticipate eleventh-hour adventurism on the part of some agency staff, such as demanding more land or money when the parties have already made a package deal.

Achievements and Unresolved Issues

What are the achievements? First, the blending of science with the value systems of the stakeholders has occurred without too much stress, although it is fragile. Second, arguably, the extreme left- and right-wing positions, once capable of stonewalling the process, have effectively been marginalized. Outlandish positions tend to be dismissed and even ridiculed in the deliberation process of the working group as a whole. Third, long-term exposure among all of the various interest groups has led to a level of trust and comfort that could have arisen only over time. A strangely warm camaraderie has even formed among groups who would be considered adversaries in other less successful planning efforts.

Fourth, and most significantly, the ambitious preserve system is taking shape, bit by bit, jurisdiction by jurisdiction. The wildlife agencies' brilliant strategic move of permitting interim losses of gnatcatcher habitat as long as the larger preserve planning effort is ongoing has kept the parties at the table, under the threat of an overall development shutdown if NCCP planning stops. Agency personnel at management levels have been present to ensure the process keeps moving ahead and to mediate disputes.

A number of issues are still unresolved after years of work. If they are not resolved, the program will fail in its goal to be a final solution.

Foremost, the use of science as the basis for land-use decision making, especially by the environmental community, is too often abused. No-growth advocates are particularly comfortable using science to support their cause. This needs to be repudiated by the agencies, or they will suffer loss of their credibility, one of their biggest assets. (To an extent, lower-level staffers within the wildlife agencies are guilty of some of the same antics.) In the biological world, where there are few certainties, the NCCP needs to have a concomitant level of flexibility. Clearly, there will have to be aggressive adaptive management of the preserve.

Major policy and technical issues with broad implications have yet to be resolved. Mitigation ratios for impacts within and outside of the preserve system have not been determined. The question of who will provide and fund ongoing preserve management and maintenance lingers. Fire management and brush control within the preserve must be organized and funded. Contractual assurances on the finality of the program (a line is a line) have yet to solidify. And, the most intractable, the funding of the local share, remains a mystery.

Concluding Thoughts

Science is the yardstick by which overall progress will be measured. Yet the biological quality of a preserve system as large as is proposed can never be stated with certainty due to financial, legal, and logistical shortcomings, so why suggest otherwise and why delay putting together the reserve? Programs such as NCCP are usually triggered by some kind of crisis, such as the listing of an endangered species. Yet, as stated elsewhere in these reviews, one preserve size and planning approach does not fit all. There is no way a generalized model can be applicable across the country, even when triggered by similar crises. What we can do is offer our experiences to others in hope that the common themes will emerge and be available to plug into a new milieu.

Timing is everything. Only the government can take an idea whose time has come and delay until its time has passed. In the case of MSCP, management personnel of the state and federal agencies played a pivotal role in the successes that were achieved and spent a lot of time to get the program set up in a workable manner for business.

Questions need to be clearly articulated so that polarizing influences can be dealt with quickly.

Efforts need to be inclusive of all stakeholders, no matter how initially repugnant the viewpoints might be. Marginalization of the extremist views will occur naturally.

Trust is essential for the process to move ahead. Strong and democratic leadership from the major participating groups and a sense of humor are critical.

Risk, uncertainty, and variability should be incorporated into the plan and schedule. Integration and synthesis should be allowed to occur among the stakeholders. Opportunities for adaptive management and adjustment should be built in; they will be needed. One of the NCCP's possibly huge successes is the harnessing of variability and uncertainty to allow for the biologically legitimate extension of the assurance of "no surprise" to landowners and local governments.

Collaboration and cooperation are nice words, but one should not forget that this is a negotiation after all. A certain toughness is required among the participants or the decisions will not be valued.

If there is adequate funding to buy land and manage it, almost all of the problems diminish markedly. Since this is rarely the case, funding needs to be considered before all else, as well as public education.

Was there too much science in the NCCP, or not enough? Like many of the NCCP issues, the question seems answerable, but the facts will not be known for some time. Therefore, only prudent judgments backed by the best available science can meet the fundamental need to proceed with implementing the NCCP plans and programs while the time, political interest, and money are there. To delay would have certainly led to the failure of the program.

Now, several years into the process, most of the policy decisions have been reached among the stakeholders, and the consensus continues to survive. The extremists on both sides have filed the expected lawsuits, and the supporters are busy buttressing the plan for the legal fights, which doubtless will go on for some time. Funding is still the bugaboo it has always been, but signs point somewhat optimistically to public acceptance of responsibility for bearing a share of the costs. None of the key participants are sure they have the will to repeat the experience, but, so far, it has been worth it.

Policy Review

Ron Rempel, Andrew H. McLeod, and Marc Luesebrink

In California, an ambitious public-private conservation partnership is fundamentally changing the manner by which the state and federal governments provide for the long-term needs of plants, animals, and other wildlife. In so doing, the state desires to involve in conservation to a greater extent than ever before the "stakeholders"—private landowners, developers, environmentalists, and local governments. This transition to "ecosystem planning" has necessitated many changes in policies and practices, including the use of sound science and the involvement of all stakeholders in this ambitious enterprise.

This review, however, focuses on two other essential components of the NCCP program: *incentives*—creating, offering, and delivering tangible benefits for long-term cooperative participation in the program to those from the landowner-development and environmental sectors—and *assurances*—establishing and providing assurances to those stakeholders that the program's guarantees of preserve creation and implementation, regulatory streamlining, and state and federal wildlife agencies' long-term commitment and responsiveness are fulfilled.

Unprecedented Habitat Planning

Owing to the large size and number of jurisdictions in this region, the planning area is broken down into eight subregions in order to simplify planning. Two of the subregional plans, the Orange County Central/Coastal NCCP and the San Diego Multiple Species Conservation Program Plan, have been completed and approved. The Orange County Central/Coastal NCCP plan, which creates a 37,380-acre reserve system that will protect the habitats of thirty-nine plant and animal species, received final approvals from wildlife agencies and the County Board of Supervisors in 1996. San Diego's Multiple Species Conservation Program plan is an NCCP plan that encompasses 582,000 acres and establishes a 172,000-acre preserve system in southwestern San Diego County. It covers eighty-five species and twenty-three vegetation types. It received approval from the San Diego City Council in March 1997 and by the County of San Diego in November 1997. (Taken together with two other plans being prepared in San Diego County, a total of 1.3 million acres are being addressed through a long-term conservation strategy that balances preservation with economic development.) Other plans are also being created and are in various stages of completion in San Diego County (Multiple Habitat Conservation Program), Orange County (Southern NCCP), western Riverside County, Coachella Valley, Los Angeles County (Palos Verdes Peninsula NCCP), and western San Bernardino County.

From the outset of the program, the California Resources Agency and the California Department of Fish and Game (DFG), which jointly administer the program, have recognized that the ambitious goals of NCCP require innovative solutions to problems not previously confronted by wildlife agencies under the ESA. These problems include preservation of ecosystem function in an urbanizing landscape and maintaining habitat values through adaptive management of large preserve systems. NCCP represents a fundamental change in the orientation of wildlife agencies and the way they do business. In 1993, a major step was taken when the new Clinton administration promulgated a specially crafted 4(d) rule under the federal ESA authorizing the limited incidental take of the listed California gnatcatcher consistent with the preparation of a state NCCP plan. This breakthrough created the partnership that continues today between the state's Resources Agency and Department of Fish and Game and the federal government's Interior Department and Fish and Wildlife Service.

NCCP plans have resulted in revolutionary changes in conservation of species and habitats through integrating conservation into the local land-use planning process in much the same way as infrastructure—roads, water and sewer systems, parks, and schools. This integration of NCCP plans into community land-use processes allows the community to participate in preserving community character, open space, and quality of life while at the same time conserving species and habitats.

Incentives

Key to successfully applying the state authority for NCCP—and to realizing the potential of the state–federal partnership—is the ability to create incentives for all stakeholders to participate in a new, voluntary, and improved form of conservation planning.

The NCCP act cited specifically the major incentives envisioned under the new law: cooperation and collaboration toward an ecosystem approach to conservation, and a streamlined regulatory process for compatible and appropriate development. The wildlife agencies also understood that additional incentives would be needed to generate the critical mass of support required for successful planning and that new incentives would be designed for different subregions based on the characteristics of their plans.

For the Orange County Central/Coastal plan, the wildlife agencies were able to create an NCCP working predominantly with a single, dominant landowner—a circumstance that greatly simplified the process. Since the major land developer, the Irvine Company, desired the benefits of streamlined regulatory permitting, a one-time planning process, and a secure future business environment, it supported the program. Because the Irvine Company could commit most of the needed lands to the preserve system, the need for extensive landowner education on the benefits of the program was not needed. The major incentives provided by the state and federal governments were the commitment to participate in the costs of planning the Central/Coastal NCCP and an agreement to manage public lands consistent with the goals of NCCP.

For San Diego's MSCP, the complexity associated with the large number and variety of landowners within the planning area necessitated greater attention to incentives. Many landowners who did not plan to develop for years to come viewed the MSCP as an unnecessary burden and expense or mistakenly anticipated that overall mitigation requirements on landowners would be increased. To gain support of these important stakeholders, the wildlife agencies had to demonstrate that MSCP would improve the regulatory process without placing an inordinate expense upon private land developers and would provide an impetus to landowners to become engaged in the planning process.

To this end, the City of San Diego, the state, and the Department of Interior agreed to several formal elements intended to encourage landowners and jurisdictions to participate. Foremost among these incentives, the wildlife agencies agreed to contribute a portion of the acquisition, management, and monitoring needs of the program. Together, the federal and state governments will acquire up to 13,500 acres over thirty years, half of the lands to be acquired by public means. The wildlife agencies also have accepted responsibility for managing lands that are contributed to the MSCP by the state and federal governments. The local jurisdictions have committed to the acquisition of 13,500 acres, with the remainder of the private lands needed to establish the preserve being conserved through mitigation of new development. The state and federal commitment to share acquisition and management responsibilities among local jurisdictions and new development was an important aspect of the agreement that guided finalization of the MSCP and demonstrated the critical role that incentives can play in encouraging participation and finalizing plans.

As the NCCP program grows, the wildlife agencies and local jurisdictions will face two major issues related to incentives.

First, the agencies must continue to offer incentives (broad assurance regarding species coverages and assistance with habitat acquisitions) that are designed for individual subregions, recognizing that an incentive utilized in one subregion may not be appropriate for a plan being developed for another. The determination by the wildlife agencies as to the level of incentives also cannot precede plan preparation but must be decided as details of the conservation plan emerge.

Second, local, state, and federal agencies need to explore new opportunities for creating incentives in order to diversify the sources of funds and lands. Proposition 218, which was passed by California voters in 1996 and which requires a two-thirds vote for approval of new taxes and fees, increases the importance of creativity in this area because of the expected impact on the ability of local governments to raise revenue. In recognition of this, the agencies are exploring new options such as estate tax credits for land dedications and creative uses of existing incentives such as conservation banking. By expanding the sources of funding and utilizing incentives wisely, the wildlife agencies expect to continue to offer reasonable and fair incentives to all stakeholders.

Assurances

The ability to provide sufficient assurances to both environmental and economic stakeholders has been perhaps the most important element of the NCCP process. In many respects, the benefits conveyed by an NCCP plan are only as valuable as the assurances that glue them together through time.

At the outset of the NCCP program, the conservation community desired two major assurances: that the conservation described in the NCCP plans for future development had to be secure, and that the lands contemplated for inclusion in the preserve system had to be managed appropriately. To guarantee mitigation for future development, the program integrated conservation into local planning processes to an unprecedented level. This was achieved, in part, by issuing the necessary state and federal incidental take permits to local jurisdictions (rather than project proponents), which then have the responsibility to comply with the specifics of the NCCP plan and implementation agreement. This, in essence, granted ownership and responsibility to the locals.

The second method of assuring future mitigation was to revise local planning ordinances to include the provisions of the MSCP plan. Mitigation and avoidance elements of the MSCP, for example, will be implemented in part through the City of San Diego's Environmentally Sensitive Lands Ordinance and the County of San Diego's Biological Mitigation Ordinance. A major benefit of this approach is that conservation becomes a part of ongoing planning processes and decision making, rather than a marginal consideration that enters into the thinking of local decision makers at the tail end of the planning process.

The conservation community also looked to NCCP to provide appropriate management of preserve lands. The Orange County Central/Coastal plan addressed this issue largely through the creation of a nonprofit organization with responsibility for the management of the preserve system. Funding for the nonprofit board comes largely from the county's highway building agency through the establishment of a nonwasting endowment as part of its mitigation for new toll roads and through nonparticipating landowners' mitigation fees. Participating landowners agree to manage lands to maintain biological values until dedicating them to the preserve system. Together, the nonprofit board and the interim management provisions form the foundation for effective long-term preserve management.

The process developed for MSCP preserve management in San Diego utilizes a strategy based on phased preparation of management plans as preserve lands are dedicated as part of development or are acquired using local, state, and federal funds. Jurisdictions are required to prepare framework management plans within six months of the jurisdictions' approval of the MSCP plan. The MSCP also requires that the framework management plans be finalized by the jurisdictions and approved by the wildlife agencies within an additional three months. Area-specific management directives will be prepared for distinct sections of the MSCP as they are contributed to the preserve system. The preparation of jurisdiction-specific management plans is anticipated to take two to three years.

For both the Orange County Central/Coastal plan and the MSCP, the initial management plans represent only the beginning of a longer process. The implementation phase of the NCCP plans, during which the area-specific management directives will be prepared, will generate new questions about appropriate uses within preserves, management of boundaries and "ecotones," management prescriptions, coordination among preserve components (both within subareas and throughout the entire coastal sage scrub planning area), and funding. Each of these issues will be addressed through an adaptive management approach wherein the results of ongoing monitoring and new information will be used to inform future management decisions.

Preserve management issues will be a focal point of the NCCP program and a challenging issue for policy makers to address. In recognition of this, the wildlife agencies and local jurisdictions have taken steps to ensure that preserve management receives adequate funding and attention. First, funding has been provided through a state grant to assist with preserve management activities. In 1996–97, $600,000 was disbursed by the Department of Fish and Game for ten different projects in San Diego, Orange, and Los Angeles counties.

Second, the City of San Diego has proposed annual workshops to review management and assembly of its section of the preserve. These workshops will follow the release of the annual report on preserve assembly and management required of participating jurisdictions of the MSCP. Developers and landowners also sought assurances from the NCCP program in the form of a predictable regulatory landscape. This certainty was achieved largely by providing the details of what requirements a project will have and backing up the requirements with the application of the "no surprises" policy, which provides that future listings under state and federal ESAs of covered species will not require additional private mitigation or a delay in project permitting. In such instances, any additional cost for species recovery and planning must be borne by public agencies. For the MSCP, the "no surprises" policy was extended to reflect "significant" and "sufficient" conservation of habitat types within the planning area. Sufficiently conserved habitats are those that are afforded the greatest level of protection, and significantly conserved habitats receive high levels of protection. For those habitats determined to be significantly conserved, the state, federal, and local governments agree to share with landowners the responsibility for protecting (either through modified management practices or additional acquisition) uncovered species dependent on that habitat. For sufficiently conserved habitats, the state and federal governments commit to provide for necessary additional conservation either through modified land management practices or through additional land acquisitions.

These assurances, however, are not without controversy. The "no surprises" policy, which is being formalized by the Clinton administration with regard to the federal ESA nationwide, has been criticized because the land conservation provided for in the NCCPs may not be adequate over the long term, it is argued. Many also view the potential responsibility of state and federal governments to fund needed adjustments for covered species and conserved habitats with skepticism due to limited budgets for wildlife protection.

The wildlife agencies recognize the uncertainty associated with protections provided by NCCP plans. This uncertainty is due, in part, to the fact that the scientific study of ecosystems is inexact and relatively new. Scientists and planners have very little empirical data to draw upon to validate that protections are sufficient—particularly in coastal sage scrub, on which ecological investigation began only recently and data are sparse. As a result, the agencies have sought to develop conservation plans that protect the highest-quality habitat within the coastal sage scrub ecosystem, regardless of whether sage scrub and the vegetation communities are "occupied" or otherwise "protected" by the federal or state ESA. To accomplish this goal, the cooperation of landowners and developers is essential. Many have indicated that they would not participate in a program that addressed unlisted species and unoccupied habitat without the assurance that additional mitigation would not be demanded of them if the species were subsequently listed—hence, the importance of the various assurances, including the "no surprises" policy. Put simply, the ability to practice ecosystem management is not possible without state and federal governments providing security to landowners and accepting some risk.

The counterbalance to this uncertainty is the ability to make preserve modifications and management adjustments in the future to reflect new scientific information and understandings. With this in mind, the wildlife agencies have emphasized the importance of applied research and have been working closely with the scientific community to better coordinate the efforts of researchers and land managers. During the implementation phase, the wildlife agencies must continue this effort to better incorporate science into preserve management and to remain flexible and creative in utilizing adaptive management. To ensure that necessary adaptive management occurs, NCCP plans incorporate requirements for adequate long-term funding of preserve monitoring and management.

Conclusion: The Future of NCCP

As the NCCP program grows and expands, its partners will face new and challenging policy issues.

First, there are a series of issues that relate to the precedential nature of the program, including the level of incentives to be provided by the state and federal governments, the level of coverage (number of species) to be provided in an incidental taking permit, the mitigation ratios for various habitat types, and the process of preserve assembly. These questions must be addressed based on the

specifics of individual subregional and subarea plans. To be effective, the wildlife
agencies must seek to maintain the core elements of the NCCP program without
establishing rigid regulations for plan development that will hinder creativity,
and local jurisdictions must be willing to start developing plans without demanding specific commitments prior to the plan's preparation.

Second, the state and federal wildlife agencies will have to work with the U.S.
Army Corps of Engineers and the U.S. EPA to fully address the management of a
key habitat type, wetlands, in the NCCP process. A true ecosystem effort at
regional planning must seek, in the long term, to include wetlands. Landowners
and developers have expressed concern that the promise of streamlining offered
by the NCCP program will not be fully realized unless the traditionally cumber-
some process of obtaining permits for impacts to wetlands is integrated. In order
to maintain the benefits of a streamlined permit process, it has been suggested
that a comprehensive approach to protecting wetlands be designed consistent
with the underlying NCCP plans. The wildlife agencies are currently exploring
this issue to evaluate potential approaches to undertaking this task.

Finally, the expansion of the NCCP program beyond the coastal sage scrub
ecosystem of Southern California will be an important step in demonstrating the
broad benefits of ecosystem management. This was begun with the promulgation
of the Draft NCCP General Process Guidelines, which provide a framework for
NCCP planning throughout the state. The challenge of applying NCCP in other
ecosystems will be to develop new tools and mechanisms appropriate for new circumstances that uphold the high conservation standards of NCCP.

Important lessons of the NCCP pilot program in Southern California are
already being applied to comparable conservation strategies undertaken by the
Wilson administration around California. Under an unprecedented state–federal
partnership launched in 1994, policy makers are working to develop a restoration
and conservation strategy for the San Francisco Bay–Sacramento/San Joaquin
Delta estuary, the linchpin of the state's water delivery system. This is a unique
program involving the formal participation of agricultural, urban, and environmental stakeholders. It seeks long-term water sufficiency for California and environmental benefits through a combination of habitat restoration, physical
improvements, and flow requirements. Similarly, California has undertaken a
comprehensive effort (yet to be fully integrated into NCCP, as noted earlier) to
increase and protect wetlands in the Central Valley, San Francisco Bay Area, and
coastal Southern California.

We are confident that the underlying principles of the NCCP program—collaboration with stakeholders, ecosystem management, sound science, and regulatory efficiency—provide the best promise for resolving conflicts between conservation and development throughout the state and offer valuable lessons for comprehensive natural resource management elsewhere.

CHAPTER 11

Interior Columbia Basin Ecosystem Management Project

Case Study
Thomas M. Quigley, Russell T. Graham, and Richard W. Haynes

Science Review
James K. Agee

Management Review
Martha G. Hahn

Policy Review
John Howard

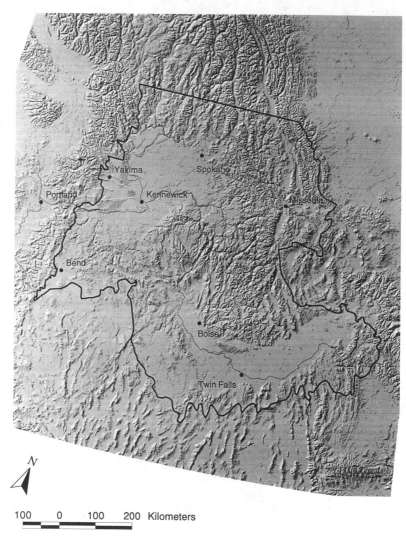

N

100   0   100   200   Kilometers

Interior Columbia Basin Management Project Assessment

Case Study

Thomas M. Quigley, Russell T. Graham, and Richard W. Haynes

The scientific assessment of the interior Columbia Basin is one outcome of the continuing debate over the management of Forest Service– and Bureau of Land Management–administered lands within the basin. To fully understand this debate, and the implications associated with it, requires an understanding of the biophysical, social, and economic components of the ecosystems for the total land area—public and private. Such an examination is designed to develop an understanding that will disclose conditions, trends, and potential outcomes associated with natural resource management in the basin.

The Political and Geographical Context of the Assessment

The assessment area in the interior Columbia Basin includes approximately 145 million acres of land in portions of Idaho, Montana, Nevada, Oregon, Utah, Washington, and Wyoming. The FS and BLM manage approximately 75 million acres (over half of the area). The majority of the people in the basin live in urban settings (even though there are only six metropolitan areas), but the majority of the area is rural. Agriculture is a dominant industry in the rural settings. The proportion of the population making its living directly from natural resources within the basin has been declining. Because the region has an increasing population and the larger economies of the basin are growing, the majority of the population is probably not aware of the changing circumstances regarding management of natural resources within the basin. Conflict over natural resource management has been largely debated among publics favoring either resource protection or commodity extraction.

The conflict over resource management in the basin intensified in the late 1970s and early 1980s as concerns grew about timber harvest from public lands, forest health concerns (defined at that time primarily by widespread tree mortality caused by insects, diseases and large wildfires), declines in wildlife species associated with old forest structures, degradation of rangeland, and dwindling populations of anadromous fish. National forest and BLM resource plans within the basin were completed during the 1980s and 1990s. These plans addressed issues on individual planning units with only limited focus on larger geographic areas. These plans were appealed, and lawsuits made implementation difficult at best and infeasible at worst. This experience is similar to events that occurred on the west side of Oregon and Washington.

Fisheries issues and concerns intensified during the late 1980s and early 1990s as anadromous fish populations declined. The listing of the Snake River salmon as endangered under provisions of the Endangered Species Act and continued declines in other anadromous fish populations precipitated several policies regarding management of habitat that culminated in the Anadromous Fish Habitat and Watershed Conservation Strategy (PACFISH), which established interim standards and guidelines for anadromous fish habitat occurring on FS- and BLM-administered lands in the Northwest.

National forests in the upper Columbia Basin (Idaho and Montana) remained relatively unaffected by these events until the Pacific Rivers Council successfully argued a lawsuit against the Umatilla and Wallowa-Whitman National Forests of Oregon and Washington. The injunction that followed necessitated consultation on Forest Plans by the FS and the National Marine Fisheries Service (NMFS) because of the endangered Snake River salmon. After the lawsuit in Washington and Oregon, the Pacific Rivers Council initiated a similar suit naming seven national forests in Idaho. This suit threatened to substantially reduce timber harvest, livestock grazing, mining, and recreation on FS-administered lands.

At the same time, the FS and BLM were being pressured to deal with the high tree mortality (caused by insects and diseases) through timber harvest. These issues gained national attention by the early 1990s (Quigley 1992; Sampson et al. 1994; O'Laughlin et al. 1993; Everett et al. 1994; Jaindl and Quigley 1996). Coincidentally during this period, large wildfires punctuated eastern Oregon, eastern Washington, and southern Idaho, intensifying the issue. Because of these issues, U.S. Senator Mark Hatfield and Speaker of the House Thomas Foley requested that the FS assess the forest health conditions in eastern Oregon and Washington. The subsequent report recommended actively managing forest conditions within an ecosystem-based approach (Everett et al. 1994).

Several members of Congress requested specific scientific societies to form a panel to examine conditions of the forests in eastern Oregon and Washington. The scientific societies report provided a set of recommendations, including a recommendation that a more in-depth assessment be initiated (Henjum et al. 1994). Until such an examination could be completed, the report recommended a "time-out" in harvesting old growth, cutting large trees, logging or road building in riparian and unroaded areas, and grazing in degraded areas. In Idaho, studies by the University of Idaho and American Forests concluded that conditions in forested environments were out of balance and that aggressive action was warranted (O'Laughlin et al. 1993; Sampson et al. 1994).

In 1993, the Pacific Northwest Region (R-6) of the FS, motivated by concerns over riparian conditions, old-growth forests, wildlife associated with old growth, and a petition from the Natural Resources Defense Council concerning these issues, initiated a screening process (screens) to preserve options for decisions that might result from additional planning. In addition to old forests, the screens addressed riparian areas and fish habitat. When applied to planned timber sales, the screens immediately decreased timber available (in some instances over 50 percent). Coupled with the almost nonexistent timber harvest on the west side of

Oregon and Washington due to injunctions over spotted owls and old growth, this screening action heightened concern over "gridlock" in forest management. These actions occurred at the same time as forest mortality and fires were increasing throughout the basin.

At this time, too, increased understanding of ecosystems, public pressure to manage federal lands more "holistically," and increasingly restrictive measures to manage on a species-by-species basis led the chief of the Forest Service to announce that the agency would adopt an ecosystem approach to the management of national forests. The BLM, shortly thereafter, announced its intentions to use an ecosystem approach as its primary means of managing public land. Each agency initiated efforts to develop ecosystem management approaches addressing a multitude of issues and geographic areas.

These events provided some of the context for the president's Forest Conference, held in Portland, Oregon, in 1993. The conference was designed specifically to address issues concerning forest management gridlock within the range of the northern spotted owl. When announcing the outcome of the FEMAT process, the Northwest Forest Plan, in July 1993, President Clinton directed the Forest Service to take the lead in developing "a scientifically sound, ecosystem-based strategy" for managing the forests east of the Cascade crest. The process that was initiated to implement this charge was chartered in January 1994 to include an assessment of the interior Columbia Basin, with the portions of the Klamath and Great Basins within Oregon included, and two environmental impact statements (figure 11.1). The total assessment area included approximately 145 million acres and portions of seven states. The environmental impact statements were to cover FS- and BLM-administered lands within the assessment area (approximately 75 million acres).

Policy Questions

The chief of the FS, Jack Ward Thomas, and the director of the BLM signed the charter establishing the Interior Columbia Basin Ecosystem Management Project (ICBEMP) in January 1994. The charter provided direction for the preparation of environmental impact statements (one for the Upper Columbia River Basin [UCRB] and one for eastern Oregon and Washington [Eastside]), a scientific framework for ecosystem management, a scientific assessment of the basin, and a scientific evaluation of the EIS alternatives (table 11.1). The charter was interpreted to include a minimum set of policy questions the project was to address. Agency leadership finalized these questions after interaction with the public and agency line officers.

The charter did not explicitly enumerate the policy questions for the assessment. This lack of explicit statement of policy questions caused confusion throughout the project. It was difficult to get everyone to accept the final list of policy questions that were developed, partly because they were not in the implementing charter. A shorter list, generated directly by the executives, could have simplified the assessment process and resulted in more focused, and useful, prod-

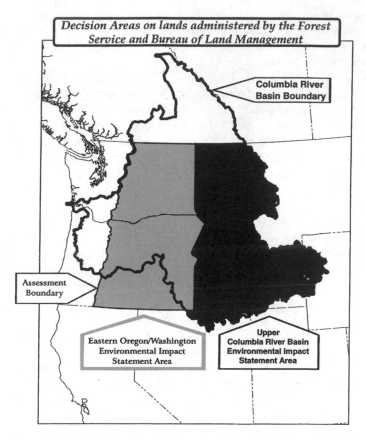

Figure 11.1. *The large region of the interior Columbia Basin was divided into two decision areas with separate environmental impact statements.*

ucts. The charter provided definitions of the project area, a description of the situation leading to the need for an assessment, a description of project expectations, a description of key participants and their roles, a definition of essential coordination, and a description of the primary products of the project. It defined each product in terms of the objective, components, and anticipated time lines. Because the policy questions were not explicitly stated in the charter, it was essential to go through a process to develop and validate the questions the Executive Steering Committee wanted addressed through the project. The science team took the lead in the process and derived policy questions from the charter that addressed major issues within the basin. They include:

- What are the effects of current and potential FS and BLM land allocations on ecological, economic, and social systems in the basin?

Table 11.1. *Products developed by the Interior Columbia Basin Ecosystem Management Project.*

Product	Responsible Team	Description
Scientific assessment	Science Integration Team	Assessment of the historic, current, and potential future status and trends for terrestrial, aquatic, landscape ecology, economic, and social systems in the basin
Framework	Science Integration Team	Description of the principles and processes appropriate for managing ecosystems within the basin
Eastside EIS	Eastside EIS Team	Proposal of new goals, objectives, standards, and guidelines applicable to national forests and BLM districts in eastern Oregon and Washington
Upper Columbia River Basin (UCRB) EIS	UCRB EIS Team	Proposal of new goals, objectives, standards, and guidelines applicable to national forests and BLM districts in the UCRB
Evaluation of alternatives	Science Integration Team	Evaluation of EIS alternatives that discloses outcomes, trade-offs, consequences, and relative risks to ecological integrity and economic resiliency

- What are the ecologic, economic, and social system outcomes associated with current (in the early 1990s) FS and BLM levels of activities?
- What is required to maintain long-term productivity (in terms of various systems)?
- What can the FS and the BLM do to mimic disturbance elements on the landscape?
- What is required to maintain sustainable and/or harvestable and/or minimum viable population levels?
- What is required to maintain and restore biological diversity (biodiversity)?
- What are the impacts of ecosystem management on major social issues and the maintenance of rural communities and economies?
- What are the impacts of ecosystem management on maintenance of late-succession and old-growth systems?
- What management actions will restore and maintain ecosystem health (forest, rangeland, riparian, and aquatic) health?

- What can the FS and the BLM do to implement adaptive management, and what is its consequence on ecologic, economic, and social systems in the basin?
- What can the FS and BLM do to protect endangered species (such as salmon, grizzly bear, gray wolf, and caribou) and insure the viability of native and desired non-native plant and animal species?

In addition to these policy questions, the assessment answered the following questions:

- What are the structure, composition, and functioning of the basin today (what is there)?
- By what developmental pathways did it get to its current conditions (how did it get there)?
- What is a plausible range of future conditions (where is it going)?

Organization

The charter provided guidance on the structure and organization related to the overall project. An interagency approach was used because it recognizes the geographic extent of, complexity of resources and issues in, and multiple jurisdictions for natural resources within the basin. The Columbia Basin Executive Steering Committee, consisting of regional foresters, research station directors, and BLM state directors for the area involved, provided oversight and direction. This committee had the responsibility for overseeing the implementation of the charter, monitoring and reporting progress, proposing amendments to the charter, ensuring internal and external participation, resolving issues related to the charter, and soliciting participation from other potential partners. As the project neared release of the draft environmental impact statements, the Executive Steering Committee expanded to include the regional directors of the National Marine Fisheries Service, Fish and Wildlife Service, and Environmental Protection Agency. This significant change resulted in new dynamics in the policy discussions.

A project leadership team, consisting of the Science Integration Team (SIT) leader, two project managers, and the BLM coordinator, provided overall direction to the specific teams, organized around the principal products of the project: assessment and EISs. The Executive Steering Committee selected the project leaders, who were given latitude to select team members. Individuals were assigned to the project through extended details or reassignments. Few of the positions were advertised; most personnel committed to the project through personal contact and negotiation with them and their supervisors. Thus, most were volunteers.

The SIT included functional staff areas for landscape ecology (physical and vegetative resources), terrestrial resources, aquatic resources, economics, and social factors. A staff of geographic information system specialists supported the spatial and data processing needs of all the science staffs. Headquartered in Walla Walla, Washington, the SIT was composed of federal employees from the FS, BLM, EPA, U.S. Geological Survey (USGS), and Bureau of Mines (BOM). Contractors were brought in for specific tasks and assignments. Detached analysis units were located elsewhere throughout the region. As a direct result of legal issues related to FEMAT, SIT was all federal employees by design.

It was first expected that SIT members would serve for nine months, remaining as employees of their current organizations and traveling to Walla Walla as needed for coordination, information sharing, and integrative discussions. It became apparent very early that this approach would not work, and each functional staff was asked to station a representative in Walla Walla. This resident group became the primary communication link among teams, within and across functional areas, and with project leadership. This group facilitated much of the integration across functional lines.

Project leadership took overall responsibility for buffering teams from political interference. For the assessment, the Science Team leader was the principal spokesperson. Through the fall of 1995, this buffering process worked effectively. However, in 1996, motivated by interest groups concerned about possible outcomes from the decision process, Congress attempted to restrict portions of the assessment. This attempt was effectively reversed through a contentious round of negotiations in the appropriation process, with this project being one of the primary reasons federal employees were furloughed while the government was shut down. The 1998 appropriation process again included language that would effectively shut down the overall project, increasing the uncertainty that the EIS process will result in a Final Record of Decision.

The overall assignment of the ICBEMP Science Integration Team included a scientific framework, detailed functional assessments, and an integrated assessment. The SIT identified the information needed, designed the process used to gather the information, monitored the information gathering (including field units submitting maps, data tables, and reports, as well as conducting workshops that resulted in additional maps, data tables, and reports), and evaluated and analyzed the information. The SIT integrated the information brought forward by the five functional staffs and described the trade-offs and potential consequences of interactions. This document, composed of detailed reports from each functional staff area, addresses the biophysical and social conditions of the basin. Another document, "An Integrated Scientific Assessment for Ecosystem Management for the Interior Columbia Basin Including Portions of the Klamath and Great Basins," links landscape, aquatic, terrestrial, social, and economic characterizations to describe the biophysical and social systems. Integration was achieved through the use of a framework developed around six goals for ecosystem management and different views of the future. In support of these documents, and as background to the conclusions, more detailed reports of databases, models, and information layers have been published (Quigley and Arbelbide 1997; Quigley, Lee, and Arbelbide 1997). These will prove useful to both public and private land managers in future decision making.

The original project time line was extended to two years when the Executive Steering Committee decided to have two broad EISs; the final Record of Decision is not expected until summer 1999. Total expenditures toward all products will approach $40 million by project closures. The FS regions, FS research stations, and state BLM offices have provided funding with help from the agencies' Washington, D.C., offices. Other cooperating agencies (NMFS, EPA, and USFWS) also contributed staff to assist in project activities and development of products. The availability of funds was not the major limiting factor in meeting project commitments. A single budget office negotiated with regions, stations, and state offices for budgets during the latter stages of the project. This greatly simplified the process. A more significant problem was obtaining the needed commitment of personnel for the various tasks needed throughout the basin. These were to be "contributed"; no central funds were made available to pay for permanent staff being tasked with duties that lasted less than forty hours. The sheer magnitude of work to complete the project made this very difficult. Participants and supervisors would have been more willing to cooperate on these tasks if money to cover salary had been provided.

Public Participation

In contrast to the traditional NEPA process, in which the public is involved in scoping and review, the Interior Columbia Basin Ecosystem Management Project attempted to keep the public involved in all phases of the science process on which the scientific framework, assessment, and evaluation of alternatives are based. The public was kept involved in the assessment through open SIT meetings and workshops and access to written material in draft form.

During the early phases of the project, regularly scheduled public meetings were held. At these sessions, each team on the project, gave a progress report, shared draft reports, and answered questions. Reports were made available to the public at a local printing-copying service and through an electronic library maintained by project personnel. Anyone with a computer modem could download material from the library through telephone access.

The SIT and the project also hosted several public workshops. This approach to "openness" required personal and financial commitment from the SIT, helped remove the potential for surprise in the end, and helped SIT members better understand public expectations. Those members of the public who tracked the process most closely received a substantial education in environmental sciences, as did the SIT members.

The SIT also made data layers and maps available to the public when the data were documented. The intent was to be as open as possible while the assessment efforts were underway. A data release policy was adopted, and several of the map layers and databases were made available during the planning phase. Most map layers and databases were releasable concurrent with the published assessment.

Governmental Coordination

Because ecosystems within the basin do not conform to jurisdictional boundaries, effective ecosystem management requires close coordination among all tribal, federal, state, and local governments. The FS and the BLM are required by law to coordinate planning activities with these governments. Coordination with other federal agencies varied in effectiveness and degree. Several federal agencies actively participated during all phases of the project—for example, the Environmental Protection Agency, Geological Survey, Bureau of Mines, and Fish and Wildlife Service. During the latter phases of the project, the National Marine Fisheries Service became actively engaged, while the Natural Resources Conservation Service involvement consisted mainly in responding to information requests.

There are one hundred counties within the assessment area, which makes working relationships difficult. Project leadership encouraged the formation of a coalition of counties that would interact with the project. The state association of counties for Washington, Oregon, Idaho, and Montana joined to form the Eastside Ecosystem Coalition of Counties (the name was patterned after the original name of the project, "Eastside Ecosystem Management Project"), with representation of three or four county commissioners from each state. A memorandum of understanding formalized the relationship between the coalition and the project. The Coalition held periodic meetings in Walla Walla to which it invited project members to provide updates and to interact on issues of common concern. The coalition developed considerable influence in the politics of the basin and helped convince Congress to continue support for the project. The coalition also effectively involved governors' offices in the project, worked with interest groups, presented material about its role in the assessment at national meetings, and worked together on several issues beyond the assessment. The working relationship between the counties and the project proved to be very effective.

Tribal Coordination

There are twenty-two recognized American Indian tribes in the basin, and tribal issues were integral concerns addressed and recognized within the project. Because tribes play an important role in ecosystem management, a tribal liaison was assigned to arrange meetings and consultations and to facilitate the sharing of information between the tribes and the project. The Executive Steering Committee consulted personally with each tribe in the project area, which required an individual approach for each. Tribes were primarily interested in decisions and future outcomes with special emphasis on treaty trust responsibilities. The tribes were particularly effective in the policy discussions regarding water quality and salmon habitat.

Tools for Cooperation and Participation

The project used a variety of techniques to identify the values people place on ecosystems and ecosystem components. It commissioned a survey of the public, analyzed current laws and policies, and held workshops throughout the basin to update the public on project status, as well as to listen to the public and try to understand its concerns. Public meetings, field trips, workshops, and written and electronic correspondence provided opportunities for substantive input to and review of written documents. The objective was to allow open and broad partic-ipation of stakeholders in development of assessment and decision documents, tempered by the need to move forward on budget and within the time line given. Project leaders shared information through various means: print, radio, and tele-vision media; satellite and computer network communications; the electronic library; a toll-free information number; personal presentations by leaders and staff of the project at meetings of various organizations; one-on-one contacts with key individuals representing interest groups; local information centers in basin communities; and stakeholder meetings held throughout the basin. As a result, fewer surprises were delivered to the public in the project results, a wide range of publics gained better understanding of the science findings and their implications for management, and project members gained an appreciation for the concerns and insights of interest groups and concerned citizens. Although it probably lengthened the total time to complete the project, the assessment and resulting policies were strengthened.

Conducting and Reviewing the Assessment

The Executive Steering Committee, in cooperation with the FS and BLM Wash-ington offices, determined the boundary for the study area. The presidential direction was for the forests and rangelands east of the crest of the Cascades. Resolving this to include portions of the Klamath and Great Basins within Oregon was basically a call by the regional forester for R-6 and OR-WA and the BLM state director for Oregon Washington, motivated mostly by the need to address forest and rangeland health issues consistently. The SIT recommended using the Continental Divide as the eastern boundary due to the hydrologic and fisheries issues. Each staff area within the SIT was given latitude to use alternate boundaries in defining systems, as long as they included the basin.

The assessment draws on information from all lands and ownerships within the basin, not just FS- and BLM-administered lands. Providing context for deci-sions and understanding that ecosystem components, structures, processes, and functions operate at multiple geographic and temporal extents requires that all lands be included in the assessment. While the rights of landowners within the basin were respected, information was gathered from both public and private sources throughout the basin. Remote sensing techniques reduced potential con-flicts. Because of the low level of data resolution used in the assessment and the large geographic extent, the assessment relied primarily on remote sensing or readily available information from third party sources. An effort was made to utilize as much of the existing information concerning the past and present condition of the basin as possible. Inconsistencies in existing information sources resulted in the need for new remote sensing efforts and an extensive subsampling process for vegetation and physical attributes of valley bottom setting characterizations. To the extent possible, the SIT relied on existing simulation models to project future conditions of the basin. Where existing models were not available, new models were constructed and simulations made to project future conditions or interpretations, and inferences were made from the information available and model results. Final publications describe assumptions, uncertainty, and variability in data elements and models.

Projections of the future were made through two approaches: scenarios and EIS alternatives. The SIT developed the scenarios as a broad description of future management approaches. They were developed as plausible, not necessarily likely or probable, futures that helped to characterize outcomes and consequences associated with different intensities of management. One example calls for aggressive thinning and prescribed fire or strong reliance on natural disturbance processes, in a very broad sense. Thus, scenarios ranged from intensive management for commodity production to passive management for a return to more natural processes. The EIS teams developed alternatives, drawing from outcomes projected by the scenarios, as specific directions the FS and the BLM could undertake to achieve objectives through establishing standards, guidelines, and processes for future management. The alternatives included a range in emphasis including continuation of current management, emphasis on commodity production, aggressive restoration, and a system of large reserves. The SIT evaluated alternatives against the likely achievement of stated goals and objectives. A team of legal advisors (from the FS and the BLM) assisted in providing legal interpretations and review.

A science review board (SRB) oversaw a modified blind review for documents developed by the SIT. The SRB facilitated the review of scientific approaches and reports of SIT. The review was broad, including diverse options and consideration for practicality and management feasibility.

Reports and projections were shared in draft form to EIS team members, agency management personnel, and executives. One-on-one interaction among team members and agency staff occurred frequently concerning the draft products. Presentations to the public were held periodically to share tentative findings and receive input. Internal meetings among SIT and EIS teams were held with the specific goal of integrating the findings and principal recommendations. The Science Team leader was responsible for resolving any differences across staff areas and concurring with final recommendations. The sharing of databases and approaches early in the formulation process resulted in few differences across staff areas. Meetings or conference calls were used to bring resolution to the few differences that were brought forward. Bringing together staffs with varying interpretations of a resource condition resulted in information sharing and agreement on interpretation.

As the comment period on the draft EISs was concluding, the SIT undertook an extensive process to examine the preferred alternative for consistency with science understanding and interpretation. This "science consistency evaluation" answered whether the preferred alternative correctly interpreted the science information, whether the risks and uncertainties were acknowledged and documented, and whether the preferred alternative elements were consistent with science. This consistency check process was initiated to address issues relative to whether scientific information was ignored, misrepresented, or misinterpreted in the development of policy and management direction. The consistency check is planned to continue during the final EIS process and development of the record of decision. It forces scientists to play out their proper role and provides better assurance that decisions are science based.

The Results

As of this writing, the project is progressing toward completion. The political processes continue to heat up, with attempts at influencing the decision-making process and the information presented. In all likelihood, the political process will get more intense as decisions become imminent. Because the project has not yet concluded, final judgments on success are not appropriate. Perhaps the best way to characterize the situation is to describe whether the project has within its grasp the ability to address the policy questions. Whether the project will realize that ability is largely dependent on political decisions.

The project has sufficient information to address the policy questions with broad and mid-levels of data resolution, but some effects and consequences at a fine level of detail may be more difficult. The project focused on the broad and mid-level resolutions. The geographic extent of the project and the charge to examine the "system" as opposed to individual species necessitated examining habitats and indicators of populations rather than strict population viability analyses. The question of sustainable, harvestable, and viable population levels has been only partially addressed.

The political process threatened to restrict SIT and EIS products. If that threat had been successful, it would have resulted in an inability to address many of the policy questions. The project weathered the political storm regarding restrictions on science reports. There were strong political pressures brought to bear on reporting all results from terrestrial species, aquatic systems, and fisheries. The final science products, however, reflect the full complement of understanding in those areas.

The science products of the project will be valuable for making broad-level programmatic decisions, setting policy priorities, and providing context for decisions within specific geographic areas (e.g., national forests and BLM districts). Circumstances, budget, and politics may overrun the assessment products, undermining their utility for current decisions. If the project had been able to meet its original nine-month time line, circumstances would have been very different; results would have been available for decisions at an earlier time, potentially avoiding issues being overtaken by political maneuvering.

The extended time line made integration and reporting difficult. The science products were authored by the SIT, using technical editors to oversee content and the writing process. Yet it was very difficult toward the end of the project to seriously engage all SIT members in discussions and writing. Concurrent and multiple products, coupled with outside pressure to address other issues, disengaged scientists as they moved to other assignments. In the end, each product was brought to conclusion by a subset of the SIT. The technical editors provided leadership; consensus was attempted through meetings and iterative reviews and were for the most part met. The technical editors resolved disagreements with the concurrence of the Science Team leader.

The primary means for sharing results is through the publication process. An extensive three-day workshop was held to share methods, findings, and recommendations. Plans are still being prepared regarding technology transfer and implementation follow-on. Several presentations by individual SIT members have been made at regional and national meetings.

Implementation

As of this writing, the implementation phase is planned to begin in 1999. National forests and BLM districts and private groups are using the assessment as they plan activities and amendments. The Oregon Department of Forestry is using the information as a key component of its statewide forest assessment. Plans are being developed to use the assessment information to amend seventy-two forest and BLM plans within the basin. The political process that actively sought to dictate implementation approaches was not enacted as law, although another attempt is being made for the 1999 appropriation cycle. This leaves the agencies the option to proceed with plan amendments. The Executive Steering Committee is currently committed to incorporating the assessment information into existing plans through an amendment process. The specific details on plan amendments are not yet finalized, and another appropriation cycle is set to begin. Politics are most likely to focus on EIS decisions and science consistency as the process continues. As decisions get closer, tribes, counties, governors, Washington office staff, congressional subcommittees, the Department of Justice, and the Council on Environmental Policy all are stepping up their engagement. The science findings have elevated the policy debate above arguments of site-level considerations. The resolution of providing assurance of outcomes in the absence of fine-level standards remains an issue.

The SIT has proposed that the assessment information be considered a part of a dynamic assessment that includes models, databases, and analyses updated through monitoring, inventory, and analysis processes (figure 11.2). This could occur through a partnership of science and management in an adaptive ecosystem-based approach that potentially could address cumulative effects at geographic extents above individual national forests or BLM districts. This requires the commitment to maintain databases consistently across multiple jurisdictions, framing monitoring and inventory within a consistent hierarchy, and periodically validating the critical questions needing resolution in the broad and mid-level of detail.

Retrospective

The capacity to accomplish the charter objectives is within reach. Political and budget realities will be the final deciding factors in the extent to which the assessment results in substantive changes in management. Some of the political concerns are general in that they originate from issues broader than the basin (e.g., the concern over private-property rights). While some of the political concerns are specific to the basin (e.g., local rather than programmatic decisions, gridlock in forest management, active management to address forest health, and fisheries

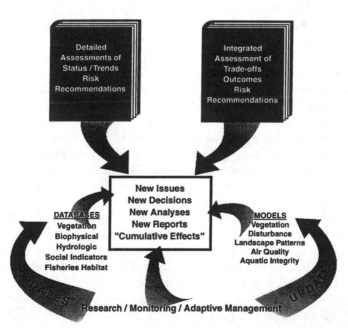

Figure 11.2. *The dynamic assessment model developed by the Science Team in the interior Columbia River Basin assessment.*

habitat management). The convergence of these concerns has resulted in attempts to use the political process to influence outcomes.

If undertaken again, an assessment of the interior Columbia Basin could be improved by taking the following steps:

- Complete the assessment and framework before proceeding to organize and develop EISs.
- Center the study in an area served by several commercial airlines to facilitate team interactions.
- Base the assessment on existing information to a greater extent—with no additional data collection.
- Assign team members to project work as their sole priority and station members in a common location for the duration of the assessment.
- Rapidly get commitment from executives on the specific policy questions being addressed.
- Commit scientists to the assessment who are integrative, are comfortable in the policy arena, understand the broad issues and concerns, and can work under tight time frames and limited budgets.
- Keep the time line condensed so that outside circumstances do not overrun the assessment.
- Identify project goals early—in the ICBEMP, it would mean to identify the goals of ecosystem management early in the project.
- Focus data collection, analysis, and writing on the specific risks to achieving the identified goals.
- Insure library and computing resources are readily available to team members.

In the ICBEMP case, there was a lack of consistent information across the basin that could be used to address the policy questions. Filling this void to the degree we could prevented us from completing the assessment in six to nine months, as originally requested. If the SIT had proceeded with the best available information, outlining processes that could be used to fill identified knowledge gaps and addressing the major policy questions identified, the likelihood of being overrun by circumstances beyond control of the project would have been lessened.

The issue of assessments and the role of research in accomplishing them should be addressed specifically through a deliberative decision-making process. The agencies need to develop the institutional capacity to conduct assessments through approaches other than a crisis. Assigning staff, committing resources, and establishing dynamic assessments (linking research, inventory, monitoring, and adaptive management) are the means by which cumulative effects and policy issues can be addressed efficiently and effectively.

Many of the problems faced in this assessment draw from issues bigger than those in the Pacific Northwest. The struggle to settle on specific goals and policy questions stems from a lack of clear direction and purpose for the management of public lands. National-level direction, in terms of law, executive mandates, or agency policy, is strong on process and low on vision or direction. For what basic purposes do the national forests and BLM districts exist? Is providing commodities a primary purpose or a by-product of meeting biodiversity goals? Are resource-dependent communities to receive priority in access to natural resources from public lands? Should the provision of viable populations of native species be a goal or a constraint? National direction continues to push these issues to lower and lower levels by providing laws and policies that establish more analysis and planning processes.

The overall assignment to develop a scientifically sound, ecosystem-based strategy for management of FS and BLM lands proved to be a task that might very well require several additional years before it will be reality on the ground. Indeed, the writings on ecosystem management have been filled with platitudes that provide no real priorities or ranking of actions to proceed with implementation. The roots go deep in this approach, as Congress has provided no clear statement of goals regarding natural resource management. Rather, recent laws (for example, NEPA, NFMA, and ESA) have been focused on pushing more processes for planning to lower levels in the FS and the BLM.

Literature Cited

Everett, R. L., P. Hessburg, M. Jensen, and B. Borman. 1994. *Eastside Forest Ecosystem Health Assessment*, PNW-GTR-317. Portland, OR: USDA Forest Service, Pacific Northwest Research Station.

Henjum, Mark G., J. R. Karr, D. L. Bottom, D. A. Perry, J. C. Bednarz, S. G. Wright, S. A. Beckwitt, and E. Beckwitt. 1994. *Interim Protection for late-successional forests, fisheries, and watersheds: National forests east of the Cascade crest, Oregon and Washington.* Bethesda, MD: The Wildlife Society.

Jaindl, R. G., and T. M., Quigley, eds. 1996. *Search for a solution: Sustaining the land, people and economy of the Blue Mountains—A synthesis of our knowledge.* Washington, DC: An American Forests Publication. 316 pp.

NAS (National Academy of Sciences). 1997. *Forested landscapes in perspective: Prospects and opportunities for sustainable management of America's nonfederal forests.* Washington, DC: National Academy Press.

O'Laughlin, J., J. G. MacCracken, D. L. Adams, et al. 1993. *Forest health conditions in Idaho.* Policy Analysis Group Report 11. Moscow, ID: University of Idaho, Idaho Forest, Wildlife and Range Experiment Station. 244 pp.

Quigley, T. M. 1992. *Forest health in the Blue Mountains: Social and economic perspectives.* General Technical Report PNW-GTR-296. Portland, OR: U.S. Department of Agriculture, Forest Service, Pacific Northwest Research Station, 9p. (Quigley, Thomas M., tech. ed., *Forest health in the Blue Mountains: Science perspectives.*)

Quigley, T. M., and S. J. Arbelbide, tech. eds. 1997. *An assessment of ecosystem components in the interior Columbia Basin and portions of the Klamath and Great basins,* PNW-GTR-405. Portland, OR: USDA Forest Service, Pacific Northwest Research Station.

Quigley, T. M., and H. Bigler Cole. 1997. *Highlighted scientific findings of the Interior Columbia Basin Ecosystem Management Project.* General Technical Report PNW-GTR-404. Portland, OR: USDA Forest Service, Pacific Northwest Research Station.

Quigley, T. M., K. M. Lee, and A. J. Arbelbide, tech. eds. 1997. *Evaluation of EIS alternatives by the Science Integration Team,* PNW-GTR-06. Portland, OR: USDA Forest Service, Pacific Northwest Research Station.

Sampson, N. R., D. L. Adams, S. S. Hamilton, et al. 1994. Assessing forest ecosystems health in the inland west. In *Assessing forest ecosystem health in the inland west*, edited by Neil R. Sampson and David L. Adams. New York: Haworth Press.

USDA Forest Service. 1996. *Status of the Interior Columbia Basin: Summary of scientific findings*. General Technical Report PNW-GTR-385. Portland, OR: USDA Forest Service, Pacific Northwest Research Station.

Science Review

James K. Agee

Four years after its initiation, the ICBEMP has released a series of reports that builds hierarchically on years of intensive effort to describe a vast, multistate river basin. At a cost of well over $30 million, the project represents the integration of massive data sets and an enormous amount of public interaction, which continues with the drafting of two environmental impact statements currently in review and revision. The scientific reports alone total more than two thousands pages. A number of issues related to the role of science in development and implementation of these large-scale assessments exist. Because of the large scope of the project, this review has to be considered one critique and not the final word.

The Issue of Scale

Natural resources science has rarely operated at the scale of an entire multistate river basin. Scientists are usually much more comfortable working at larger scales (considering scale in the cartographic sense, that is, larger scale might be 1:1,000, and small scale might be 1:1,000,000). Reductionist science is an important part of science, but to some extent it is of limited utility at smaller scales of the real world, where complex interactions limit generalizations. One of the strengths of the Columbia River basin assessment was to break the 145-million-acre basin into roughly million-acre sub-basins defined by watershed boundaries, a more workable strategy. Integrating the sub-basin results appears to be a black box, at least at this stage of disclosure; how it is done and whether or not it works are yet to be seen.

The Integrated Scientific Assessment for Ecosystem Management in the Interior Columbia Basin (Quigley and Cole 1997), called here the ISA, has utilized the sub-basin approach to show regional trends, often by use of shaded maps, allowing the reader to synthesize ecological change at a regional scale, be it fire regimes or hydrologic integrity, and to identify portions of the region that are most at risk. While it is not altogether clear that the sub-basin summations revealed emergent properties, they effectively summarize issues at smaller scales, and some creative work was done in ecological scaling. A creative use of cluster analysis was used to define "forest" and "range" sub-basins by a variety of characteristics, such as vegetation conditions, hydrologic sensitivity, and human-caused disturbances. Sub-basins with similar characteristics were then mapped as one of six potential clusters for either forest or range ecological integrity. Basin and sub-basin scales of terrestrial and aquatic resources are simultaneously addressed by this technique. The summaries of these data suggest that federally managed lands have a composite ecological integrity much higher than that of the intermingled state and private lands (the forest cluster with 100 percent "high" composite integrity has 80 percent FS and BLM ownership, while the cluster with 100 percent "low" ecological integrity has only 35 percent FS and BLM ownership; the rangeland cluster data are similarly skewed).

The intent of the process is to supplement or modify existing Forest Service and Bureau of Land Management plans, not to replace them, so at implementation the process of scaling will move down to the lower level again. The Forest Service and Bureau of Land Management administrative units will have the benefit of planning with a small-scale context in place, a context essential when the administrative units intersect and overlap ecological unit boundaries. This approach will likely improve federal land management plans, but it does nothing to address generally lower ecological integrity on private lands. The assessment does little to provide direction on the scaling-down process even on federal lands.

The Baseline

The temporal baseline for much of the assessment is "the last hundred years." Realistically, there exists almost no detail concerning resources before that time. The ISA does not make the mistake of defining this earlier period as necessarily the most desirable end point for future management, but it does indicate that the interaction of some processes in earlier times resulted in much more sustainable outcomes. Some analyses, such as rangeland condition, begin after the most significant change had already occurred, and the ISA clearly recognizes this, as in its description of nineteenth-century introductions of livestock.

In other databases, the change that occurred early due to European settlement is not always recognized. For example, much of the vegetation database used to interpret ecological change comes from early aerial photos. This restricts the temporal period to fifty years or less; in the earlier Eastside assessment, a predecessor to ICBEMP, the average "early" photo date was 1948. Much of the ecological shift in species composition and forest structure was well in place by that time. Graphs in the ISA that show changes in fire regimes after 1950 significantly underestimate the change that had occurred before that time.

What Is Quality of the Baseline?

Questioning the quality of baseline data strikes to the core of how to do such broad assessments. Should we concentrate on already published information plus well-documented unpublished data or rely on collection and interpretation of new data? For example, significant amounts of data were generated from old aerial photographs, and the quality of the early photos is generally poor (small-scale and fuzzy). Therefore, the basic data going into the secondary analyses were somewhat questionable, and their weaknesses may be compounded as several of these databases are merged and synthesized. The quality of the basic data becomes more easily ignored in each successive, higher-level assessment.

Similarly, a number of major assumptions in other parts of the assessment are not clearly reiterated at subsequently higher levels of analysis. The nature of any specific set of assumptions is less relevant to this criticism than the fact that the results of analyses are highly influenced by the assumptions. The assumptions are often quite broad in these science-based assessments, as they need to be, yet the necessary caveats fade at each successive level of integration, with the emerging conclusions appearing to be factual rather than the result of sets of assumptions made at the primary level of data analysis.

The ISA discussion on science gaps largely ignores the effect of these gaps on the assessment itself. The ISA concludes, "We found it difficult as scientists to accept that existing information presented in a timely fashion had more influence than detailed data brought forth later." I would conclude that existing information of known validity has more influence than detailed data of questionable validity, whether brought forward now or later.

The scientists involved in the ICBEMP are currently engaged in a "science consistency check," evaluating the use of the scientific data provided to the EIS Team in the two environmental statements. Given the uncertainty of some of the data provided to the EIS Team, and the evaluation of data incorporation into the environmental statements by the same individuals who provided them, this appears to be an exercise that focuses on precision rather than accuracy. The EIS may precisely incorporate inaccurate information, so that the "science consistency" review has a limited potential to scientifically "bulletproof" the two environmental statements. Even with these caveats, it is a refreshing instance of managers asking scientists "How did we do?" in incorporating science in decision documents.

Does FACA Suppress Science?

The ICBEMP process has appeared to be a close-to-the-chest process to most of us in academia, primarily due to the Federal Advisory Committee Act (FACA). At the time the ICBEMP Team was being developed, FEMAT had just been declared in violation of FACA because of the incorporation of selected nonfederal scientists in the process and the general exclusion of other nonfederal people. According to the law, when assessments are treated as decision documents, interaction with nonfederal scientists and others must be closely controlled unless the process is totally open. For some assessments, this has not been a problem in practice: Sierra Nevada Ecosystem Project (SNEP) was not a decision-based assessment, so FACA was not directly applicable. Almost all of the nonfederal input in FEMAT was complete before FACA violations were established in court. In the ICBEMP process, lack of outside scientist interaction, except in special situations such as limited peer review, clearly limited the possibility for intermediate checkpoints. While the project provided tremendous opportunities for public input and access to support documents, including access for outside scientists who wished to review materials, this is a far cry from insuring adequate, independent scientific input into the documents and their review. Did a lack of interaction affect the outcome of the process? I don't think that question can be answered yet, and it cannot be answered by the "science consistency" review now underway. Realistically, given the penchant for new data collection in ICBEMP and short time lines for the process, many academic scientists would not have been able to play large roles in the process anyway. My guess is that many participants in ICBEMP would have felt more comfortable with fewer restrictions on interactions with nonfederal scientists, and I know that some nonfederal scientists active in the region, when asked about ICBEMP, are uncomfortable responding to questions about the science involved in ICBEMP.

Too Much or Too Little Time?

The case study by Quigley, Graham, and Haynes in this chapter suggests that if undertaken again, less time should be allotted for the assessment and less data collection should be undertaken. In general, I agree, possibly for different reasons. I think science has served management best by being a unbiased source of reliable information. From what I have seen of these bioregional assessments (including SNEP and FEMAT, as well as the original Eastside assessment), data that were already available were generally more reliable (but still very incomplete) than data generated as part of the project (see discussion of baseline quality earlier). Analyses funded as part of these assessments are rushed and expensive (>$30 million in this case); secondary and tertiary analysts have little interaction with the primary analysts; and to some extent a can-do attitude will result in analyses being attempted that are too ambitious, even if time and money constraints are absent. However, analyses that are done as part of a regional assessment can be tailored to questions being addressed and do have unique value in that regard. There is probably a project-specific balance between using existing data and generating new data that would mitigate the problems of lacking project-specific data (the existing database) and lacking reliability (the new database), but finding that balance may be difficult. Clearly, time- and money-constrained assessments will be forced to rely more on existing databases, while assessments without these constraints will be more free to generate new data as a dominant source of information.

Emphasize What We Don't Know

Most scientists with interests in drier forests and rangelands, not only in the Columbia Basin but elsewhere, have been disappointed by the lack of research attention given these wildlands in past years. The preexisting database from which the assessment was drawn was not as strong as the preexisting database, for example, used with FEMAT. In extrapolating from this limited database, an important element is to identify major holes or gaps in the database. To some extent, this can be addressed through the adaptive management approach, and the ISA has a detailed section dealing with science gaps that, if filled, would improve future management. It can supply a blueprint for future research, both in terms of disciplinary areas and the scale at which it should be conducted. After all, none of these assessments are destinations; they are temporary rest areas along the uncertain highway of ecosystem management.

Is Adaptive Management the Cure?

Adaptive management is certainly a defensible approach in a world of ecological uncertainty. But it can work, and be defensible, only if simultaneously adopted in the sociopolitical world. In my view, that means it must be integrated in the budgetary process from top to bottom. Adaptive management cannot work if money is allotted for roading and timber harvest, but not for monitoring. The political world does not have to embrace uncertainty itself, but it must fund activities that reduce or define uncertainty if it continues to fund those that cause it. Otherwise, feedback to management cannot occur, and goals, objectives, and strategies cannot be altered intelligently based on such feedback.

The scientific documents suggest that the major ecological problems in the interior Columbia Basin are not with federal land management, but are with management on state and private lands within the planning area. The documents carefully avoid direct statements implicating state and private lands, but the data summaries speak clearly to this issue. How can ecological problems in the interior Columbia Basin be comprehensively addressed when only half of the landscape is addressed? The stated answer in the *Summary of Scientific Findings* (USDA Forest Service 1996) is that the Forest Service and the Bureau of Land Management will need "help from others" to make ecosystem management work. Perhaps the clearest implication from these assessments is that ecological science has defined a pathway that is strewn with sociopolitical obstacles. The federal agencies in the executive branch are powerless to solve these problems, many of which involve constitutional issues of states' rights and private-property rights. The federal role may be a strengthened technical assistance and education role, discussed in the 1997 National Academy of Sciences report on nonfederal forests (NAS 1997). That may be viewed as a solution with a low probability of success, but I think it recognizes that federal land managers do not shoulder all of the responsibility for sustainable regional landscapes—we all do.

Management Review

Martha G. Hahn

A brief review doesn't allow a thorough discussion of a multiyear, $40 million project that has involved hundreds of local governments, dozens of agencies, and a cast of thousands. Rather than spend a lot of time on process, management, and implications of the scientific assessment of the Columbia Basin ecosystem, I'd like to cut quickly to the chase and consider what we learned from the experience. T. S. Eliot wrote, "We had the experience but missed the meaning." Writing this review has given me the chance to think about not only the experience but also the meaning. And I will reflect on the meaning from two perspectives: as the chairperson for the project's Executive Steering Committee and as the BLM Idaho state director.

The Interior Columbia Basin Ecosystem Management Project, or ICBEMP, is an attempt to assess ecological conditions on almost 145 million acres in seven western states and to provide guidance and recommendations to managers about how to improve and maintain ecological integrity. Part of the ICBEMP mission was to produce several documents, which I will lump together for purposes of this review and call "the scientific assessment," which were published in the fall of 1996. We hope the scientific assessment gives an accurate, science-based view of the state of the region. Ultimately, the recommendations of ICBEMP will provide guidance and management coordination for the 72 million acres of federal land covered as part of the project.

Before I plow ahead, allow me to detour slightly and give a tip of the hat to those who worked long, worked hard, and worked often in adverse—if not occasionally outright hostile—conditions. The people involved with ICBEMP were given an assignment that would have been difficult to complete on time and in a credible manner under the best of conditions. That they have been able to accomplish as much as they have is a tribute to perseverance, dedication, professionalism, flexibility, and a healthy appreciation of the bizarre.

Size and Scope

ICBEMP is a huge undertaking in several respects. First, consider the size of the project. The scientific assessment area looks at almost 145 million acres of public, state, private, and tribal land, covering portions of Oregon, Washington, Idaho, Nevada, Montana, Wyoming, and Utah. One hundred counties are included in the assessment's purview, along with twenty-two tribal governments. The assessment area contains some of the most diverse and complex ecosystems in America. Up to seventy-four individual land-use plans will be amended by the final record of decision.

The scope of the project grew rapidly in the early stages, and it became apparent that the resources needed to keep pace with the project—people, time, and money—would need to be expanded to keep up with the effort's raw ambition. The project was originally scheduled to last nine months, then was extended to a couple of years.

Complicated? Yes. But there are good reasons for allowing the project to grow. BLM's national director and the chief of the Forest Service made the call to expand ICBEMP. Their reasoning was, "Why stop at the Idaho border, when we're placing more emphasis on landscapes and basins? Migrating salmon don't recognize state boundary lines and national forest jurisdictions. Devastating wildfires don't stop burning when they cross from BLM to private or national forest lands. Many of the regional economies are tied to what happens on the other side of the next mountain range over. Let's do this assessment based on what the geography tells us is smart."

The burgeoning size of the project was a two-edged sword. It caused mixed feelings among managers. On the one hand, the project would provide much more information and help ensure a more consistent management approach on a bigger chunk of ground when ICBEMP is completed. On the other hand, our existing workloads did not go away. While our workforce was shrinking (and still is, in many places), we were asked increasingly to provide more people and more money for longer stretches to continue the project. Not that BLM employees were in the Walla Walla or Boise project offices every day, but the frustration level among our managers soared in direct proportion to the number of times they heard employees announce, "I'm off to Walla Walla for a few weeks."

I give credit to our managers and field employees for making good calls on what projects were allowed to slide somewhat because of the exodus of key specialists yet still accomplishing the essential work. We learned to pick up the big pieces first, as we juggled our day-to-day management responsibilities. If a project like this were again undertaken, there needs to be a commitment for employees to work full-time on the project, unencumbered by other duties to the extent possible, and return to their jobs when the work is completed. Yet, with our shrinking budgets, we may not have the luxury of doing things the "right" way. The practicality of this issue is that we may not be able to allow employees to work on long-term, wide-ranging projects, no matter how valuable the results might be.

Scale

The scientific assessment and other documents produced by ICBEMP are designed to give a big picture of the region. I am optimistic that this big picture will help us make decisions and design management in individual watersheds and local communities. But there is always a danger that the size and figurative weight of the product will be too much for local managers to work with effectively. They need a local road map, and we could be dropping a world atlas on them.

Our managers need a snapshot—clear, concise, and accessible. Yet what we see coming out of the effort is a product of epic scale—a mural. Can our local managers get the clear, "backyard" picture they need from such an expansive mural? The answer, I think, is yes, but there is a risk that critical information will be lacking when we ratchet down as much as needed with ICBEMP. A good part of the whole project's success will depend on its usefulness, not only on a regional scale, also but in how it is applied on the local level.

Resource Issues

Early in the process, forest health became a dominant issue. Certainly, it is an important issue, but there was a tendency to look less critically at some of the other issues because of forest health's dominance. For example, more consideration of rangeland health would have been useful. We are only now coming to an understanding of how rangeland ecosystems function; our research and knowledge of how rangelands function is on a sharp, upward curve. And about 43 percent—almost 31 million acres—of the federal land covered in the ICBEMP is rangeland. Yet many of us are a little disappointed that rangeland issues were not given proportional attention.

Part of the fault lies with BLM. We did not have all the data and information needed to complete the rangeland picture. This is not a situation unique to ICBEMP or the scientific assessment in particular. No, it is a fact of life that the information data managers and specialists need is not always at the tips of their fingers. We need to mitigate our lack of information with more inventorying and monitoring, more research—and more pressure on our paper-thin budgets.

BLM primarily manages rangelands. Since rangelands were not given the attention we had hoped for, we felt our agency often took the backseat. We felt a little bit like the kid who shows up on the playground with a basketball in hand, while everyone else brought baseball bats and gloves. If one resource or issue dominates in a scientific assessment process, then it comes at the expense of other resources. Jockeying for attention may be a part of scientific assessments, but it should be minimized.

Public Participation

The ICBEMP is an outgrowth of the President's Forest Ecosystem Management Assessment Team effort, which was viewed with skepticism by those who saw that decisions would be made by the White House. ICBEMP was guilty by association.

People in the interior Columbia Basin seemed to be thinking, "Here we go again; the shots will be called from Washington, D.C." The primary public concern about ICBEMP was that decisions affecting Murphy, Idaho, or John Day, Oregon, would be made at a Pennsylvania Avenue address.

This skepticism found root in Congress. At one point in 1995, it looked as though Congress would place severe constraints on the project, allowing the scientific assessment to be produced but not a record of decision. That would be roughly equivalent to a physician diagnosing an illness and then not prescribing a treatment to help you get better. Fortunately, key members of the local public believed in the value of the project and trusted that decisions would be made at home. We have taken extraordinary steps to involve counties, elected officials, tribal governments, interest groups, industry, conservationists, and just about anyone else who showed the slightest expression of interest. We have tried continually to reassure the public that, in the end, local people working with local managers will be the primary decision makers.

We still have a bumpy road ahead. The cost of implementing ICBEMP recommendations is high, about $267 million for the first year alone. Although the majority of this is already funded, the sticker shock of the project has raised some eyebrows and cost us support among elected officials. But we feel it is an investment, not just an expense. As the old commercial says, "Pay me now, or pay me later."

Process

Organizationally, the project has suffered from a lack of consistency. The Executive Steering Committee, made up of three BLM state directors, three regional foresters, and two Forest Service research station directors, changed five of its eight members during the first two years of the project. Other scientists, employees, and contacts changed, as well, making it difficult to track the work's status. New layers and players were added, and by necessity, lots of improvising took place. The more we learned, the more questions arose. The more questions, the more we invited others to participate. Regulatory agencies, tribal governments, county executives, and others were brought into the circle.

This does not necessarily suggest a bad situation, although it did complicate our work. Some innovative and extraordinary steps were taken to involve organizations as they became part of the widening circle of interest groups. For instance, we worked closely with a council of county officials to ensure good communication with local elected officials. We have made it a point of being accessible and explaining each step of the process to interested groups and individuals. And I can't even guess the number of miles employees assigned to ICBEMP have driven to hold informational sessions, public hearings, and other meetings, not only in Portland, Spokane, and Boise but also in Grangeville, Burns, and Elko. Some of our staunchest support and best information came from groups that were on the periphery in the beginning.

Implementation

Implementation is going to be a challenge for many reasons beyond its high cost. Congress and key constituencies will be watching implementation. They may accept that the science is pretty good, but they question if the decisions will be made locally or somewhere else. Implementation will give us the chance to back up our promises and allow local managers, working in concert with residents and others who have local affiliations, to make the decisions. The credibility of the project hangs in the balance.

We will amend seventy-four land-use plans for BLM and the Forest Service after the record of decision is issued. It is difficult to know if we can really do what the recommendations suggest. For example, the prescribed burning recommendations in the preferred alternative are ambitious and aggressive and involve some risk, but provide few practical guidelines.

It should not come as a blinding revelation to anyone: how the science and recommendations are implemented will determine if we got our money's worth out of ICBEMP.

This raises the question of scientists' role in implementation. We believe we need a partnership between managers and scientists in the nuts and bolts of making it all work on the ground. We don't want the scientists to think they are finished because the assessment is complete. There is more to do. Monitoring will be critical to the implementation, and we will need scientists to continuously assess the monitoring data. On the basis of these continuous assessments, we will update and fine-tune our management practices, which will require more monitoring to start the cycle anew. It is a dynamic process, this marriage of science and management.

Lessons

First, bioregional assessments need the appropriate time, funding, and personnel to complete the work. Managers need to make it as easy as they can for specialists to complete reviews and attend important meetings, even when it means that others must pick up the big pieces left behind.

Critical participation of nonfederal entities needs to be fully thought out at the scoping stage of the project—how do you get busy people to engage in reviews and meetings when they have only so much time and money? This is a long and often tedious process. Anyone who has stuck with it throughout the process deserves a healthy pat on the back, as well as our thanks. As managers, we need to commit to making the process as simple, clear, and accessible as possible for the public.

Set clear goals concerning what is needed from the project and set them early. The scope needs to be carefully planned; firm conclusions need to be reached early about what can be accomplished and what cannot be accomplished. Scientific assessment work and moving targets are not a good combination.

In something as far ranging as the scientific assessment of the ICBEMP, one issue should not dominate; the needs of all participants must be considered, and ideally, the information generated and recommendations developed will have high applicability. I like the analogy that large-scale projects such as ICBEMP allow us to first look at the ground around us, then climb the mountain for a broad-scale view, and then return to the ground with both perspectives to make the decision. We understand how the part fits into the whole and how the whole fits with the part. In the best of all worlds, political influence is not allowed to dictate what will be done with the information and conclusions developed by science. Obviously, we do not live in the best of all worlds, because politics can influence these kinds of projects. Political influence can be lessened, however, by executives and managers who agree to allow science to function independently and shield it from political pressure to the degree possible. Internally, blind reviews by peers can help keep science from being tainted.

Next, projects of similar scope and ambition must be simple, straightforward, and service-oriented. Leaders of them must never lose sight of what should be achieved.

Science and management must work together. When we base our management decisions on science, our positions should be unassailable. At least, that is the assumption. Unfortunately, it is not the case—yet. We live in a complex world, and the union between science and management is too often a rocky one, influenced by factors including politics, organizational needs, honest differences of opinion, and personal views and agendas. Science and management are joined at the hip, however, for better and worse.

Last, I believe sharing experiences with practitioners of other assessments in this book offers us an opportunity to learn from each other, and perhaps even steal a few good ideas to use back home. We may become a little closer to realizing a mutually productive relationship between science and management, and with eyes wide open to the pitfalls and controversy that await any science-based policy assessment in today's political climate.

We've all had the experience. Let's share the meaning.

Policy Review

John Howard

As a county commissioner viewing the resource issues that surrounded the interior Columbia Basin prior to the planning initiative, I saw many complex ecological and social problems that plagued the basin. These problems had not occurred just in the past few years, but resulted from public and private land management decisions made throughout this century. The condition of the basin's forests and grasslands today is a result of these management decisions, decisions based primarily on a need for natural resources (lumber for housing, grazing lands for cattle) and on a policy of wildfire suppression. Along with the effects of these management decisions is an increase in recreation use of the basin's forest and grassland ecosystems.

Policy Background

The ecological conditions of the basin and the social and economic well-being of its communities are intertwined. Prior to the assessment, during the late 1980s and mid-1990s, large areas of forest were burned, partially as a result of past management practices. Many forests also suffered from insect infestations. As a consequence, many communities saw a severe reduction in lumber manufacturing, resulting in local economic hardships. Local governing bodies, in order to provide services for the public, could no longer ignore the important connection between the ecological condition of the ecosystem and a stable, healthy economy.

This link between the health of the ecosystem and the economic and social values that exist in each community within the project area acted as a magnet to draw county governing bodies together at the beginning phases of the ICBEMP. The individual state associations of county governing bodies chose in 1994 to form the Eastside Ecosystem Coalition of Counties (EECC). This coalition represented 108 counties from Montana, Idaho, Washington, and Oregon and provided a unified effort to coordinate input from county governments into the planning process of the project. This approach to public participation in a federal planning process is the first of its kind to involve local partners from the beginning, and it hopefully sets precedence in resolving management issues in the future.

The Counties' Objective

The EECC's objective for the project was and is today to ensure that the public of the basin supports a management alternative that addresses social and economic needs and puts the basin back on track toward restoration of a balanced and healthy ecosystem. The science information from the assessment describes several major ecosystem problems that need addressing in this restoration effort—forest health and noxious weeds. The composition of the forests, in the Blue Mountains, in particular, has changed dramatically over the last one hundred years. Forest management practices and fire suppression have changed tree composition and stand density, resulting in an increase in epidemic insect outbreaks and an increased risk of lethal fires. Current forest conditions impact the ability of the forest to provide resource needs for humans and to sustain fish and wildlife populations.

The science work has also been of great assistance in mapping areas of noxious weeds in the Columbia Basin. The problem of noxious weeds has been a shadow on the horizon, receiving little attention from policy makers or managers. The amount of area taken over by noxious weeds during this century has been phenomenal. If the continued rate of spread of noxious weeds goes unchecked within the basin, it will have a direct impact on the basin's ability to provide adequate habitat for fish, wildlife, and livestock on both private and public lands. Loss of plant root structure to hold the soil together often accompanies the invasions of noxious weeds, resulting in erosion. Erosion of soil, especially into streams, will affect water quality in the streams and cover spawning beds for anadromous fish. Clearly, a coordinated noxious weed initiative with partners from federal, state, and local governments and the private sector is needed to reduce the level of weed expansion in the Columbia Basin.

There are different alternatives that point toward ecosystem recovery; however, the counties are in support of an accelerated active management plan that quickens the pace of recovery.

Public Participation

Public participation has been a large part of the development of the ICBEMP. There have been numerous town hall meetings throughout the region and larger meetings held in different stages of the development of the project. The EECC has participated in most of the public forums. These forums and the interaction between the coalition and the federal team during the ICBEMP project have created a close working relationship. The EECC has gained a great deal of knowledge that can be applied in the future in the implementation of a public-supported alternative.

All along, the EECC has shared input and suggestions with the federal agencies in order to make improvements to the draft EIS. In the implementation stages of the project, the coalition's intent is to have each county work collabora-tively with the local federal districts and resource areas to achieve the goals and objectives of restoring forest and rangeland health. Finding common ground on resource issues is extremely difficult but is ultimately necessary in order to im-prove resource health.

Usefulness of the Assessment

The science assessment has been the major contribution to the EIS. The science provides valuable information for policy makers to help determine future courses of action and evaluate the impacts that these courses of action will have on their communities. The maps that describe the changes in fire regimes and forest structure over time have been most useful.

The social and economic information provided by the science assessment is also useful to the counties. In this area, the counties are not necessarily in agree-ment with the assessment. The assessment found that most service jobs in the counties are in the recreation sector. The counties believe that this suggests that the highest and best use of the federal land is for recreational value. While we realize that a proportion of the service jobs relates to recreation, it is our opinion that the majority of the service jobs should be attributed to activities in the wood-products industry and the government/service sector, jobs that con-tribute to the local economy throughout the year. The emphasis on recreation jobs at the expense of these other kinds of jobs, such as wood-products and cat-tle industries, will have an unfortunate effect on how policy makers look at the social and economic needs of the communities of the Columbia Basin. We fully intend to work with the project team to correct this piece of the social and eco-nomic assessment.

The counties are also concerned with the predominance of standards that the draft EIS proposes for public land managers to follow. The draft standards are not always internally consistent or clear, and they frequently cannot accom-modate site-specific needs. These faults will unnecessarily delay critical, active restoration.

Measure of Success

The true measure of success of the assessment will be the recovery achieved for the ecosystem. Establishing benchmarks to measure reductions in the threat of lethal fires and recovery of forest and rangeland health, and monitoring the region's economy, would go a long way toward measuring the results of the cho-sen alternative.

My advice to other local governments involved in bioregional assessments would be to engage yourself in a partnership process as we have done. From my observation, one can achieve more in working together from the very beginning of a large land-planning process than you can when you are asked to choose an alternative that has had no local input into its development. There is more work involved in engaging in the planning process early; however, the outcomes of knowledge gained and improvements to the work products can be of significant use in dealing with future issues.

CHAPTER 12

Sierra Nevada Ecosystem Project

Case Study
Don C. Erman

Science Review
Robert R. Curry

Management Review
Jon D. Kennedy

Policy Review
Dennis T. Machida

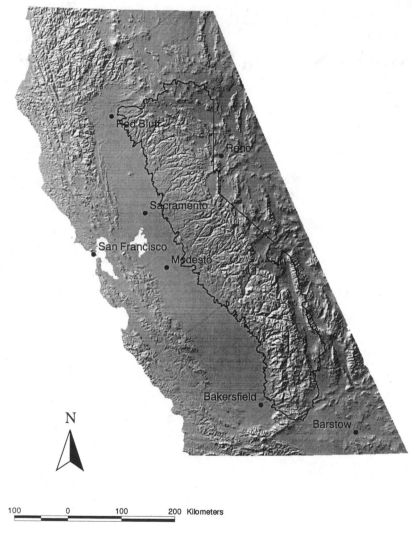

N

100   0   100   200 Kilometers

Sierra Nevada Ecosystem Project

Case Study

Don C. Erman

Mountain ranges hold a special place in many societies. High peaks often are considered sacred by native populations. The Sierra Nevada, the great "snowy range," in the words of the first Spanish missionaries, or "the range of light," in the words of John Muir, is certainly special and sacred. The collision of cultures once gold was discovered in 1849 on the western Sierra foothills brought almost instant annihilation to one of the most dense and diverse Native American populations on the continent and a pillaging of natural resources unparalleled in its speed and sweep. Areas outside the devastation of the immediate gold fields were used for fuelwood, lumber, water to power the mills and workings, and forage for the swarms of livestock turned loose everywhere. But the wounds and scars from that age were not the reason for concern in 1992.

At present, nearly 22 percent of the 20.7 million acres of the Sierra Nevada are in designated biological reserves (table 12.1). As in much of the West, most of the land—nearly two-thirds—is held in public ownership (figure 12.1). Large reserve areas and a majority of public ownership, some would argue, ought to confer protection in balance with other uses and values. From the perspective of many scientists, however, neither seemed to be working in the Sierra. That was the message of a five-part series of in-depth, front-page stories in the *Sacramento Bee*: "The Sierra in Peril." Writer and Pulitzer Prize–winner Tom Knudson's story brought to a head the concerns of many that management of land and natural resources was not leading in the right direction.

But what does a journalist know? How could one man know enough about the sciences, the history, the individual places of such an enormous area to tell it right? Other versions followed. California Governor Pete Wilson, at the prodding of some of his advisors, called for a "Sierra Summit" at which scientists could give their views on conditions to a select audience of the public. Regional public workshops organized by the state following the summit were rancorous. Environmental organizations called their own conference, "Sierra Now" (Environment Now 1992), to assess the status of conditions and to look to the future, and business and industry interests held a "Sierra Economic Summit." The swirling controversy in the Pacific Northwest over clear-cutting, loss of old-growth forests, and damaged rivers had already engaged the courts and the personal involvement of presidential candidate Bill Clinton. Many of the concerns were the same in the Sierra Nevada. The pressure for the Sierra Nevada Ecosystem Project (SNEP), a comprehensive scientific assessment, came from many directions. An independent study was needed to resolve the controversy about the condition of the range.

Table 12.1. *Areas of designated biological reserves
in the SNEP core study area.*

Public/Private Ownership	Acres (subtotal)	Acres (total)
Private Reserves		31,340
Nature conservancy reserves	31,340	
State of California		144,675
Ecological reserves	2,090	
State parks and reserves	28,837	
Wildlife areas	113,748	
Federal		4,282,204
Bureau of Land Management		
Areas of Critical Environmental Concern and Wild and Scenic Rivers	208,550	
Wilderness Area	306,535	
Fish and Wildlife Service	1,129	
National Park Service		
Devils Postpile National Monument	806	
Lassen Volcanic National Park	37,979	
Sequoia and Kings Canyon National Parks	861,077	
Yosemite National Park	746,121	
Forest Service		
Research natural area	45,617	
Special interest area	54,916	
Wild and scenic rivers	34,055	
Wilderness areas	1,985,419	
TOTAL RESERVE AREAS		4,485,219

Congress Acts

Late in 1992, led by California Representatives George Miller and Leon Panetta, Congress began work on action related to the Sierra Nevada. The Conference Report for Interior and Related Agencies 1993 Appropriations Act (HR 5503) authorized $150,000 for a "scientific review of the remaining old growth in the national forests of the Sierra Nevada in California, and for a study of the entire Sierra Nevada ecosystem by an independent panel of scientists, with expertise in diverse areas related to this issue."

A bill introduced by Representative Panetta (HR 6013) spelled out the details of how the study was to be conducted, but the bill fell victim to adjournment. Although Panetta returned to Congress in the next session to read the text of the bill into the *Congressional Record* as guidance for the study, failure to complete passage immediately clouded the legal authority, the source of funding, the charge, and the many other details of the study. For example, even though the

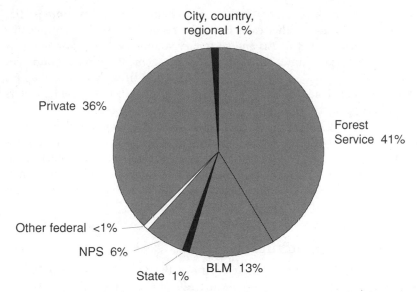

Figure 12.1. *The diagram of ownership and land management within the study area of the Sierra Nevada Ecosystem Project illustrates the predominance of federal land (approximately 61%) in the region. (BLM = Bureau of Land Management; NPS = National Park Service.)*

Appropriations Act funded a small part of the study, and language in the Panetta bill stipulated exemption from the Federal Advisory Committee Act, a later suit by the California Forestry Association against the USDA Forest Service eventually won in federal appeals court, declaring that the project was not exempt from the FACA. (No remedy was imposed by the lower district court, however, as it had initially concluded that the study had fulfilled most of the intent of the FACA and, by then, had been completed and published.)

Nevertheless, it was clear from the Appropriations Act language and the Panetta bill that the composition of the team members should be independent scientists largely from outside the management agencies. This aspect made the SNEP at once a departure from some other large federally dominated ecosystem studies. A federal agency would not be in charge (the entire funding was eventually managed by the University of California through agreement with the Forest Service), but federal scientists could be science members of the team (although they would not represent their agencies). The largest federal agency in the Sierra, the USDA Forest Service, which manages 41 percent of the Sierra, after various letters from members of Congress and its own internal discussions, decided to take on the major obligation for funding a study, at a cost of nearly $7 million during the proposed two and a half years.

The Sierra Nevada Ecosystem Project report was eventually delivered to Congress in June 1996, six months after the deadline. Nearly 150 scientists par-

ticipated as authors or coauthors of the report chapters and many more participated as advisors and in other capacities. The report consisted of an executive
summary (22 pages), a synthesis and summary volume (volume I, twelve chapters, 209 pages); two volumes of assessments, management options, commission
reports, and background information (volumes II and III, eighty-three chapters,
2,629 pages); and an addendum of late chapters printed in 1997 (eight chapters,
328 pages) of policy background, models, and case studies (SNEP 1996). Now, a
year and a half later, the project seems to have been largely successful, in that it is
being studied, discussed, and debated. Interest groups have used information and
are preparing their own plans for future management on the basis of the findings.
The reports will continue to educate scholars and the public about the human
and natural resources of the Sierra Nevada. But the path to completion was not
straight or simple, as the following pages will attest.

Creating the Structure

The failure to pass the authorizing element of the SNEP science team created
enough uncertainty to dog the process thereafter. The Forest Service asked the
National Academy of Sciences (in accordance with the Panetta bill) to draw up
names for the team and serve in a peer review capacity. The academy, however,
declined to become involved in such a "regional" issue. The Forest Service then
formed a steering committee to select potential team members. The Steering
Committee was composed of the director of the USDA Forest Service Pacific
Southwest Research Station (first Dr. Barbara C. Weber, later Jim Space; both were
elected as chair), the deputy chief for research of the Forest Service (Jerry A.
Sesco), a research administrator from the National Park Service (Dr. Bruce M.
Kilgore), representatives from the University of California (first me, later Jeffrey
Romm) and the California Academy of Sciences (Dr. Dennis Breedlove), and a
scientist of high standing from California and a member of the National Academy
of Sciences (UCLA Professor Emeritus George Bartholomew). This committee
suggested names for the team as well as its science team leader. Who selected the
team? Partly, the first team leader, Professor John Gordon of Yale, decided. Most
of the names of the initial team were not on the list developed by the Steering
Committee, including the name of John Gordon. When the Forest Service picked
up the full cost of the study, the Forest Service, not surprisingly, took the lead in
the process that followed. In the language from the charge for the study, "The
Forest Service's recommended approach is to develop a study based on achieving
the general requirements of HR 5503 and attempt to meet intent of the ecosystem study established in HR 6013."

The Steering Committee framed a charge for itself and for the Science Team
for the conduct of the study out of the two congressional bills (one legitimate,
one not) and from letters from various congressional representatives to the chief
of the Forest Service. The subtle differences among all these documents were
striking and left an impression on those who were involved in the political efforts

to establish the study that at various times the SNEP was not following the intent of Congress (or the law). For example, the size and disciplines of the team, the duration of the study, and the areas of emphasis were inconsistent. The first letter from the key House committee chairs (as well as from Representative Panetta) to Forest Service Chief Robertson requested a six-person panel to submit a map of the existing old growth within six months. These were not guidelines from either HR 5503 or HR 6013 (text of the congressional bills, the letters from members of Congress, and the SNEP charge were included in the Sierra Nevada Ecosystem Project Progress Report, May 1994).

Three elements of the eventual charge from the Steering Committee were significant. First was the requirement to assemble and assess *existing* information on conditions of the entire Sierra Nevada ecosystem, and, by inference, without regard to ownership. This analysis would assist Congress in making future policy decisions, particularly regarding the condition of old-growth forests. The second element was to examine alternative management strategies for the resources of the Sierra Nevada, including the need for old-growth reserves. And the third (created out of the interpretation of all of the various congressional background) was to frame the alternative management strategies "to maintain the health and sustainability of this ecosystem while providing resources to meet human needs." These words, not expressed directly in any of the language from Congress, became the goal for the study.

Within a few months, the first team leader resigned. The majority of the Science Team of nine had previous experience in the preceding assessments of the Pacific Northwest but little recent contact (or residence) in California. Whether these factors were significant can only be surmised. But in California, the political undercurrent regarding the composition of the team from different poles of the environmental spectrum was one of dismay. A new team leader, University of California at Davis Professor Deborah Elliott-Fisk, took over, and the contract for management switched from Yale to the Wildland Resources Center of the University of California. Six months later the second team leader also stepped down, and I became leader of a team in disarray and a study on the brink of collapse. The small team assembled by John Gordon had grown to nineteen Science Team members and an almost equal number of special consultants with additional expertise or agency liaison responsibilities. The Science Team was still predominantly made up of academic scientists, although special consultants were mostly from agencies or other institutions. Even with such a sizable group, many areas of specialized knowledge were lacking. Team members were directed to hold workshops and engage others with expertise to strengthen the assessments (Fleming 1996). Exclusion of some scientists from the team remained a criticism of the team's operation.

In the first, slow year, a major step had been taken in the assessment stage of old growth (Franklin and Fites-Kaufmann 1996), but no other funds had yet been allocated for assessments by the team members. Many ideas were considered; discussions about direction and charge, and meetings with other scientists occupied the team. Students, scientists outside the team, and even the public thought, incorrectly, that proposals could be submitted to SNEP for funding, like grants to the National Science Foundation. In addition to a third team leader, the Science Team concluded that it needed a small group, a coordinating committee, to meet regularly to help plan and coordinate the various study areas. Members were Constance Millar, U.S. Forest Service Pacific Southwest Research Station; Rowan Rowntree, U.S. Forest Service Northeastern Forest Experiment Station; David Graber, U.S. Geological Survey, National Biological Service; Douglas Leisz, Consultant and U.S. Forest Service retired; and Deborah Elliott-Fisk, University of California. With the team leader, this group of five other team members met with increasing frequency to manage every stage of the project. One of its first tasks was to budget the funds for various subprojects. Additional team members met with the coordinating committee during reviews and for special planning and especially when the summary report was drafted. The continuous review of progress and planning by the coordinating committee was essential, and it was instrumental in completing a synthesis. A preponderance of agency members on the coordinating committee (who could devote the time required), however, gave an impression of potential lack of independence to the study. The time taken in the first year, although hampered by the changes in team makeup and leadership, was probably necessary. A large group of independent scientists, brought together for the first time on a project with enormous dimensions and consequences, needs time to consider and weigh its direction as a team.

The team also drew up its own five questions to guide the study. These gave the general direction for all assessment chapters: (1) What were historic ecological, social, and economic conditions, trends, and variability? (2) What are current ecological, social, and economic conditions? (3) What are trends and risks under current policies and management? (4) What policy choices will achieve ecological sustainability consistent with social well-being? (5) What are the implications of these choices?

The beginning of SNEP came under the unusual circumstance of a Congress and a presidential administration controlled by Democrats with a strong interest in reshaping natural resource management. Before the project stabilized, however, the elections of November 1994 swept in a new political majority to both houses, and national political leadership interest in SNEP was gone. Support for the study changed to efforts to eliminate all funding, but only a portion was eventually withdrawn. A clear line of congressional direction and interest was lost when Congress changed.

The Public in the Process

The initial SNEP policy was that no one outside the science team should participate in the process. When I became team leader, this position changed to a formal, well-structured contact with the public on a regular basis (SNEP 1996; Kusel et al. 1996). A group of about ninety people were identified as key contacts to meet with the team, discuss preliminary ideas and findings, and review early drafts of the report.

Planned and announced meetings with the public and key contacts occurred quarterly. We soon learned an important lesson. Part of each meeting included formal presentations by team members, followed by questions from the audience. The greater part of each meeting, however, was similar to a scientific poster session. Team members sat at tables arranged by project themes (e.g., old growth, aquatic/riparian, fire) with data, interactive computer terminals, and posters. Participants at the meeting circulated among the tables and had intense exchanges with team members and others at a table. Every twenty minutes, we rang a bell to signal people to change tables. This device kept a few individuals from dominating a session and gave everyone an opportunity to make contact with a variety of topics and team members. These less formal and unstructured exchanges were invaluable in reducing the skepticism of the public participants about the purpose of the study, and they increased the level of respect for the scientific process. Questions and feedback were rapid. Team members were educated by the knowledge and insight of the public, and the public became educated in the openness of the scientists to their questions, concerns, and interest.

Following the delivery of the report to Congress, a final public conference was held at the El Dorado County Fairgrounds in Placerville, California. This event provided a last formal exchange between the entire team and the interested public. As in our previous meetings, more than half of the two-day conference was devoted to discussions around tables among the team members and the public. By this time, the team enjoyed a high degree of acceptance for its process and praise for its method of operating.

Substantial time and resources were spent managing public outreach and involvement. In addition to the monthly Science Team meetings and the quarterly public workshops, team members attended frequent and innumerable smaller meetings by groups and as individual science team members throughout the state. We created a newsletter and established a web page on the Internet, which included the newsletter, plans for public meetings, team member biographies, and other information. These expanded the outreach of the project (SNEP 1996).

The core team was selected to conduct the assessments and consider alternative management strategies. Many performed double or triple duty in the public outreach and administrative tasks of such a large project. Team members from academic institutions generally found it difficult to devote the full time demanded of these multiple responsibilities. Although monthly team meetings remained closed, the frequency and extent of our public contacts were essential in defining the purpose of our work in the minds of the public, and in many cases public questions helped sharpen the focus of individual studies. What began with misgivings by many of the team grew into welcomed working relationships at public sessions. Nevertheless, because team members were involved in both the scientific assessments and public involvement, their energies were taxed and the project resources were stretched.

Workings of the Team

The personal interactions of the team members often threatened to dominate the scientific. Few of us know all the human dimensions of our colleagues, least of all by the creative works they may have published. The tensions that inevitably emerge in a team probably can never be avoided. At best, they can be somewhat managed. During the course of the project, strongly held differences in data interpretation emerged, antagonisms of particular analytical approaches developed, and suspicions about motivations (e.g., was there a controlling function of Forest Service members of the team?) frayed relationships. Seldom did these issues come out directly; they came through secondary sources, often from people outside the team who had heard about the problems. Problems were relayed to the team leader. Occasionally, other scientists were invited to bridge some of the conflicts. When a team is selected through some kind of formal process (and has a shaky beginning), there is little leeway to reassign weak performers or difficult personalities. In SNEP, I could only add additional scientists, shift responsibilities around, and let time pass.

The separation of public sessions from working team meetings provided a reasonable balance between the need to inform society and the need for the team to debate information and interpretation. Requests from those outside to attend or become part of the team continued for some time. But except for formal presentations by specialists and announced public meetings, the team met alone. An emphasis on independent scientists in all the congressional language was interpreted by the Steering Committee and the Science Team to mean that managers from federal or state agencies, or county supervisors and planners, were excluded from team membership. However, during public meetings we were urged to work with managers so that results of the project would have a better chance of being understood and used. The team made a point of including managers on the list of key contacts and of inviting them to public meetings. We requested analysis from managers to gain their perspective, and managers eventually contributed chapters to the report.

In the early stages of the study, we concentrated on finding sources of information, both published and unpublished, and we evaluated quality and suitability of information. Other scientists with detailed knowledge of some aspect of the Sierra were brought in for conferences and workshops or as authors of chapters. Over one hundred scientists in addition to the Science Team and special consultants contributed chapters. Specialists and managers within the major federal agencies were repeatedly asked for records, data summaries, and other assistance. The cost for this crucial assistance, although not fully charged, was nonetheless not free. Roughly 25 percent of the project funding was used to cover personnel costs of the Forest Service (including some of those who were team members) and other agencies.

For some time there was an unstated assumption that the project would result in a consensus report. Disagreements were strong, but perhaps muted, by this unspoken assumption. The team eventually reached the position that disagreement was not only acceptable but was also valuable to the final report. Further, all the basic chapters of the report would be authored and hence carry with them the direct source of responsibility and authority. Thus, chapter writers were encouraged to make their own conclusions, and the summary volume drew attention specifically to several of the more contentious areas—historical forest structure, fire patterns, structural standards for old growth, and aquatic protection strategies. Readers of the report find this aspect surprising, and it is often commented that the report is "inconsistent" and "contradictory." Most of these differences were not from lack of editing; they were well-known and intentional. Where broad general agreement did occur, then, it emerged as a consequence of the analysis rather than some deliberate process to find common ground. These conclusions, such as of the dire condition of the aquatic system (SNEP 1996, volume I, chapter 8) and the small amount of high-quality old growth in some forest types (SNEP 1996, volume I, chapter 6), are all the more powerful.

Sadly, the debate within the team was often less than fully engaged. One might assume that with a group of largely academic scientists and no managers, policy makers, or interest groups, an atmosphere of probing questions and reasoned argument would prevail. Such was not the case. Perhaps a gathering of specialists in the same discipline behave so. But with only one or two scientists from a discipline, team members seldom challenged the authority of each other. G. B. Shaw wrote that "all professions are a conspiracy against the laity." The conspiracy is just as great against other professions. What can be more embarrassing for an expert in his or her field than to ask a dumb question of another expert?

And yet the frequency and duration of meetings of the full team were considered essential by the team. Possibly the deemphasis on reaching a consensus also promoted a feeling of lessened concern about the reports of others. However, this lack of communication is not specific to the SNEP team and is a growing problem in much of science (Greene 1997).

Staffing

A well-organized and professional staff was necessary to the project. Two types of staff assistance were assembled specifically for the project: an administrative group (who worked at the SNEP offices at UC Davis) and a technical group (for geographic information systems) who worked out of state offices in Sacramento. Another team member (Professor Frank Davis, UC Santa Barbara) directed the highly competent GIS staff and met with them regularly. Their presence in offices shared by state and federal specialists in forest GIS assessment allowed them to take advantage of experience near at hand. This sharing of knowledge and data was a significant contribution to the project.

The administrative staff, with little previous experience in such a project, worked under pressure from people inside and outside the team and generally maintained professional working relationships during the project. But overwork, personal conflicts, and probably inexperience in this group resulted in demands on me to reinforce schedule priorities, check on follow-through, and deal with personnel problems. The most critical area where inexperience showed was in the complicated management of chapters through the revising, editing, and layout process. However, demands to plan events, meetings, and arrangements; coordinate schedules of team members; and schedule public meetings continued throughout the project, when undivided staff attention to the reports was needed. A larger, more experienced staff in this area would have improved efficiency and prevented errors and confusion. At the time, the trade-off between shifting more resources away from the Science Team or delaying the schedule did not seem justified. In retrospect, the project needed an equivalent level of funding and skill for staff, as for scientists. The cost for developing this kind of support staff must be borne directly by an independent project and paid out of the study funds.

Assembling the Pieces

The team argued two issues at great lengths. One was the delineation of the core study area, the Sierra Nevada. Congress had stipulated several areas clearly outside the Sierra Nevada proper (e.g., Lassen Volcanic National Park, Modoc National Forest), and this larger study area was considered to the extent that time and data were available. The source of disagreement about the boundaries of the core area rested on the sources of data for various disciplinary analyses. For example, much social and demographic data and distributions of biota are given by county, yet many of the counties extend well beyond the conventional boundary of the Sierra. Firm delineation created a tidy map and simplified some analyses; lines were ignored when data and analyses dictated.

The second issue discussed at length was What constitutes an ecosystem for purposes of analysis, and does it include humans? The explicit inclusion of human dimensions, including institutions, in the assessment expanded the scope of the analysis and strengthened the report.

Overlapping both issues was the question of whether to exclude private lands in the assessments. Some team members argued that the intent of Congress was to assess only the federal lands and hence avoid questions of private-property rights and other complications. The team, however, concluded that when the bill stated "the entire Sierra Nevada ecosystem," it clearly applied to all lands, as necessary, to conduct an ecosystem study.

The study eventually found that current conditions were severely impaired in the foothill region, especially riparian habitats, where private lands predominate (Kattelmann 1996). Of equal concern, the scientists forecast major continuing degradation of the many distinct natural community types and habitats in the region (Davis and Stoms 1996; Graber 1996; Shevock 1996). Precisely because a majority of the land in the lower elevation zone is private and currently undergoing rapid change, there is little area in official biological reserve status.

Alternatives for how this region may be further developed, based on county general plans and population projections, showed that a wide range of possibilities exist, depending on how future management contends with increases in population and settlement patterns (Duane 1996). Potential impacts from high population growth and sprawl-type development would also reverberate into the adjacent public lands and higher elevations through increased roads, traffic, air pollution, and water demands and fragmented wildlife corridors. Ignoring the private lands in the Sierra Nevada would have been a serious omission. Here, again, findings from SNEP are being examined by county planners, charitable foundations, and others as they consider future conservation strategies for this vulnerable area of the Sierra Nevada.

The shear magnitude of the area made it impossible to bring in many of the fragments of data from private holdings on 34 percent of the land. But information available from public sources (e.g., county general plans, human population density and building parcels, dams and water diversions, road networks, vegetation cover) helped fill out a picture of the entire range.

All authors were instructed to use the five questions as a guide and to be explicit about the sources of information, assumptions used, inferences made, and logic used to make conclusions. They were also asked to write for a nontechnical audience and to define terms or use alternative plain language.

The original guidance from Congress and the Steering Committee, while stipulating a range-wide assessment and consideration of alternative management strategies, did not request a plan for the Sierra Nevada with options such as those constructed for an environmental impact statement. Thus, most authored chapters gave ideas for alternative management. These were synthesized by a subset of the entire team in the overview report (SNEP 1996, volume I). We organized the synthesis chapters around broad components of the ecosystem (e.g., fire, air quality, old growth) and concluded each chapter with conceptual strategies to meet specific goals and their implications. The conceptual strategies lacked explicit location detail, which would be included in a true planning process. They also lacked a full integration of the individual components and strategies across the entire Sierra Nevada. In future ecosystem studies, a greater level of synthesis might be obtained if a small subset of the team were brought back after several months of reflection to integrate the completed chapters and assess the overall study.

The team rejected an approach to synthesis that used quantitative models to incorporate the many dimensions of the entire ecosystem or to project change under alternative management. Many of the component parts could be described only in qualitative terms. Nevertheless, two models were noteworthy products of the study: a system for selecting biodiversity management areas (Davis et al. 1996), and a simulation model that focused on alternative forest management strategies that included several components developed by others on the team (Johnson, Sessions, and Franklin 1997). Both of these models are being considered at present for future management in the Sierra Nevada.

The Role of Review

The Steering Committee assumed the role of managing an anonymous peer review process for the report. The team constructed a much more involved process. As chapters were developed in early rough form, they were reviewed by colleagues for clarity and accuracy either inside or outside the team. A more polished draft then became available to any of the ninety key contacts (representing a broad public) who wished to comment, read, or review. Simultaneously, the draft was distributed to five team members both for critical review and to help integration and cross-referencing of related chapters. When these comments were returned to the author, another draft was submitted to the Steering Committee for distribution to anonymous peer reviewers. Delays in manuscript preparation and errors in the management of papers by the staff meant that not every chapter passed through all stages of review, except the final peer review. (Some commissioned papers were not intended for chapters but as sources for team members. For completeness, these were published but without peer review.)

Although the thoroughness of this review process was admirable, the consequence for some authors was bewildering. As many as a dozen or more independent reviews were common (exclusive of detailed editing of the "final" version). Consolidating a version that satisfied all critics was usually impossible. The greatest care was taken in response to the anonymous peer reviews (each chapter was sent out to three reviewers). Once responses were returned from the Steering Committee, they were given to two individuals (one staff, another a team member) who formed an "editorial board." They summarized the critical comments from the reviewers and indicated which points should be addressed in a revision (if necessary). Authors were required to respond in writing to the editors, much as one would to a journal, indicating how they had made changes or defending their positions. And as in professional journals, final responsibility for the content of the chapter rested with the author.

As one might expect, many of the reviews (peer and nonpeer) were calm critiques that caught slips and helped the writers improve their chapters. What was surprising, on the other hand, were vitriolic reviews by some and what seemed to be blatant attempts at censorship. One senior scientist with a long career of journal publishing commented, "I have never experienced such viciousness on the part of reviewers." During and after the study, various individuals and organizations attempted to stop the study and attacked team members. Letters to state and federal legislators and to newspapers claimed that the project was illegal or trampled on private-property rights. Some complained to top University of California administrators that faculty on the team were biased or unethical or that the project should not be supported by the university. Such is the terrain when research scientists address problems that are relevant to the policy arena.

Was the complexity and extent of the review necessary and useful? The three elements—internal team review, "public" review, and anonymous peer review—each had a different and useful purpose. Review by the team allowed members to understand and integrate the work of an allied scientist into his or her assessment and allowed much cross-referencing among chapters. Public members gave a reading that challenged the way scientists in their disciplines ordinarily write and identified loose connections in logic and construction. The peer reviewers examined papers for scientific merit. The trade-off for the project was time. Each step of the review process was a loop. In retrospect, I think the review process was probably too complicated, but I am unsure what step could be eliminated in the future.

Before leaving the subject of review, I must make a few remarks on the use of reviews. It is interesting to read again the proposed legislation (HR 6013) and how Congress intended review to be handled. The context for this discussion is the increasingly frequent request that all work destined for science-based policy undergo "peer review." The Panetta bill stated that both draft and final reports should be submitted to the National Academy of Sciences "for technical peer review" and that comments from the peer reviewers and detailed responses to those comments be included in the final report. Thus, the intent of Congress was that the peer review process be available to the public. Such a requirement is a significant departure from the accepted peer review process of independent scientific studies and scientific journals.

Only in the last months of 1997 was the academy itself specifically exempted from the requirement of the Federal Advisory Committee Act following legal challenge. The "internal workings" of the academy, such as peer review, officially will remain closed to public scrutiny. Academy president Bruce Alberts testified before the House on November 5, 1997, saying, "We believe that keeping the committee deliberations and our review process closed and confidential is fundamental for ensuring the independence of our studies and the scientific quality of our reports, enabling our recommendations and findings to be based on science rather than politics." The National Academy of Sciences and the Congress acted quickly once challenged about public openness. And perhaps for good reason. I know of no journal that opens its process to the public inspection of comments by reviewer and author. The culture of science assumes that the social contract among scientists ensures the general probity of the process. The political framework of independent, scientific ecosystem assessments (such as SNEP) asks for the same process but demands a public accounting. A requirement for public distribution of critiques and responses will likely change the purpose of review and the way both reviewers and reviewees couch their remarks. After completion of the entire study and as a result of the legal challenge to SNEP by the California Forestry Association, the Forest Service decided that all peer reviews would be opened to the public (but with reviewers' names removed). This decision was made by the Forest Service alone without consulting the Steering Committee. The only direct response to the SNEP report from any member of Congress to date was an earlier request for the peer reviews of a few chapters.

No review system is without flaws, as anyone who has published knows. The minimum requirement should be that a third party select reviewers and act as intermediary between authors and reviewers. Whether this process is confidential should depend on the agreed upon uses of the science but, in any event, should be made clear to all parties at the outset.

Making a Report Accessible

The strongest advice SNEP received as it prepared the final report was to make the report readable to a general audience. We hired a team of professional editors unconnected to the science team to make stylistic changes to each chapter and provide the report layout and artistic design. We were also advised to make the report short or summarize the contents in a few statements. There is a conflict between making a comprehensive assessment of a large, diverse ecosystem and writing a short report. There is also a problem with reducing complex scientific relationships to simple, unqualified statements. But the work to reduce 2,900 pages of technical assessments and case studies into a 209-page readable summary was worth the considerable effort. However, the report was so large and detailed that it will not be completely read until scientists and the public use the information in summaries, conferences, and other sources and as it is cited and used in the literature. Now, a year and half after release of the report, continuing demands are made on many team members and other authors to present some of the findings.

Expanding Outreach

The technology of the Internet provided access to the final report and to much of the basic data (http://ceres.ca.gov/snep). Following SNEP, we spent additional time and resources, from the state and university, to format the data used in GIS analyses for easy searching of files. Besides the printed report, a CD-ROM version was created. Advantages of the computer version include low cost and a capacity to search all volumes quickly for keywords and print out maps and sections from any chapter. We need a commitment for teaching, training, and demonstrations at the local level in most areas of the Sierra, however, to bring the findings and analytical tools developed by the project into general use. The cost to create and maintain such an educational program is not trivial. This dimension was not anticipated by Congress, and neither funding nor planning was adequate to continue responsibility for managing the data once SNEP concluded.

Some have said that the major contribution of the National Environmental Policy Act was the opening up of federal decision making to the public (Diamond and Noonan 1996). Major ecosystem assessment studies can do more than allow the public to comment formally on proposed projects as in NEPA. By revealing the basic data as well as the tools used in analysis, we empower citizens to check the results or conduct their own analyses. Thus, expanding the availability of information and analysis methods should be as important as the conclusions that individual scientists reach as part of a study.

Uses of the Report

The fact that political interest from Congress shifted does not mean that the shift was permanent or that Congress is the only body to act on many of the SNEP findings. The findings have been used to assess the Revised Draft Environmental Impact Statement (RDEIS), "Managing California Spotted Owl Habitat in the Sierra Nevada National Forests of California: An Ecosystem Approach." The RDEIS was to be issued after completion of SNEP in order to take account of relevant findings. When the Clinton adminstration decided to withhold release of the RDEIS (based on comments from scientists about its shortcomings), two science panels were formed—one under FACA by the administration and another under a joint Senate–House Republican committee. Both panels were asked to review the adequacy of the RDEIS and to consider it in light of the SNEP report to judge whether the RDEIS should be released to the public. The FACA panel found the RDEIS was inadequate, and the Forest Service is recommending another revision. The Senate–House panel did not reach the same conclusion and used the occasion to argue against some of the findings of SNEP. The final report of this panel has not been completed as of this writing.

Other groups are also working with the SNEP data. The Wilderness Society has created maps and additional analyses from the original SNEP data and has prepared new GIS maps from combining SNEP data with other sources. Private foundations are considering SNEP models for identifying priorities in land conservation easements and purchase. Several other environmental groups are using the report as a basis for proposing new policies for the Sierra Nevada (e.g., Pacific Rivers Council, Forest Service Employees for Ethics in the Environment), and some Sierra counties are using the information for general county planning or to refine restoration projects. The pattern of use will more likely continue to be one of changes at project and regional scales rather than some grand sweeping new policy engendered by the SNEP report. These changes may be improvements because of the use of better information for management; however, a program for managing the Sierra Nevada as an entire system of interconnected parts remains elusive.

Conclusion

In science, as in evolution, what begins with one purpose often changes with time into another use. Because SNEP was not initiated to solve a specific problem or to develop alternative management plans, the enduring value of the study might be easily underestimated. The Congress that requested the report to guide its decisions was not the one that received the report. The full value of the SNEP report may not be known immediately. The assessments identified critical gaps in some of our most basic knowledge about components and processes in the Sierra Nevada. The report will serve as a benchmark for assessment of conditions, trends, and problems in many aspects of the ecosystem; as a guide to improving management; and as an educational source for those who will examine related questions in the future. It is already being referred to by local and regional planners, interest groups, and public agencies. But as the SNEP report concluded:

> Left unresolved is the question of whether our society has the will and the capability to correct such problems. Implementation of new approaches or possible solutions is the responsibility of the public and its institutions. The beginning is to acknowledge that problems exist: willing minds and able hands can find solutions. (SNEP 1996, volume I, p. 174)

ACKNOWLEDGMENTS

I thank Nancy A. Erman for help with and editing of this chapter and the SNEP team members for their many contributions to the study. This chapter was supported in part by the Sierra Nevada Ecosystem Project through a cost-reimbursable agreement, No. PSW-93-001-CRA, between the U.S. Forest Service, the Pacific Southwest Research Station, and the regents of the University of California, Wildland Resources Center.

LITERATURE CITED

California Spotted Owl Federal Advisory Committee. 1997. *Final Report of the California Spotted Owl Federal Advisory Committee, December.* Portland, OR: U.S. Department of Agriculture.

Davis, F. W., and D. M. Stoms. 1996. Sierran vegetation: A Gap analysis. In *SNEP*, volume II, pp. 671–689.

Davis, F. W., D. M. Stoms, R. L. Church, W. J. Okin, and K. N. Johnson. 1996. Selecting biodiversity management areas. In *SNEP*, volume II, pp. 1503–1523.

Diamond, H. J., and P. F. Noonan. 1996. *Land use in America.* Washington, DC: Island Press.

Duane, T. P. 1996. Human settlement, 1850–2040. In *SNEP*, volume II, pp. 235–360.

Environment Now. 1992. *Sierra now: A vision for the future.* Malibu, CA: Environment Now. 50 pp.

Fleming, E. 1996. Compilation of workshops contributing to Sierra Nevada assessments. In *SNEP*, volume III, pp. 1073–1101.

Franklin, J. F., and J. A. Fites-Kaufmann. 1996. Assessment of late successional forests of the Sierra Nevada. In *SNEP*, volume II, pp. 627–661.

Graber, D. M. 1996. Status of terrestrial vertebrates. In *SNEP*, volume II, pp. 709–734.

Greene, M. T. 1997. What cannot be said in science. *Nature* 388:619–620.

Johnson, K. N., J. Sessions, and J. Franklin. 1997. Initial results from simulation of alternative forest management strategies for two national forests of the Sierra Nevada. In *SNEP*, addendum, pp. 175–216.

Kattelmann, R. 1996. Hydrology and water resources. In *SNEP*, volume II, pp. 855–920.

Kusel, J., S. C. Doak, S. Carpenter, and V. E. Sturtevant. 1996. The role of the public in adaptive ecosystem management. In *SNEP*, volume II, pp. 611–622.

Shevock, J. R. 1996. Status of rare and endemic plants. In *SNEP*, volume II, pp. 691–707.

SNEP (Sierra Nevada Ecosystem Project). 1996. *Sierra Nevada Ecosystem Project (SNEP): Final report to Congress,* volumes I–III, plus summary and addendum. Davis: University of California, Centers for Water and Wildland Resources, Wildland Resources Center Report 37–40.

Science Review

Robert R. Curry

From the standpoint of a research scientist, the SNEP report is a marvelous document. It supplies an enormous amount of information in a single place, synthesizing much of what is known about the ecosystems of an area that has been the subject of avid academic study for over 120 years. The task put to the authors was overwhelming from the standpoint of objective science because so many have written so much about the area, from so many different perspectives. It is in the Sierra Nevada that much of our understandings of geologic, biogeographic, limnologic, hydrographic, meteorologic, and climatic science have originated, because the area has been so attractive to world scientists. The Sierra has given the world critical understandings of principles of mountain building, glacial history, genetic evolution of plants and fish, pollen rain dynamics, volcanic processes, soil science, and even the origin of the earth's magnetic field, to name but a few basic areas of natural history. The task to undertake a "study of the entire Sierra Nevada ecosystem," as mandated by Congress, is as daunting as summarizing what is known about the human genome.

Physically, the SNEP report is also enormously impressive. On a single CD-ROM, the summary report of nearly three-thousand pages has been organized to permit rapid, although complexly nested, access through a conventional table of contents that spans multiple printed volumes, with search capabilities and ability to view and print all the diverse full-color maps, photos, and figures. Further, the background documentation that could be made public is linked through two primary California database Web sites to provide the GIS, metadata, and tabular basic data, so that the "reader" has access to a fundamental compendium that is nowhere else available publicly in a single repository.

The report is fundamentally a science effort, of 107 carefully selected senior scientists and collaborators, as well as many additional peer reviewers. It attempts to take a whole-systems approach, looking at the integrated interactions between ecosystem elements, including very significant inputs and assessment by social scientists and economists. But this bioregional assessment had an initial agenda. It was to focus on remaining old-growth forest ecosystems in the over-twenty-thousand-acre core area, where only 36 percent of the lands are in private ownership. Overall, the SNEP report succeeds admirably in providing an unbiased critical assessment of fundamental contemporary questions in ecosystem science. It contributes at a critical time, when political actions are being proposed that are based on pseudoscience. The careful reasoning of the acknowledged scientific leaders should make this a timely document to assist in land management decisions over much of United States for at least the next twenty years.

For example, the public contention that there is a "crisis in National Forest health" that has generated self-serving legislation in the name of science is clearly shown to be false by careful and well-documented analyses by specialists in forest insects, drought, basic forest ecology, and historical analysis. Management questions dealing with conflicts between introduced fish species and amphibian populations are assessed using clear, well-documented and well-reasoned data and analysis. Gap analyses of plant populations show how past public land management decisions have often been completely contrary to avowed goals or justifications for long-range planning. Areas of science with too little data for valid conclusions, such as assessments of spotted owl, wolverine, fisher, and marten and reintroduced large mammals, are shown to be weak links in land management planning.

A fundamental strength of the analysis and findings of the SNEP report is that its conclusions were often reached by researchers within the agencies who have been responsible for past mistakes. In essence, this bioregional assessment has provided a forum in which the agency specialists have been free to provide opinions and information highly critical of their own agencies or corporations. To this end, the independence of congressionally mandated scientific review serves its highest and best purpose. No such report could originate from the State of California or the U.S. Forest Service because its findings tend to contradict the avowed management goals of those entities, despite their supposed bases of ecosystem values and science.

A primary case in point is the issue of fire management. In the Sierra Nevada, the U.S. Forest Service and other agencies have been eager to feed the flames of public fear of fire to develop management plans that call for road building and forest cutting in areas where past misguided fire management has contributed to present high fuel loading. There is little question that construction of second homes and population expansion into the Sierra has increased public risks associated with fire. The SNEP team members could not agree about the historical character of closed-canopy, mixed-forest ecosystems in the western Sierra, and thus could not fully agree on the effect of fire suppression on present fuel loading, although apparently all agreed on the ecological role of fire in maintaining the diversity and stability of forest ecosystems.

The report thus presents alternative views but quite clearly concludes that management alternatives are available to counter the idea that trees that are dead and dying need to be cut down to save the forest. A defensible fuel-zone concept is proposed that would use fire management to return to a more natural fire frequency and mosaic.

On the equally controversial issues of range management, the SNEP report team attempted to use comparative range assessments made by the National Forest range conservationists over the past four decades to determine trends. As with most of the SNEP ecosystem component assessments, such an approach was qualified with caveats about the changing assessment methods and noncomparability of successive assessments, but those were the only data presented.

The general issue of data uniformity and availability must have plagued the SNEP team. A major shortcoming of the entire SNEP report is the very dissimilar data available for different National Forests at different times within the core study area. A most fundamental failing is the fact that two different regional offices of the U.S. Forest Service have responsibility for the SNEP area. Most of California is in Region V, headquartered in San Francisco, but the Toiyabe National Forest is managed out of Ogden, Utah, where very different standards for data collection are evident in the SNEP databases. This difficulty may have compounded the most serious shortcoming of the SNEP assessment—the difficulty it had assessing old-growth forests. Old-growth assessment was the primary charge of both of the pieces of legislation that mandated the SNEP study, and, in the opinion of this reviewer, that task was not fully accomplished.

Late-Serial Stage and Old-Growth Assessment

What is unquestionably the oldest median-aged forest in the world, the Bristlecone Pine stands in the White Mountains of the Inyo National Forest, within the study area of the SNEP review, were not identified as late-serial or old growth (LSOG). These forests were ranked as class 3 ("selectively logged or burned but retaining significant numbers of large trees or snags, or second-growth forests approaching maturity"). Because they are not characterized by canopy closure and other complexity criteria established to evaluate the small remnants of old-growth mixed-conifer stands in the western Sierra, these and probably most of the other old-growth stands in the Sierra were not classified as LSOG.

It can be argued that the impetus for the SNEP study was the encroachment of private and public forest cutting on the remaining LSOG in the El Dorado National Forest and surrounding areas where historic railroad land grants created mixed land ownership patterns in sites of highest forest productivity in the Sierra. In can also be argued that most of the unassessed old growth is presently in designated wilderness or other protected status in National Parks and National Forests, and thus did not demand SNEP review. But the maps were produced, and the criteria established to develop them simply do not reflect the many mature and old-growth stands that may be critical to wildlife and plant communities and that require preplanned management strategies for fire suppression, insect or disease infestation, and other long-range planning.

The SNEP team has carefully pointed out that its criteria exclude many small patches and open-stand species that comprise old-growth or late-mature forest communities. Among these are the many eastside Sierra types such as lodgepole, red fir, and Jeffrey pine assemblages. In the Inyo National Forest, these are the commercial species that are being cut on public lands today, and the natural communities surrounding these forest blocks are among the most diverse in the entire Sierra. In the Toiyabe National Forest, late-serial stage and old-growth mixed-conifer red fir, white fir, lodgepole, Jeffrey pine, and disjunct populations of several white pine group species were nowhere identified as LSOG, even where these stands may have included the most valuable and threatened commercial species in the Sierra—sugar pine.

The reasons that the very comprehensive analysis of LSOG failed to identify the open eastside forest communities are because of a ranking artifact. A six-point scale of structural complexity that served to proxy for habitat support capability in the authors' opinions failed to rank the old-growth mixed-conifer stands with open, parklike structures as complexity class 5 or 6, which they equated to LSOG. In part, the decision to equate stand complexity with the values of "old-growth" forests predicted that much old growth would not be classed as LSOG. While the justification for such an approach seems plausible for the mid-elevation forest communities of the western Sierra, it is ecologically untenable if applied throughout the SNEP region. What was tested and found to work well in controlled case studies in the West is precisely contrary to standards for habitat diversity and support capability in areas with patchy soils and/or lesser precipitation.

This inconsistency is clearly stated by the authors (cf. volume II, chapter 21), but the volume I primary assessment (chapter 6) presents simplifications (cf plate 6.2, "Late Successional Forests Ranked 4 and 5") that lead most readers to conclude that there is no such forest type anywhere in the entire Toiyabe and Inyo national forests. Caveats and qualifications notwithstanding, the primary purpose of an independent scientific analysis is to present clear data that, in this case, can guide land management activities. While the presentations on individual groups of organisms and habitats, such as birds, riparian habitat, fish, and amphibians, are of the highest caliber and are very clear and useful, the systems approach that should have integrated those findings and opinions with the basic ecology of diversity in patchy parklike or pocket stands of less than one thousand acres of LSOG seems to be missing. This is probably an almost predetermined effect of the magnitude of the task requested by Congress. Coordinating a group of senior specialists as large as that required for SNEP and trying to come up with a consistent finding over so large and diverse a geographic area was probably impossible. That so much agreement does come from so many authors provides a measure of the extremely competent leadership and publication effort that the SNEP report represents.

Conclusions

As a land management scientist, I found great value in the numbing array of tables and figures. The GIS and map presentations are state of the art and really constitute the core of this work. As many will doubtless perceive, this study is not simply a bioregional assessment but is the core of a virtual database. Because so much is based on GIS-compatible data sources, such as the California Natural Diversity Database and the PRIME Forest Inventory, and because these must be constantly updated, the real core value of this work will be that it can be used as a framework for a continuous on-line reference publication.

Like much that is overwhelming, this initial three-thousand-page printed report will be out of date by the time it is routinely incorporated into day-to-day land management science. I have met no one who claims to have read far beyond his or her own limited interests. Indeed, I have been asked to excerpt sections, for example, on insect threats or presettlement fire frequencies for congresspersons who would not even accept copies of the entire printed report. It is a primarily a reference volume and an assessment framework. Much of it is available on-line, and that effort deserves continuing support. The policy decisions that created the need for the SNEP report will not be answered simply from the report in a one-time effort. Good science must guide policy, and that is a continuing responsibility.

This study uses the scientific method to test hypotheses and validate professional scientific opinions. Within the budget and time constraints, it could not do this with equal effort throughout the entire geographic region. Thus, the authors chose to focus on a western, mid-elevation, central Sierra subregion and to develop sound methodology to evaluate ecosystems in that rapidly growing area, where sociopolitical conflicts seem boundless.

The report is being sought and used by policy makers. It serves as a template for future science. Now, can Congress come up with a line-item budget to keep the framework open to continuing input and additional evolving inquiries?

Management Review

Jon D. Kennedy

The Sierra Nevada Ecosystem Project was commissioned by Congress after strong lobbying by many environmentalist interest groups. Congressman Leon Panetta was responsible for the House action that authorized the SNEP effort and with the original allocation of $150,000 to fund the effort. This amount was based on the reported cost of the Gang of Four report completed for the northern spotted owl in the Pacific Northwest, an early model for SNEP.

The Managers

The direction for the SNEP effort was to prepare an assessment of the current condition of the Sierra Nevada and to develop management scenarios that might guide policy for future direction for managing these lands. This was to be accomplished by an independent scientific panel overseen by a small steering group with minimal agency representation. The Forest Service testified against the enactment of this project, because the agency was already embarked on an environmental impact statement for the California spotted owl covering the entire Sierra Nevada, which would address many of the elements that SNEP would cover. Local governments and municipalities within the affected area were never invited to be a part of the process leading up to the approval of SNEP.

At the time this direction from Congress was developed, the Congress was controlled by the Democratic Party, as previously mentioned. Congressman George Miller was chairman of the Natural Resources Committee and supported the SNEP proposal. It seems apparent that the Congress considered itself to be one of the managers who was anxious to see the results of this effort. However, in 1996, when the Republican party gained the majority in the Congress, scientists were given the charge to develop alternative strategies to guide the management of the public, and in some cases, the private lands. The scientists became another one of the managers concerned with this effort.

However, groups who were much closer to the day-to-day management of the Sierra were not directly involved in SNEP. Local governments, counties and municipalities were not represented on the assessment team but are critical in regulating private lands to meet the objectives of future management. The federal land management agencies were represented by agency scientists but not by managers, who were provided only remote access to the process and were not involved in the determination of alternative management strategies. Those who had the responsibility to implement the policies developed through this process were very seldomly involved.

Scientists as Managers

The charge to the Science Team included not only the assessment of the condition of the Sierra Nevada but also the development of alternate management scenarios that would result in improvement of the Sierra Nevada ecosystem(s). As one of the members said, in an early meeting of the SNEP scientists, "This is heady stuff." While there is no question about the importance of the scientific data that have been accumulated in the SNEP report, it is also clear that some scientists have certain biases that go beyond their findings. These biases are demonstrated in the proposed management options, including the effects of management on water quantity production and the management for "old-growth values."

It is important to note that there is not total scientific agreement on all parts of the SNEP report. Rather, there are findings in the SNEP report that can be used to support many sides of a debate.

Local Government Managers

Local governments (counties and municipalities) have the responsibility to plan and regulate private land use within their area of jurisdiction. This includes land-use zoning, permit requirements, and fee determinations. In addition, local governments are charged with coordination with land managers of lands within their jurisdictions that they do not regulate, such as state and federal lands. While the SNEP effort was focused on federal land management, issues such as water quality, air quality, wildlife habitat and migrations, fuel treatment, transportation, and recreation cannot be limited to the public lands. The linkages to state and private land are critical in addressing these issues. Local governments are likewise very aware of the integrity of private-property rights. The balance between the larger public need and private landowners' rights is a continuing debate at the local government level.

However, the SNEP program had very little involvement of local governments and decision makers, except during key contact meetings held after the SNEP effort was well underway. No local government representatives served on the SNEP team, nor were they asked to prepare papers for consideration in the final documents. These local managers were asked to accept what was being given to them and then to make the best use of the final product.

Federal and State Land Managers

The managers mostly responsible for implementing on-the-ground activities to respond to the agency mandates are the federal and state land managers. While their involvement in the SNEP process would seem important, the direction from Congress was to make the scientific panel an independent group not affected by agency persuasions. While there were some federal and state scientists deeply involved with the SNEP effort, they were selected for their scientific contributions, rather than any agency affiliation.

As a result, state and federal land managers had little influence over the development of management scenarios and strategies for implementation.

Congress and the Administration

The House of Representatives, on the basis of results of the FEMAT assessment, decided to apply the same process to the Sierra Nevada. There was heavy lobbying by the environment community to take this approach, even though the Forest Service had committed to embark on an environmental impact statement for management of the California spotted owl. While no legislation was passed, the direction from the House was sufficient to cause the process to proceed. It became apparent that the Congress would be asked by the environmental community to develop management strategies and policy for actions in the Sierra Nevada, and the SNEP reports would form the basis for those actions.

The administration, committed to a course of action to attempt to resolve the impasse in the Pacific Northwest over the northern spotted owl, also supported a similar approach for the Sierra Nevada. And, in addition to the original $160,000 directed by Congress for the SNEP effort, the administration authorized over $7 million to complete the SNEP reports.

It also seems clear that Congress and the administration were using science to serve as a surrogate for solutions to social–political issues. Since direct political response to the questions of old growth, threatened and endangered species, fish and wildlife habitat, and private-property rights was likely to result in months, and even years, of debate, scientists were being asked to provide a basis for answering those questions. Both the Congress and the administration were, in a sense, managers, using the science to direct management of the federal lands in the Sierra Nevada.

The Application

There is little question from any level of government that the SNEP reports offer a tremendous amount of information that is, and will be, useful for future decision making. Many counties are already accessing the SNEP databases for information needed for development of land-use plans. Federal and state agencies are likewise taking advantage of the information developed by SNEP in program and project planning, including fuel treatment projects, recreation development priorities, and revision to forest management plans. Decision makers at all levels are asking for assistance and definition from the scientists involved in developing the SNEP reports. At this level, the SNEP program has been a valuable step forward in our knowledge about the Sierra Nevada.

Yet, since neither local government nor federal and state interests had been deeply involved in the formulation of the SNEP effort, there is little acceptance of the scenarios or the premises behind the SNEP reports. This is one of the major shortcomings of this effort.

At the congressional level, there has been a good deal of interest in the results, but, at this time, little evidence exists indicating that Congress is willing to debate legislation that would address the issues of public land management, commodity production versus preservation of federal lands, or the relationship of intermingled private lands.

However, the administration has used the SNEP reports, and other assessments directed by Congress, as a basis to defer actions by the federal agencies to proceed with environmental impact statements that would address the Sierra Nevada, the Interior Columbia Basin, and elsewhere. Many of these decisions, which historically would have been made by the agencies, have been elevated to the Secretary of Agriculture, or even the White House.

Summary

As in several of the other assessments included in this book, scientists were asked to go beyond the presentation of scientifically based information, to propose how this information could (or should) be used. Management options developed by scientists are not necessarily implementable, either politically or socially.

The value of the SNEP project has yet to be determined. Much of the data is useful for local decision-making processes, and many are now using it. However, it is clear that local decision makers and the general public must be more involved in the process, or it will become another top-down, unfunded mandate that local interests must implement.

Policy Review

Dennis T. Machida

The Sierra Nevada promises to be the next opportunity, among California's great regions, for government, the private sector, and the public at large to address natural resource policy issues in a comprehensive fashion. The sheer size of the Sierra region, and the scope and complexity of the issues facing it, are daunting, but they are matched by the significance and beauty of its mountains, rivers, forests, and oak woodlands. As California's population growth continues to put pressure on environmental resources in the region, and as resource management decisions by all levels of government are increasingly being challenged, there is a growing need for society and its institutions to grapple with these issues if the Sierra's resources are to be preserved in the future.

The purpose of the following discussion is to report on the role that the Sierra Nevada Ecosystem Project is playing in the incorporation of ecosystem concepts in the protection, management, and use of this region by public and private institutions—what the SNEP report refers to as "closing the loop." This is a status report, because it is really too early to fully assess the effectiveness of SNEP in achieving this objective. Given the sheer size and complexity of the SNEP effort, it is understandable that institutions and individuals will need time to absorb its results. This discussion will focus on the institutional responses, because SNEP was largely targeted at institutions, and also because institutional responses are the easiest to discern at this time. But it should be recognized that there may be other important and unforeseen ramifications of SNEP now that it is in the public domain. The scale of public dissemination is illustrated by the fact that the California Environmental Resources Evaluation System (CERES) reports that over 194,000 inquiries concerning SNEP (both entries to the Web home page and requests for related documents accessible through the system) have been made since the release of the SNEP report in June 1996.

Context

In evaluating the impact of SNEP, it is important to understand the context in which it was undertaken, and the original scope and purposes of the study. As discussed in earlier sections of this chapter, the SNEP report was requested by and prepared for Congress. However, over the course of the study, control shifted from Democrats to Republicans, and by the time the SNEP report was finally completed, there was less interest in SNEP at the congressional leadership level than when the effort was begun. Consequently, it is unlikely that Congress will respond to the SNEP report in the foreseeable future. Second, at the same time, there was a sense in many quarters that changes in policy must emerge from a "bottom-up" process involving local government, business, and environmental groups, rather than being imposed at the State and national levels, in a top-down approach. (For those concerned with the Sierra, the lesson had already been well learned from the public response and level of concern expressed during the California Resources Agency–sponsored Sierra Summit process in 1991–92, which involved six conferences and workshops with over fifteen hundred people in attendance.) Third, the focus of SNEP was a scientific assessment. Consequently, it was expected that implementation issues would be dealt with at a later time by the implementing agencies. Fourth, the field of vision in SNEP was on a relatively large scale; the report was not intended to provide information on a project level. Indeed, to have done that would have greatly exceeded the available time and resources. Finally, the authors of the SNEP report were not prepared to support any particular decision-making process. As the report states, "The team was not asked to prepare a single plan, a range of options for implementation, or preferred alternatives as in an environmental statement process required by the National Environmental Policy Act." The team did not view its charge as being to select "goals for society or the Sierra Nevada Ecosystem." It is important to note that, at about the same time, the environmental impact statement that the U.S. Forest Service was undertaking could have had a dramatic impact on a resource management policy in the Sierra Nevada. In other words, there was already a decision-making process underway at the time the SNEP report was being prepared. Consequently, the EIS was viewed as the key mechanism for dealing with many of the issues facing the Sierra Nevada.

Given this context, the SNEP team defined its purpose as being to "advise Congress on existing and possible future conditions of old growth and late successional forests and ecosystems of the Sierra Nevada." The objectives of the study were to assemble and integrate available information in the form of: (1) multidimensional assessments of the ecosystem (with an emphasis on late-successional-stage old growth) and (2) alternative resource management strategies. This information when assembled would guide "a mutual search for strategies for improved management." Essentially, SNEP represented a scientific effort to provide information and management approaches for future policy decisions. This objective did not have the galvanizing effect of a decision or controversy that required the establishment of policy and the institutional alignments needed to implement the policy. At this time, SNEP is an example of an effort to provide information to society; it was left to society as a whole to respond.

How Well Did SNEP Do Its Job?

Certainly, there have been a wide range of responses to the SNEP report, ranging from enthusiasm to claims of bias by some scientists to disappointment with the lack of project-level information in the report.

Generally, the responses have arranged themselves into two categories. First, there is wide recognition that the assessments were overall well done. Although SNEP was largely based on existing data, the study did achieve notable breadth— for example, by including people and institutions within the scope of the ecosystem analysis. In large part, the breadth of the SNEP report reflected a decision to include a wide variety of academic disciplines on the SNEP team. Additionally, SNEP helped to better define the linkages among various components of the ecosystem. The SNEP report also provided new ways to view riparian protection (e.g., calculation of buffer areas); fire processes (e.g., incorporation of climatic and silvicultural information in fire-spread models); old-growth protection (e.g., managing on both the patch and the matrix levels); community well-being; the need to include both public and private lands in planning efforts in order to achieve ecosystem goals; the development of models to help set goals (e.g., biodiversity management areas); and the need for institutional processes (e.g., collaborative planning efforts) to achieve ecosystem objectives.

In addition, SNEP's public participation and outreach efforts were considered to have been successful, in that they balanced the SNEP team's need for independence with the need to answer the public's questions and to gather public comment. Further, public input assisted the SNEP team in framing the management strategies. These efforts generated a widespread sense, long before the SNEP report was released, of where SNEP was heading and why, greatly reducing the level of trepidation that had existed in some quarters. The public process employed by SNEP should certainly be looked to as a model for future regional assessment efforts in California and elsewhere.

Several aspects of the study left room for improvement. It is generally accepted that the alternative management scenarios should have been further developed and analyzed by the full Science Team. Preferably, this process would have drawn more on the experience of resource managers at some point, as was done in other subjects covered by the SNEP report. Additionally, the usefulness of the report would have been greatly enhanced by integrating the various strategies and approaches on a landscape scale. Although there were several efforts to incorporate various concepts on a subregional scale, there was no integrated vision of how the use of the different strategies would affect the landscape of the Sierra Nevada. The enormous and complex demands of simply completing the assessments left insufficient time and resources to carry the study to this further level.

Outcomes of the SNEP Process

It is probably premature to try to evaluate the overall success of SNEP. Not enough time has elapsed for the report to be (in its own words) "absorbed, adapted, and implemented." As yet, there has been no fundamental shift of national, state, or local policy or significant change in operations, because of the SNEP report. Nevertheless, the report has resulted in "outcomes" involving various levels of government and the private sector. The following is a partial discussion of these outcomes.

At the federal level, Congress has not as yet responded to the report. The Forest Service has recognized the value of SNEP, but has not fully integrated the results of SNEP into its policies and operations. This reflects, in part, the need to further develop SNEP data and models. It also reflects that the Forest Service was focused on the preparation of the EIS on the California spotted owl during the same time period that SNEP was being prepared. Further, a Federal Advisory Committee Act challenge to SNEP by the California Forestry Association put in question the usability of SNEP information in the California spotted owl environmental documents. Additionally, the U.S. Forest Service resource managers were not directly involved with SNEP during the preparation of the report because of the scientific orientation of the effort. These factors, in the view of some, created at least a temporary barrier between the EIS and the SNEP efforts. Ultimately, there was interaction between the two processes when a FACA committee (which included former SNEP team members) was created by the Forest Service to evaluate whether the EIS contained a "complete and integrated analysis of all significant information" (including SNEP information) related to the management of Forest Service lands. It was the conclusion of the FACA committee that the EIS, in part, did not adequately address the larger ecosystem concerns. There is a great deal of speculation as to why this FACA review was undertaken. However, it is clear that there are larger questions (other than the specific issues generated by the SNEP report) underlying this process, including how broad the scope of these efforts should be (e.g., focus on a particular species or the entire ecosystem) and how resource management policies should be developed (e.g., the extent of collaboration with interested groups in both assessments and decision making). These are fundamental questions that are inherent in the development and implementation of ecosystem approaches.

At this time, the Forest Service is implementing the Sierra Nevada Conservation Framework process to try to arrive at a process for answering these questions in a collaborative manner with other public agencies and to improve communication and trust among the various stakeholders (e.g., business interests, environmental groups) in the Sierra Nevada. The EIS experience seems to reflect the transition from a National Environment Policy Act–oriented process involving a set of species to a broader participatory process involving an entire ecosystem. However, it is not clear how this process will relate to recently initiated NEPA-based efforts to revise forest plans in the Sierra during the next year. The Forest Service is also consulting SNEP data for a watershed assessment of the Lake Tahoe Basin, which is a product of the Presidential Forum at Lake Tahoe that occurred in July 1997.

The State of California is involved in a number of bioregional efforts— including the National Communities Conservation Program and the Cal-Fed process involving the San Francisco Bay Delta—that incorporate some of the same types of thinking that went into SNEP. The state has also incorporated SNEP principles into the State Fire Plan. However, SNEP has not yet resulted in new comprehensive state policies or programs focused on the Sierra Nevada.

Consistent with its emphasis on "bottom-up" and collaborative approaches, the state is providing, through the Regional Council of Rural Counties (RCRC), funding for various local government efforts to incorporate or build upon SNEP information and approaches in a number of areas in the Sierra. Among them are: El Dorado County (development and implementation of a plan for a network of Defensible Fuel Profile Zones throughout the Cosumnes River and South Fork of the American River watersheds); Mono County (utilization of SNEP information on recreational open-space and habitat conservation to update a General Plan); Plumas County (documenting the process to reach agreement on baseline monitoring parameters and protocols); and Tuolumne County (action plan for watershed management and timber harvesting activities). The state is also expending funds through the U.C. Wildland Resources Center for the transfer, documenting, and storage of SNEP data and to make such information more available to and usable by the public and institutions, through CERES and other channels, and for the provision of technical assistance to local governments and other public agencies and to watershed groups.

On a regional scale, there has been an effort by the California Biodiversity Council to hold forums in the Sierra Nevada for local resource managers to collaboratively discuss and address regional issues that were illuminated by SNEP. Through a grant from the Forest Service, faculty at the University of California at Davis are utilizing SNEP information in the analysis of regional growth issues, in part, related to central Sierran counties.

At the local government level, SNEP information has yet to be extensively utilized. However, SNEP helped to generate local government support for the reinvestment of revenues generated by water development in Sierran watersheds in watershed restoration and resource management. RCRC is actively promoting this policy as part of the Cal-Fed deliberations. SNEP information has also been used in various local economic development efforts.

At this time, the most wide-scale use of the SNEP information is occurring among environmental and nonprofit organizations in the private sector. The Sierra Nevada Alliance is using SNEP information to formulate a watershed-level approach in dealing with ecosystem issues, evaluating forest plans, and increasing public awareness of biodiversity issues on private lands. The Wilderness Society has undertaken a number of such activities, which include publishing a citizen guide to SNEP, conducting workshops, and reproducing and distributing maps at a more useful scale. A number of forest protection groups (e.g., the Sierra Nevada Forest Protection Campaign) are calling for the use of SNEP information in developing a strategy to protect old growth, riparian habitats, watershed restoration, and reintroduction of fire into the ecosystem.

The Sierra Business Council used SNEP data extensively for its Sierra Nevada Wealth Index, which measured the socioeconomic and environmental health of the region. This was one of the first private-sector efforts to define and evaluate the Sierra Nevada as an environmental and economic region. A number of foundations are utilizing SNEP information to develop a biodiversity preservation strategy for the Sierra Nevada to guide their grant programs. SNEP information is being used to identify areas that contain significant plant communities.

In summary, SNEP is being used by both the public and the private sectors, and undoubtedly, these activities will lead to greater consideration of ecosystem objectives in both sectors. However, it is difficult to predict whether these activities will coalesce into new regional programs or institutions, which will be needed to sustain the effort.

Constraints on Using the Results

Although a great deal of activity is starting to take place, it is clear that the SNEP's impact has been constrained by a number of factors. First, there is lack of broad-based political support of ecosystem strategies within the region. This lack of public support translates into wariness of the ecosystem strategies in Washington, D.C., in Sacramento, and in county seats. However, there seems to be support for some initiatives if they are funded and involve public and private collaborative efforts. Second, there is a lack of institutional capacity to deal with the complex substantive and public process demands of ecosystem management. There appears to be greater acceptance of ecosystem principles among senior administrative policy makers; however, it is difficult to apply these principles when resource managers in the field are being asked to do more with less staff and less money. Third, SNEP data are not easily usable in their current form. Certainly, a great deal has been done to make the data accessible to the public. However, additional effort is needed to help the public and organizations figure out how to use the data and models.

Future Needs

In order to realize the full potential of SNEP to benefit institutional planning and decision-making processes, the following actions should be undertaken to develop further data and strategies and to increase the capacity of institutions to use the information.

Development and Refinement of Data and Strategies

There is a great deal more that should be done to add value to SNEP's efforts by developing and refining data; developing and testing its various models; and improving the usability of data by the public, including:

- Additional research and monitoring to refine and to add to the SNEP database. SNEP, like the Sierra Summit before it, identified a number of research needs to be addressed. Monitoring is essential in determining baseline conditions and changes in the ecosystem. This feedback is important for "adaptive management."
- Support of scientific efforts, in consultation with resource managers, to develop and integrate the strategies on a landscape and watershed scale. This

need was recognized by former SNEP team members, resource managers, and private-sector groups. Such an effort would add a great deal to the acceptance and use of the strategies proposed by SNEP.

- The development of maps and other information at a scale that is useful for project development and implementation activities. This is the next logical step for making SNEP data and models more useful to resource planners and managers.
- The development of materials that would show what the integrated strategies would look like in applicable landscapes or how features would be distributed throughout the range. Such an effort would help the public to understand the implications of SNEP and visualize the use of the various strategies in their communities and in natural areas.

Development of Institutional Capacities

At present, there is generally a lack of institutional capacity to make full use of SNEP information and to generate ecosystem perspectives at various scales. This capacity could be increased by:

- The provision of additional staff, expertise, and training for public agencies to handle the substantive and process-oriented demands of ecosystem planning (e.g., collaborative planning), development of projects, and ongoing management. There is a lack of trained staff and sufficient resources to deal with the demands and complexity of incorporating ecosystem considerations in existing planning processes at every level of government. Without sufficient resources to undertake data gathering and analysis and the necessary public participation processes, the "adequacy" of such efforts will likely be questioned on a continual basis. The issue is one of trust and communication as public agencies try to reconcile competing demands in a fair and reasonable manner.
- The establishment of a center or other institutional arrangement to conduct monitoring and research, to share information, and to provide technical assistance for both the private and the public sectors. There is also a need for a center or other institutional arrangement to sponsor research and monitoring activities, to serve as an information clearinghouse, and to provide technical assistance with regard to "on-the-ground" application of approaches. A wide range of activities is being undertaken in the region. However, there is currently no easy way to share information or to gain advice. All of these activities would benefit from an institutional (e.g., creation of a center, greater utilization of University of California Extension, etc.) and programmatic focus on the Sierra Nevada.
- The establishment of institutions or institutional arrangements to provide an ecosystem context at various levels. There is also a need for various levels of context for various levels of activity. For example, watershed-level activity would benefit from a perspective of what is happening in other related watersheds. Similarly, each subregion would benefit from an understanding of what is happening in other subregions. Further, it is important to place individual activities within a context of sustainability. Certain short-term project bene-

fits may not be sustainable over time. Currently, there are no mechanisms to provide such contexts. The California Biodiversity Council may be one avenue to provide such context.

• The establishment of a state institution or institutional arrangement to focus state resource management policies (which cross regions and watersheds) and additional resources to the region. There is also a need to align various state programs to achieve greater consistency with ecosystem objectives in the region. The state may provide a means to link various levels of activity because of its special responsibilities with regard to land-use planning and private lands; regulation of air, water, wildlife and water resources; and implementation of restoration programs.

• The provision of support to increase the capacities of nonprofit organizations, such as land trusts to develop and implement programs and projects. The establishment of watershed groups and land trusts is a growing trend in the Sierra Nevada. They could serve as important partners in achieving ecosystem objectives and serve a link between communities and public agencies. Currently, they would benefit from assistance in establishing their organizations and programs.

Commitment of Funds

The utilization of an ecosystem approach would be greatly accelerated through the commitment of additional public and private funds to implement the approach. Such a commitment of funds would be consistent with a bottom-up collaborative approach because it would support locally sponsored efforts. It would also result in a voluntary and nonregulatory alignment of public and private institutional programs. The fiscal needs of this approach transcend institutional responses to a controversial issue (e.g., the California spotted owl) and should be based on sustaining the effort over the long term. This objective could be achieved by:

• Commitment of a dedicated source of funding for all ecosystem activities in the Sierra Nevada (e.g., research and monitoring, planning, restoration, and management). Currently, there is a lack of financial resources to achieve ecosystem objectives at every level of government and in the nongovernmental organization sector. The development of committed funding sources (congressional authorization, state bond, foundation grants, etc.) would be useful in developing and implementing cooperative and collaborative public- and private-sector efforts to achieve ecosystem objectives. It is encouraging to note that the State of California is currently in the Sierra as part of the Cal-Fed process.

• Distribution of funds on a comprehensive program basis, based on a "sense of place," and involving all levels of government and appropriate private-sector entities. There is a need to develop and fund comprehensive programs (i.e., including all institutional functions) based on a "sense of place." Place-based programs would help ensure that the specific needs of the area are met and that they receive local and statewide support. They should be inclusive of both the public and the private sectors and should cross public jurisdictional lines.

Conclusion

SNEP was a remarkable effort. It is one of the most comprehensive resource assessments ever undertaken; indeed, it demonstrated what can be done in this form, and on this scale. Certainly, it has done much to define the Sierra Nevada region at the end of the twentieth century. However, realizing the ultimate benefits of SNEP will depend on increasing the capacity and financial resources of public and private institutions to take advantage of this important work.

ACKNOWLEDGMENTS

I wish to acknowledge the contributions of (in alphabetical order) Laurel Ames, Phil Aune, Lucy Blake, Lewis Blumberg, Katherine Clement, Tim Duane, Don Erman, Jim Gaither, Greg Greenwood, John Gussman, Patricia Kelly, John Sheehan, Lynn Sprague, Bill Stewart, and Douglas P. Wheeler to this chapter. Any errors of interpretation are, of course, my responsibility.

PART FOUR

Synthesis

CHAPTER 13

Understanding Bioregional Assessments

K. Norman Johnson and Margaret Herring

At their best, bioregional assessments build understanding about the condition of a bioregion and the consequences of particular actions; they provide principles that can be integrated into future management action; and they help solve problems. The seven assessments described in this book accomplish these purposes to a degree. In this chapter, we consider these case studies, individually and as a set, to learn how bioregional assessments can be used to help shape policy and improve our understanding and management of bioregions.

Most of the assessments described in the preceding chapters began in response to a call from a policy maker to help solve a particular problem. These assessments do not follow a single prescription for regional-based problem solving because there was no ready-made process available for them to follow. The people involved in these assessments navigated by the seat of their pants, using scientific information and scientists to help inform decisions. Thus, these case studies illustrate a process that is developing across the country not through explicit design but through experience and innovation.

The assessments described in this book represent a variety of natural resource issues throughout the country. The assessments invite comparison but defy reduction to a standard methodology. Our purpose, therefore, is not to prescribe steps for others to follow, but rather to review the steps that some have taken in order to learn how bioregional assessments can be used to improve our understanding of the condition and potential of large regions and to assist resource management and policy formation within them.

Our synthesis examines four stages within the assessment process. The first stage (bioregional context) explores the conditions existing in the ecosystem and the social system prior to each assessment and what problems precipitated the need for an assessment. The next stage (conducting an assessment) considers how the assessments were conducted and the contributions of those involved. The following stage (results and outcomes) explores what each assessment produced and how those results were used. The final stage (building capacity for understanding and change) considers the effect these assessments have had on our understanding of bioregions and their management and policy.

Finally, we summarize the lessons we have learned from these seven assessments, and other assessments we have encountered, in twenty-five points, building on all that preceded them in this book.

Bioregional Context: Existing Conditions, Precipitating Issues, and the Questions to Be Answered

Bioregional assessments are strongly conditioned by their context. The dominant theories in science, current resource management practices, and the values embedded in public policy combine to create the context for natural resource issues. Add to the mix the condition of the regional ecosystem, existing environmental policies, local concerns, and regional and national interests, and you have the ingredients to propel an issue to the forefront of public concern. All you need is a spark.

A Common Concern: Long-Term Sustainability of Resources, Economies, and Communities

In these case studies, sparks flared from the tension between short-term development and long-term conservation of resources. Battles erupted over consumptive water use and ecological values in the Everglades, old-growth forest habitat and logging in the Pacific Northwest, and traditional lifestyle and potential development in the north woods of New England. These conflicts reflected a shrinking base of resources—a scarcity—and a growing list of the values society has placed on those resources.

To complicate matters, there seemed to be disagreement among experts about the condition of the ecosystem and the implications of different policies, as well as a variety of values that the public wished to emphasize. Uncertainty about facts and values in many of these case studies made the job of decision making doubly punishing. In such a situation, decision makers can become paralyzed, unable to make a choice, not because they lack formal authority, but because no available alternative is politically preferable to continued delay (Lee 1993). Increasingly, established planning systems had failed to deal with these conflicts. Planning at the scale of forest units, county ordinances, and point-source controls provided guidelines with little connection beyond their specific jurisdictions. Decisions made at the scale of single timber sales, for example, had little ability to consider the cumulative effects of multiple timber sales on watershed integrity or the range of critical species. Environmental decision making was often a patchwork of local initiatives, uncoordinated and unable to respond to large-scale changes occurring over time. Planning processes tended to invite public review and interagency consultation only after plans were drafted, which often resulted in challenges, delays, and continued conflict rather than resolution. When established processes failed, the job of decision making often passed to the courts, and management of natural resources slipped into receivership.

In the Everglades, a lawsuit brought by the federal government forced the issue of water quality but may have delayed efforts to deal with additional problems of water quantity and timing of release that were necessary for the wetlands restoration. Multiple lawsuits in Southern California, brought against the building industry, temporarily halted development on critical coastal habitat but provided no long-term guarantee of conservation. And a series of court injunctions in the Pacific Northwest stopped much of the timber harvest in federal forests.

An alternative to court-ordered, crisis-driven decision making was needed. Policy makers, torn between constituents with conflicting interests, needed a credible process to help them make decisions that would stand up in the courts and provide a measured response to the many dissenting voices. And they wanted some assurance that the decisions they made would not come back to haunt them.

The Classic Policy Makers' Response

With so much controversy and confusion before them, policy makers often resorted to a time-honored tradition—sending the problem to a committee. It just so happened that in these cases, the committees often included scientists. While scientists have a long history of advising policy makers (see chapter 3), a new depth of scientific involvement was triggered in 1989 with the designation of the Interagency Scientific Committee for conservation of the northern spotted owl. With lawsuits on the horizon over protection of the northern spotted owl, members of the Northwest's congressional delegation sought to insulate the existing timber sale program from appeal while searching for a long-term solution. They called for a review of the existing owl protection plan, and Jack Ward Thomas was asked to head a team of scientists to study the issue and make recommendations.

Thomas was wise enough to assist with drafting the instructions to his team. Its charge became one of developing a "scientifically credible plan for conservation of the northern spotted owl." With these few words, Thomas moved scientists from bystanders to central players in federal forest management. The call for "scientific credibility" fundamentally changed the criteria for acceptability of forest policies and gave scientists a central role in policy formation. Where previously managers and policy makers could wheel and deal with fair abandon in choosing what sources of information to use and what risks to take in conservation of species, they now had to cope with a judge and jury uninterested in political compromise.

After six months' work, the Thomas Report, which resulted from the ISC analysis, presented one of the first regional conservation strategies in the West based on conservation biology, overturned forest plans that had taken ten years to construct, and broke the hold of the forest industry on forest policy of the Northwest. In addition, the Thomas Report firmly established scientists as new players on the policy block, holding trump in any card game that policy makers wanted to play. Thinking that if a little science was good, then more science was better, policy makers called for a series of reports that ultimately led to FEMAT and the President's Plan for Northwest Forests.

With the drama of a warm, fuzzy owl pitted against the continued harvest of old-growth forests, FEMAT gained enormous publicity. Other policy makers in other regions were soon coming forth with their own science committees, and new mandates for assessments were showing up, from northern New England to the Sierra Nevada. Not surprisingly, many of these assessments referenced FEMAT, some in admiration, some in disdain. But in many ways, the work that began with the Thomas Report brought bioregional analysis and the role of scientists to center stage in addressing regional resource problems throughout the country.

Defining the Questions to Be Answered

Most of the bioregional assessments in this review began with questions posed from policy makers to a committee of experts. Assessments were a pragmatic response, enlisted to answer these questions. The request from the policy maker was generally: We have a problem. How do we fix it?

In retrospect, it must be asked if there was a valid role for expert analysis in the solution of these problems, or if the assessment was being used as a tactic for delaying decisions. The searchlight of public concern does not stay in one place indefinitely. A lengthy assessment could be seen as taking action without the responsibility of making any decisions, allowing public concern to shift to other issues. It would be very difficult to know the motives of each mandate, but the results show that these assessments took on lives of their own, perhaps beyond the policy makers' intention. The problems in each of these case studies were undeniable, and public interest in finding a solution was keen. Problem definitions varied; some were posed as broad questions of bioregion condition, while others were stated in terms of a specific question. Some assessments were closely bound in an attempt to avoid further conflict; others seemed to be lightning rods for increased confrontation.

Not surprisingly, the way the assessment objectives were stated at the beginning affected the focus and outcome of the assessment. We briefly summarize below the questions that triggered the assessment along with the general approach taken to set the stage for discussion throughout this chapter.

In the Great Lakes, a series of questions posed over the course of many years focused on several nearly catastrophic environmental disasters. Policy makers wanted to know the cause of nearshore eutrophication and collapsing lake trout fisheries. The regional scientific community was able to unequivocally link cause and effect and recommended effective solutions. However, more recent problems in the region have had more diffuse effects, with far less certain links to cause. Direct application of empirical science no longer has been able to predict the outcome of any particular management action. Bioregional assessment has become a long-term, recurring process driven by evolving questions, rather than a discrete, one-time activity.

In contrast, in the Southern California NCCP, one very distinct question was posed: How much coastal sage scrub is necessary to sustain the species dependent on that habitat? Policy makers needed specific information to make specific decisions about land use. The first thing everyone learned was that the necessary information was not known. As a result, research has been pursued in tandem with policy development, and land-use decisions are being made based on the professional judgment of scientists as they gain more understanding of this deceptively simple question.

In FEMAT, it took time to get the policy questions aligned with evolving public concern. The fate of old-growth forests was at first narrowly focused on the legal debate of a single endangered species. New assessments followed each iteration of the policy question as FEMAT's concern grew from spotted owls to multiple species to an entire bioregion, until at last the public concern for the forest's natural and social economies was addressed. The request to FEMAT from policy makers was clear: Provide a credible management alternative for federal forests of the spotted owl region that will pass muster in court. The use of expert knowledge was critical to this process because it lent credibility to the results. Science was equated with credibility, and finding a credible defense in court was paramount to policy makers. The question asked and the information provided were a good fit, although managers have had a devil of a time implementing the resulting plan.

While FEMAT provided a court-approved framework for managing the national forests of the spotted owl region, the question did not stop growing. It topped the crest of the Cascade Mountains, spread across the interior Columbia Basin, and became a suite of questions aimed at managing a vast intermountain region. Significantly, questions were not explicitly stated in the interior Columbia Basin assessment charter; it was left to the assessment team to formulate them. Instead, the assessment was chartered to develop specific reports and linked databases that would be used to advance ecosystem-based management on federal lands within the basin. Making the transition from assessment to plans has proven to be difficult in this case.

The policy question in the Northern Forest Land assessments focused on concern not over the biological or physical environment but rather over a way of life. How could the pressures for development of the northern forest be resisted and the traditional way of life maintained without a strong regulatory intrusion? The question of public rights on private property was assessed in a very public process.

SNEP originated as a one-line request in a congressional budget bill, asking for an assessment of the old growth and ecosystems of the Sierra Nevada, along with a modest amount of money to address those issues. It was left up to the scientists, various steering committees, and the Forest Service to further define the scope of the study and to find the resources that would be needed. Not surprisingly, given its amorphous beginnings, SNEP mushroomed into a multidimensional, somewhat unwieldy study of this celebrated mountain range.

During the long history of engineered water control in the Everglades, new institutions have been continually created to address different problems. As a result, a jigsaw puzzle of jurisdictions and mandates overlays the South Florida ecosystem. Although the call was clear from the governor to save the Everglades, the agencies concerned all functioned under very different mandates and sources of authority. Within this jurisdictional jigsaw resides a consortium of scientists and engineers, from many different organizations in the Everglades, who saw that despite support to save the Everglades, there was no synthesis of understanding that could provide guidelines for drafting an ecosystem-based plan. This self-organizing collaboration of scientists undertook their own assessment of the bioregion, adding the voice of scientific consensus to the political melee. But because they had no institutional backing for the assessment, the scientists have found it difficult to gain a hearing for their ideas and solutions.

Conducting Bioregional Assessments

A Search for Greater Understanding

The assessments in this review generally got their start through pragmatic questions from policy makers. Generally, the questioners have little patience with long-term studies or lengthy tomes to answer their questions. For many policy makers, three months is a long time, and two years could find them out of office.

Scientists, on the other hand, recoil from quick answers. They are used to contributing a few drops to the sea of knowledge, carefully conditioning any of their results in a myriad of qualifiers. Three months is a very short time, hardly enough time to write the research plan for the studies that will be needed.

Not surprisingly, if given the chance, scientists will turn the pragmatic questions of policy makers into a search for understanding about the bioregion and the complex web of life within it. It takes tremendous discipline for the scientists ever to return to the basic questions that triggered the assessment in the first place. When successful, these assessments answer the questions posed and contribute to broader regional understanding, while the policy makers are still in power and can use the information and before the managers must take action.

Generally, in these case studies, assessment teams attempted to answer the following questions:

- What do we have (in terms of resources, economics, and societies of interest within the region)?
- What is happening to what we have? (What are the trends of these things we have?)

- What might we do to fix any apparent problems, and what are the effects of the alternatives? (Develop options and evaluate their consequences.)

In addition, some of the studies attempted to answer two more questions:

- What do we want to happen? (Set goals; choose options.)
- How can we make it happen? (Develop implementation plans and measures of success in implementation.)

Time Frames and Timing

Assessments conducted to answer questions amid crisis do not have the luxury of attenuated response. Policy makers need solutions, and in such cases, decisions will have to be made with or without the framework of credible scientific information. As Jack Ward Thomas points out in chapter 1, avoiding decisions will result only in fewer, more restrictive management options later.

Time frames for assessments usually do not allow for extensive new research. Generally, less time and information are needed to reach closure for policy makers than are needed to reach closure for scientists. Understanding this basic difference is important to finding closure in the assessment itself. With few exceptions, experts were not able (or willing) to give up their careers for assessments. The work of the assessment had to be of limited duration, even though implementation of assessment results may be protracted over many years. FEMAT and NCCP are good examples of this approach. In these assessments, there was rarely enough information to achieve scientific certainty, but often there was enough information to suggest options for policy action. As Frank Mazzotti warns (chapter 8), we cannot wait for unequivocal results, for instead of gaining knowledge needed to save ecosystems, we will only have documented their destruction.

An often-cited example of where scientists missed the boat is the National Acid Precipitation Assessment Program (NAPAP), an assessment mandated by the Carter administration to determine what was known about the causes and consequences of acid precipitation and to develop options for reducing its effects (chapter 5). The assessment lasted ten years, during which policy advocates attempted to use the program to support their own efforts and scientists were pigeonholed by policy debates (Lackey and Blair 1995). The assessment report was finally presented to Congress months after the legislation was passed that the assessment had been intended to inform.

The temptation exists with some experts to study the problem until the money runs out, then hand off the report to someone else. To some degree this occurred in SNEP and ICBEMP. In both cases, the policy makers who commissioned the studies were gone from office by the time the reports were issued, and the managers had to jerry-rig interim solutions to their problems. While in some ways the ship sailed without them, both assessments developed a treasure of information and conclusions that is proving valuable for other purposes. As an example, the Forest Service moratorium on road building in roadless areas depends strongly on findings from the aquatic section of ICBEMP. Also, few plans for the Sierra Nevada will go forward, in the future, without extensive reference to SNEP.

Matters of Scale

There are many issues for which policy makers draw on expert opinion for advice, including very specific areas of product safety and regulation. Bioregional assessments are distinct in that they answer questions about very large, integrated systems of people and their environments. A bioregion, such as the Great Lakes–St. Lawrence River basin, suggests a large geographical area, eschewing state and national borders for the ecological boundaries of watersheds and biomes. It follows that an assessment of a large geographical space should involve large amounts of information and possibly large, diverse groups of people. However, the problems assessments are asked to help solve are sometimes most apparent at local scales, and different aspects may appear at different times and places. Conducting bioregional assessments involves matching the scale of the inquiry to the scale of the problems and the decisions that need to be made.

Resource managers traditionally have dealt with resource issues at relatively distinct points of time and focused on a particular locale. In recent decades, technology has given us new tools, such as GIS and remote sensing, that have improved our ability to describe natural resources over much larger geographic areas. New computer models and refined techniques for measuring long-term change allow analysis of bioregional conditions over increasingly large time scales. Still, the integration of local-scale information into large-scale databases (and back again) is not an easy transaction for managers or scientists. Maintaining connections among different scales in order to track the effect of specific actions is undoubtedly one of the most challenging aspects of assessments.

The problems addressed in these seven assessments resulted from multiple impacts over many years. Hydrologic changes have been engineered in the Everglades for over a century. Forest health problems in the interior Columbia Basin stem from ninety years of fire suppression and selective logging. After years of ecosystem alteration, effects begin to multiply and problems can manifest rather suddenly. A call for immediate response may translate into short-term solutions at small spatial scales. John Ogden (chapter 8) points out that such small improvements create the illusion that solutions are being found when actually the ecosystem continues to deteriorate.

Scientists, Policy Makers, and Managers

The division of responsibilities among scientists, policy makers, and managers for bioregional assessments has taken many different forms. Table 13.1 compares these responsibilities in different assessment models, highlighting the role of scientists in the process. We have evaluated the five different models in terms of five criteria: (1) role of scientists, (2) opportunities to examine the assumptions underlying current policy and management ("wild science"), (3) scientific credibility, (4) time needed, and (5) support from managers.

The role of scientists in the different models varies from philosopher king (moving beyond the assessment to choosing the plan) to policy analyst (evaluating the consequences of different alternatives) to selecting the alternatives to study (policy making) to standing on the sidelines and throwing rocks at the work others do. Developing alternatives for solving problems is a very important and powerful responsibility within an assessment. When scientists have that responsibility, they are being asked to go beyond their special expertise as scientists and function as a surrogate for policy makers. Only model 1 has the

Table 13.1. *Potential responsibilities of scientists, managers, and policy makers in bioregional assessments.*

Assessment Model	Role of Scientists	Chance for "Wild Science"	Scientific Credibility	Time Needed (relatively)	Support from Managers	Examples
1. Scientists assess the situation and develop a management plan	philosopher-kings	very high	high	small	low	Thomas
2. Scientists develop management alternatives and evaluate them	policy analysts, policy makers	high	high	small	low	FEMAT, SNEP, Everglades
3. Scientists evaluate current condition and trends; policy makers and managers develop management alternatives; scientists evaluate consequences	policy analysts	mod	mod	mod-large	mod-high	ICBEMP
4. Same as above, but scientists help develop alternatives	policy analysts, contributors to policy making	mod-high	mod-high	large	mod-high	NCCP, Great Lakes
5. Policy makers and managers develop alternatives and evaluate consequences	bystanders, critics	low	low	mod	high	National forest plans, CAL OWL

scientists taking it on themselves to go beyond the assessment to development of a plan. All the others leave selection of a plan to managers and policy makers at a later stage.

As managers and policy makers become more and more dominant in the assessment process, the opportunity diminishes for "wild science" that questions the foundations of current policies (see the Foreword). Without the opportunity for this work, the foundation for justifying major changes in policy (when it is needed) disappears. Managers, understandably, resist major changes that upset the plans that they have developed for an area. Also, because major change to the status quo means that some people or groups will lose their established advantage, policy makers are resistant about advocating such changes, unless they are unavoidable.

Scientific credibility rises in direct proportion to the degree that the assessment process appears free of political influence. As managers and policy makers become more dominant in the assessment, the charges of political interference likewise increase. Scientific credibility reaches its nadir when scientists are largely absent from the assessment process, which occurred during the development of the national forest plans in the 1980s. This approach was recently repeated in the development of the management alternatives for protection of the California spotted owl (Kennedy, chapter 12), in which managers developed conservation strategies and evaluated their consequences with little involvement of scientists. Not surprisingly, a technical review of this effort (California Spotted Owl Federal Advisory Committee 1997) concluded that the EIS produced by the management team was "inadequate in its current form as either an owl or ecosystem management EIS." The technical review stated, "Perhaps most important, the . . . EIS was not developed with formal collaboration with the scientific community, adjacent landowners and managers, and other federal agencies." It would seem that the lessons from past planning processes were not heeded. Increasingly, formal collaboration among many stakeholders is a required part of federal land management planning, integrated into the process and not just tacked on at the end.

The time needed for assessments rises in direct proportion to the number of interests (policy makers, managers, scientists, and the public) who have influence over the assessment process. Inevitably, the more that assessments are done in partnership with all the different players, the longer they will take and the greater the likelihood that they will not be completed on time, as evidenced by both ICBEMP and SNEP.

Managers will generally support assessments that address their problems, include their involvement, and occur quickly. Unless they occur quickly, managers must craft their own solutions to the problems at issue while guessing where the assessment and resulting management plan will come out. FEMAT, as an example, occurred quickly and attempted to address their problems but did not include them in crafting solutions. ICBEMP included managers and attempted to address their problems but took far too long to produce results. Both of those efforts would be found wanting from a management standpoint.

Scientific Credibility

In recent years, natural resource policy making has shifted to emphasize policies that are scientifically credible. But what does this notion mean in the context of a bioregional assessment?

Science, it was thought, offered information that could be trusted because it had been tested through the rigors of scientific method. Tools such as statistical tests of confidence and peer review helped verify the credibility of scientific information, which in turn helped support policies. Many questions from policy makers, however, call for experts to move beyond their traditional methods for testing knowledge. Increasingly, scientists are asked to analyze the implications of different management strategies using concepts and data that are incomplete and fragmented. The agreement among experts that comes from tested theories in narrow disciplines is more difficult to attain in analyses of broadly integrated disciplines and databases. Disagreements occur. Scientific disagreement in assessments can be subverted if it is used selectively, as bullets across the extremes of different opinions. Contrarian ideas may be essential to the advancement of science, but the magnification of scientific disagreement can disrupt policy makers' understanding of scientific findings (Ehrlich and Ehrlich 1996). The SNEP assessment team, nettled by lack of consensus, found that disagreement actually strengthened their final report by allowing discussion of contentious issues. Agreement, when it occurred, was all the more meaningful as a consequence of analysis rather than a process of negotiation.

Science and scientists play many roles in bioregional assessments (see table 13.1). In some cases, they help frame the policy questions into a conceptual framework for analysis. How questions are framed has much to do with the expertise and experience of the scientist involved. To the man who loves his hammer, every problem is a nail; so every problem could be one of simply improving silviculture, for example, were an assessment team to be composed entirely of silviculturists. Most often, though, scientists in an assessment compile and synthesize information across a broad range of disciplines and scales that greatly exceed their previous experience. This raises critical questions about the credibility of analyses aggregated over vast scales using data sets from a variety of sources and protocols.

A strong dose of professional judgment is required to answer assessment questions, even in the rare cases where relevant data are available. Problems can arise if policy makers or the public expect science to provide statistical evidence in support of a particular conclusion reached by scientists in an assessment. Much of the credibility of these efforts rests on the reputation of the scientists leading the efforts—to claim otherwise is to misunderstand the nature of the process being used. Analysis and data are valuable but do not substitute for the judgment of respected experts that is requested by policy makers.

No matter how good the quality of knowledge produced in the assessment, the chance exists that managers will misapply it in analyzing the choices before them. An innovation developed in the Tongass Forest Plan and applied in the interior Columbia Basin assessment is the scientific evaluation of the preferred alternative chosen by the decision makers (Everest et al. 1997). Because scientists are not just encyclopedias to be opened and shut on command, the assessment team has developed a list of criteria against which the decisions can be compared for consistency with the scientific information developed during the assessment. This check attempts to answer the question, How do we know if the available science has been appropriately used in the planning process and resulting plans? Potential criteria to assess the scientific basis of the planning process and plans include asking: (1) Is all relevant information used? (2) Is it interpreted correctly? (3) Is the level of risk to species and ecosystems associated with the decision displayed? (4) Is the uncertainty of our knowledge recognized?

The science consistency check can be used to achieve consistency through iterative application that may involve successive improvements in how the scientists state their finding and how the framers of management policy interpret the implications of those findings. In the case of the Tongass National Forest planning effort, the science consistency check was itself subjected to scientific peer review. Because a finding or lack of consistency can be a point of appeal or legal challenge, a thoughtful, thorough check can sidestep a potential problem. With such a process, many of the difficulties associated with the California spotted owl EIS most probably could have been avoided.

The Conundrum of Public Participation

Table 13.1 does not mention a vital fourth group of participants in assessments—the public—nor does it consider public acceptance and support as an evaluation criterion. Yet it is obvious that the public, as a whole and through interest groups, plays a key role in determining the usefulness of assessments.

Most of the assessments in this review strived to provide a forum for public participation that was different from the failed one-way methods of the past. Some (such as the northern forest) raised public hearings to a high art; others (such as the interior Columbia Basin) engaged local communities and decision makers in an exhausting, years-long seminar. This outreach was often done concurrently with the scientific synthesis, and integrating these separate parts into natural resource policy has not been easy. Many of these assessment frameworks, built to analyze scientific information, do not easily integrate social concerns beyond the more empirical measures of economic analyses and demographics. The difficulty in integrating such local understanding with expert knowledge can be summed up by Hanna Cortner et al. (chapter 5): what counts to people cannot be counted.

Many contributors echo Cortner's admonition not to ignore the people. Laura Falk McCarthy (chapter 9) describes the process adopted in the northern forestlands to engage open public discussion of concerns for the future of the north woods. Their efforts led to a clear picture of public concern about possible changes to regional traditions, but they apparently found no way to measure whether things are changing, or to what extent. Perhaps regional decision makers did not need empirical evidence of change, because decisions are being made anyway to protect particular areas from future harvest or development. However, it is still not clear if these decisions will protect the north woods from widespread changes in land use and ownership.

In contrast, the burgeoning population in South Florida seems relatively distanced from the assessment of the Everglades. Clear scientific consensus has not translated into strong public consensus. Mazzotti (chapter 8) describes the incongruity of local government taking action to protect lands against continued sprawl at the same time it is approving land use for development in the same area. Public interest at the national level for this monumental restoration is not matched by local interest.

Forgetting Neither the Science nor the People

By limiting science to a traditional role of research and monitoring, Ogden (chapter 8) sees that the application of science in the Everglades restoration has been stymied. He points out "that neither the scientific community nor the management and policy levels entered the Everglades restoration process with an overall strategy for how science and policy should be linked." Until this link was forged, the Everglades restoration was stuck in an unending loop of planning with no framework for implementation and little connection to local decision making.

Ogden reminds us not to ignore the science, and Cortner reminds us not to ignore the people. This apparent contradiction is manifest throughout our review. It may be seen as a conflict between those who believe that science can provide credible, tested, value-neutral information and those who believe that science is a social endeavor, connected to the politics and culture of its time. It suggests a struggle between protracted negotiation among stakeholders and the autonomy of expert advisors. What do they each contribute to the assessment?

There are many approaches to bioregional assessments modeled by these case studies. Some followed a more public process (NFLC), others a more expert process (FEMAT). Generally, the more public processes were longer and more expensive; the more expert processes were quicker and cost less. It is difficult for us to compare them beyond this simple accounting, because our review was not designed to ask "What was the contribution of the public?" but only "What was the contribution of science?"

The application of good science is only one ingredient in making good policy. According to Cortner, the decision to engage science as a partner to management and policy is itself a question of value. Our key questions are meant to probe ways in which this partnership can be made more effective:

- What questions from policy makers can be answered by science?
- What information from science is most useful to policy making?
- How can science be used to help guide policy and management decisions?

These questions are not meant to imply that good science is enough. Certainly, more is necessary to making good policy than good science explained well to well-intentioned decision makers (Denise Lach, policy scientist at Oregon State University, personal communication). But if the choice is made to engage science as a partner to policy and management, how can the contribution from science be more useful and effective?

Different approaches to including the public in assessments involving scientists have received analysis in the literature. In her book *The Fifth Branch*, Sheila Jasanoff (1990) describes the work of scientific advisory teams, what she dubs the fifth branch of government, created to support the fourth branch, namely regulatory and management agencies. She describes two approaches used by scientific advisors to develop and provide credible, science-based recommendations to policy makers. The technocratic approach relies exclusively on expert witnesses, in which scientists confer in isolation to assemble empirical evidence in support of one or more options (model 1). The democratic approach relies on broad participation and an open review of facts and values surrounding one or more options (models 3 and 4). The results of the two approaches may be very different. Technocratic assessments may provide an analytical rigor and sense of authority not possible with a democratic approach, but they may not satisfy all the concerns of policy makers and the public.

FEMAT seems to most closely fit Jasanoff's technocratic model. The response to heated conflict was to mandate the assessment on behalf, but not including participation, of the public after an initial conference chaired by the president of the United States. As Judy Nelson (chapter 6) points out, without the participation of managers and the public, ownership and acceptance of the policies resulting from the FEMAT assessment have been fraught with problems. Implementation has developed slowly, in part because integration with management was late in coming and budgets have dwindled, in part due to lack of public support. Paradoxically, the need for FEMAT arose because fifteen years of intense public and management participation in the development of forest plans for the region had failed to make the changes needed to protect biodiversity.

Some assessments go to great lengths to include the participation of many people and a wide range of the public concerns. Traditional New England town meetings opened public discourse on the future of the north woods before scientists were enlisted to assess the condition of the ecosystem. Throughout the interior Columbia Basin, interagency assessment teams held public meetings to hear the concerns of local people and inform them of the process. As a result in this case, county governments became partners, political allies, and, perhaps, political hostages to the process. Forest Service managers were extensively involved in the process, but, as James Agee (chapter 11) points out in his science review of the project, such openness did not extend to the academic scientists; many of the key scientific decisions were made internally without the open scientific dialogue that might have been.

Pluralistic assessments, in general, tend to take much longer than three months to complete. As participants in the Northern Forest assessments learned, public consensus takes time. But the support built from within the communities has kept the Northern Forest Lands Council recommendations moving forward long after the council officially disbanded. Also, public consensus has been building slowly in the Great Lakes for long enough to see results, including the Sustainable Great Lakes Ecosystem Initiative, which includes a growing number of agency, industry, and community signatories in voluntary compliance with the goals of ecosystem integrity and sustainability.

In the process of public consensus, it is difficult to ensure that agreement has not been reached at the price of removing substantive conclusions from the results. Reaching consensus about making major changes is not easy and may be impossible. Those who stand to lose the most in the change will stall it as long as possible. Michael Donahue (chapter 7) suggests a law of diminishing marginal utility that applies to the consensus-building process, especially when the motives of some participants may be to undermine or delay decisions.

The Case for Anticipatory Assessments

Not all questions of policy are posed in the same tempo. Some are more urgent than others, sparked from intensifying conflict. But there are other reasons to conduct an assessment without the motivation of existing crisis—that is, in order to avert future crisis.

In comparing such preventative, knowledge-driven assessments with the conflict-driven assessments of our review, Elizabeth Estill (1995) argues for assessments in advance of crisis and ecological surprise. Such anticipatory assessments, Estill asserts, would begin with questions from management not policy. Focus would be first on people as the primary source of impacts on ecosystems, analyzing current demographics and the effects of technological and cultural change on the way people use land and water. This information, superimposed with an assessment of ecological trends, would help illuminate how patterns of social change are redefining patterns of ecological change in particular parts of the bioregion. Such an analysis would help to locate growing problems and focus management action before crisis ignites, while the social and biological contexts still have enough resilience to allow creative, collaborative solutions.

For a number of reasons, anticipatory assessments are difficult to pull off. First, as John Gordon points out in chapter 3, it often takes an environmental crisis to generate research dollars and to encourage willingness of management and scientific institutions to make their staff available. Without crisis, it is difficult to demonstrate a need for such an investment of money and brainpower. Second, many interesting issues worth further study always exist in a bioregion—which will be chosen for assessment? Third, these anticipatory assessments, if successful, will point out emergent ecological, economic, and social problems. This will undoubtedly be seen by many as creating problems that did not previously exist.

Results and Outcomes

Answering the Policy Questions

Cortner et al. (chapter 5) urge that assessments begin and end with policy. Certainly, one major criterion for evaluating the results and outcomes of an assessment is whether the assessment addressed the policy question at issue. Another criterion is whether the assessment team communicated the results in a brief, nontechnical summary that addresses the policy questions in a way that policy makers could understand. Generally, the assessments we studied did better at analysis of the policy questions than in communicating the results.

"What is there" and "what is happening" are questions asked of all the assessments we reviewed. All of the assessments, to one degree or another, attempted to address these questions. In some cases, though, the findings are so voluminous (even in summary form), such as the two-thousand-page SNEP report, that they leave policy makers and their staffs scratching their heads.

"What else could happen" asks the assessment to communicate options for future actions and their consequences. Here the assessments differ widely. On one extreme, the FEMAT assessment developed an array of options and their consequences for managing the northwest federal forests, from which President Clinton selected his preference to become the basis of federal policy. On the other extreme, the Northern Forest Land Council suggested changes in tax law and other institutions to address the problem of land being converted to nontraditional uses but made little attempt to project the consequences of its proposed solutions.

Only a few of the assessments were asked to evaluate the need for new institutions to implement assessment results. The NCCP program in Southern California, for example, assumed considerable discretion in realizing the goals of federal law in ways that accommodated local needs. A key product from this assessment is a blueprint for local planning that meets federal standards through nonstandard procedures, allowing for local negotiation and market-based incentives rather than blanket regulation. One result is conservation banking, which identifies important tracts of habitat for conservation within the context of a regional conservation program. Landowners are able to market their land on the basis of on its ecological, rather than its economic, value. In exchange for contributions to the land base of the preserve system, landowners receive assurance that their development of land outside the preserve will not be interrupted.

Other kinds of market incentives have been developed in the largely private industrial forests of New England. Included in the Northern Forest Lands Council assessment was the review of several tax policies, including the ad valorem appraisal of property that assigns the highest possible land-use value for tax purposes. Following recommendations developed in the assessment, the states are now revising tax laws to provide economic incentives to maintain farms and forestlands.

Of the assessments reviewed in-depth here, only the Northern Forest Lands Council attempted to answer "what do we want" and "how do we make it happen." The first is a question of values, not science. However, the second question can be evaluated in terms of the likelihood that proposed policies will lead to the desired result. The Northern Forest Lands Council, though, relied on public opinion rather than scientific assessment to develop its recommendations and steps toward implementation. It is interesting to speculate why New England has been able to find common ground that seems so elusive in other regions. Perhaps it can be credited in part to the council's willingness to hear out the public before proceeding with any other aspect of the inquiry. Perhaps, though, they left key problems unsolved. As one critic of the process has warned, "We must be aware of the danger of legitimizing 'business as usual' under a veneer of 'consensus building' and 'participatory democracy'" (Trombulak 1994).

Decisions

Each of the bioregional assessments we reviewed provided a method to acquire enough knowledge to make credible decisions. However, it was not always clear in each case who was responsible for making those decisions and who passed judgment on their credibility. As mentioned earlier, the scientists involved in FEMAT assumed a technocratic approach. Although the president would be the one to make the final choice of options, the responsibility to develop and evaluate an array of choices was given to a team of scientists. And once the team agreed to develop an "improved" alternative, it became obvious that that option would be the president's choice. The debate about what to consider and how to consider it occurred within the closed community of FEMAT scientists, rather than with the public at large. This closed process was widely criticized and declared a violation of the Federal Advisory Committee Act (for inclusion of nonfederal scientists without following FACA procedures), but the plan that emerged passed legal muster with Judge Dwyer, and the gridlock over management of the federal forests in the Northwest was broken.

In contrast, in the Great Lakes, decisions have been made through collaboration. Assessments in this region have become an attenuated response to evolving public concerns. More than in other regions, science, management, and governance work in concert to develop and test options. What is most notable about the power relationships in this region is their devolution from federal regulatory mandates to regional, voluntary initiatives. Decision making seems to be shared among a multitude of agencies, commissions, and communities.

Michael Donahue (chapter 7) suggests that the devolution of power was greatly aided by the New Federalism philosophy ushered in by the Reagan administration during the early 1980s. This, says Donahue, thrust state and regional governments—ready or not—into the role of decision makers and inadvertently launched a more inclusive, partnership approach to planning and management.

In the Everglades, science and policy seem to be less in concert. A bioregional assessment undertaken independently by a consortium of scientists in the region resulted in remarkable consensus on the extent of the region's environmental problems and recommended solutions. Yet John Ogden suggests that the independence of this assessment from any policy maker's mandate resulted in no link to a constituency to advocate for implementation of the ideas in the assessment, and so the recommendations, considered technically credible, were slow to be adopted into policy.

The power to make and influence decisions has been closely bounded in the Northern Forest assessments. According to Perry Hagenstein, policy makers who asked for the assessments were not looking for confrontation and had no urgent need to involve scientists. The assessment searched for areas of public consensus and built from there, the very opposite approach from most of the other assessments we reviewed, many of which were driven by legal mandates. Although it can be argued that very little policy has changed in the northern forest, most participants agree that the public's understanding of forest issues has greatly changed and that the sometimes grueling public process paved the way for conservation planning and negotiations that are occurring in each state.

Underlying an examination of decision-making power in many of these case studies is the notion that multiple powers existed and that they each had the other by the tail. If one group truly held all the power for decision making, there would be no forum for disagreement, no explicit conflict, and no need for assessment. This can be seen in the application of laws whose injunctions provided the impetus for change and constraints on the solutions that could be found. The courts provided a measure of uncertainty to decision making that had to be bounded by the assessment.

Locating Opportunities and Risks

In bioregional assessments, as in real estate, location is everything. Assessments need to evaluate the spatial distribution of resources in order to be of value in making management decisions. Managers need to know where there are the greatest opportunities for solving the problem and where there are the greatest risks. Land allocation alternatives developed in FEMAT helped to locate specific areas to target for conservation and for commodities. ICBEMP developed a map of salmon strongholds that clearly identified sites that warranted extra protection. NCCP located the anchors for its suggested reserve system as a starting place for building a conservation strategy. SNEP located landscape units of significant biodiversity to provide spatial definition to the assessment.

Without locating the time and place of specific opportunities and risks, it is difficult to make decisions, because it will be unclear where the key pieces in the puzzle fit. The challenge of most bioregional assessments is to aggregate understanding developed from local experience into a big picture of regional trends and conditions. Without links back down to local applications, the assessment is not useful. However, without connections back up to bioregional condition, there is no way to assess the cumulative effect of local applications.

Integration

Bioregional assessments call for thinking on a far grander scale than is usual for scientists. Most disciplines work on a much smaller canvas, and scaling their thinking up to an entire bioregion can be difficult. In addition, finding comparable measures among diverse fields of study in order to integrate the findings of many disciplines can be overwhelming. Yet it is through this scaling and integration that we turn the impressions of the nine blind men into a picture of the whole elephant.

When we consider an integrated assessment of the consequences of alternative policies, we have reached the heart of integration for policy making. Both ICBEMP and SNEP illustrate the promise and difficulties of this kind of integration. Both uncover a number of problems in sustaining the ecosystems of the Columbia Basin and the Sierra Nevada, respectively: overstocked forests, decline in old-growth characteristics, degradation of riparian areas and watersheds, many aquatic species on the brink, and many other difficulties. Then the scientists studying each of these problems prescribe cures for that particular difficulty: extensive understory thinning and prescribed burning for overstocked forests; elimination of roads, riparian buffers, and watershed analysis to help restore riparian areas and watersheds; and many other proposals. Each scientific group believes, from the perspective of its analysis, that its strategies will improve the sustainability of the ecosystem.

By and large, though, these studies lack a portrayal of the consequences of undertaking all these corrective actions simultaneously. Most important, they lack an analysis of the degree that it is feasible to undertake them simultaneously. Is an extensive thinning operation consistent with restoration of streams and watersheds? Can we improve the structure of late-successional forests at the same time that we thin stands to reduce fire hazard? How much timber will be produced while simultaneously addressing all the problems that have been identified? The implications of these integrated strategies are vital to the policy-making process. Except for a partial analysis in SNEP done by Johnson et al. (1998), the studies, generally, lack an integration of their proposals for improving the sustainability of the ecosystems they studied.

Implementation

Although all these assessments were asked to compile information on regional resources, not all were asked to develop and analyze options that could be turned into blueprints for action. At the crossroads of science, management, and policy, it is not always clear where the lines should be drawn to influence decisions and implement them. Scientific information that suggests appropriate action is useful to policy makers, yet some assessments provide no suggested action. In some of these cases, options and their implications must be inferred from the tonnage of published reports left on the policy makers' and managers' doorsteps. With no roadmap for application, the reports stand the significant chance of not being utilized, even if they are delivered to a sympathetic policy maker.

Implementation is difficult, even when the assessment provides options aimed at providing a strategy for management. The scientists involved may have little experience in land and resource management, may give prescriptions that prove impractical upon implementation, and may provide little guidance as to how to modify their general prescriptions to the myriad of conditions encountered by managers.

The tension between managers and scientists was expressed by many assessment practitioners. Scientists must maintain a broad focus and integrate large amounts of information; managers need to narrow the focus to apply the information to local decisions. If the information is too broad, poorly linked, or irrelevant to local conditions, implementation will be difficult or impossible. Conversely, the local knowledge managers provide to scientists and policy makers regarding implementation may be too specific to neatly fit into the integrated, perhaps abstract, understanding developed in the assessment. This conflict highlights the difference between learning to adapt actions and learning to adapt understanding; both are important to building knowledge (Lee 1993). Thinking in terms of large systems while taking action in small pieces (as in the slogan "Think globally, act locally") requires an explicit strategy that is lacking in most bioregional assessments. It involves the difficulty of maintaining connections between single, small-scale actions and their cumulative, large-scale consequences. And it involves the difficulty in allowing the freedom to experiment with local solutions when the risk to further damaging the bioregional condition seems too great.

FEMAT, as an example, has been haunted by the inability of managers to adjust the prescriptions in the president's plan to their site-specific problems. As Nelson explains (chapter 6), in order to adjust riparian reserves, local biologists with generalized knowledge must certify that the change will not be detrimental to hundreds of rare species. Because few biologists are willing to take this risk, few riparian buffers have been adjusted.

One example of an explicit implementation strategy can be found in NCCP's subregional unit planning, where decision making has been left to local county and city governments, with general guidance taken from regional planning guidelines. A regional reserve designed for conservation will be adjusted to fit what is locally possible and agreed upon. The implementation strategy embraces the seemingly contradictory ideas of adaptive management—in which new information from monitoring management actions is used to modify subsequent actions—and "no surprises"—in which no additional regulatory responsibility will be heaped on landowners in the event of new discoveries.

The Role of Adaptive Management

Most assessments in this review end with a call for adaptive management, in which actions are viewed as experimental, with purposeful testing of different theories about ecological, economic, and social relationships. Yet most of these assessments occurred in times of crisis in resource management, when some species or aspect of the ecosystem was considered imperiled. A species or ecosystem on the brink of collapse does not leave many degrees of freedom for experimentation.

Therefore, in most of our case studies, adaptive management is more of an abstraction than an acceptable enterprise, and institutions still do not allow managers to risk failure. In part, this is because the experiments have been applied too late, when too much of the natural system has already been lost (see chapter 2). Fishery managers in the interior Columbia Basin have little to experiment with when only two wild sockeye salmon return to spawn in the Snake River.

The difficulties of adaptive management when species are at risk are further displayed by the adaptive management areas of the President's Forest Plan. These landscape-sized areas spread over the Pacific Northwest and Northern California were meant to be places where managers' were allowed to be bold and creative in applying innovative strategies to meet the goals reflected in the president's plan. But because these areas were not released from either the commodity targets or the standards for species protection in that plan, they have been largely hamstrung in their attempts at experimentation.

And yet adaptive management suggests a compelling mechanism for coordinating local action and bioregional effects. The difficulty described by Martha Hahn (chapter 11) in applying what is learned at a large scale to decisions made at a small scale, and coordinating those small-scale decisions in terms of large-scale goals, will require a system of checks and balances that seems to be implicit in adaptive management.

Uncertainty

Bioregional assessments attempt to consider resources at many scales of time and distance in complex relationships of cause and effect. In this way, assessments apply emerging ideas of complex systems to real-world problems, and the effort is fraught with paradox and uncertainty. Bioregional assessments attempt to apply *scientific* tools to help in the solution of *social* problems. They attempt to combine expert scientific information and nonexpert local experience to build better understanding about bioregions. And they are often conducted at large regional scales, then applied at small local scales.

Jack Thomas begins our review by describing the challenge of making difficult resource decisions amid contradictory views held by society. John Gordon explores the paradox of comparing noncomparables in science and policy. And Lance Gunderson questions how we can assess what we cannot predict in order to manage what is inherently unmanageable. Understanding where we are uncertain helps us to navigate these turbulent seas.

Uncertainty has many layers in these assessments. It was approached in some cases as the probability of something happening (species survival, for example) if particular actions were taken. At another level, it became the focus on which to apply new knowledge (where, for example, to protect the most critical remaining habitat). In many cases, the policy questions centered around the uncertainty of knowing how much is enough. Assessments were conducted to find out how much coastal sage scrub, how much old-growth forest, how much water is enough to sustain the natural system; or how much development, how much harvest, how much water is enough to sustain the social system.

How much is enough? Answers can be uncertain. Data are often lacking; public consensus is often lacking. John Gordon points out that policy consensus is possible, even with public uncertainty, if the experts seem certain; or, conversely, policy consensus is possible with scientific uncertainty if the public seems certain. But when the experts *and* the public are uncertain, policy making is especially difficult.

Uncertainty may be messy, incomplete, and frustrating, but, as Michael Jones (chapter 7) reminds us, uncertainty is not weakness. It provides management with a framework for understanding risks, and it can guide innovation in science by identifying gaps for further investigation. Such a framework for understanding risk and uncertainties may be more important than an array of facts for an assessment to provide, according to Lance Gunderson. Only by questioning the possible frailties of our understanding can we begin to make management choices that allow room for error and are robust to surprise.

Building Capacity for Understanding and Change

Building Institutional Capacity for Bioregional Assessments

Many would argue that bioregional assessments engage science in a new way. During the course of our review, we have given many names to this new use of science: integrated science, interdisciplinary science, assessment science, bioregional science. The names suggest a synthesis of new understanding, a collaborative interchange among diverse perspectives, an evaluation of conditions and options for the purposes of decision making, and a link between large-scale natural and social systems. All these different aspects of science are engaged, to some degree or another, in bioregional assessments. They represent an emerging new field of science, as well as an emerging era with new tools, techniques, and questions.

Henry Regier (chapter 7) describes an emerging era in which changes in our conceptual understanding of the world have made it possible to pose questions of evolving systems that could not be answered by the rational scientific method of defining objectives, methods, analysis, results, and conclusions. New questions are pushing many fields of science toward greater integration of disciplines and scales of inquiry. We do not yet fully understand the scope of these new questions and the challenge they pose to scientists. Through demonstration, these case studies begin to shed light on the possible contributions, and limitations, of science in answering new questions.

Frederick Swanson and Sarah Greene (chapter 4) describe new technologies in science that have made it possible to consider questions beyond the traditional bounds of a single discipline and a single scale. Innovations such as GIS and remote sensing have provided the means to collect and display multiple layers of complex information and to develop new understanding about protracted change over large areas. These tools have extended the reach of scientists to investigate changes in natural systems that otherwise might have gone unnoticed. New technologies brought advances in the studies of cumulative effects, landscape ecology, and conservation biology, which in turn brought troubling new discoveries about the galloping rate of change in some systems. New discoveries brought new concerns for managers, policy makers, and the public.

Hanna Cortner and colleagues (chapter 5) discusses the responsibility scientists have to explain the significance of their discoveries to the public, to be publicly negotiated with other public concerns. Coming from an entirely different point of view, John Ogden (chapter 8) concurs—scientists must explain the significance of their findings in order for them to be considered in decision making. Both believe (possibly for different reasons) that assessments serve a political, as well as a scientific, purpose to earn the support necessary for implementing changes in management, policy, and public understanding.

Evolving questions and technologies seem to be pushing the boundaries of disciplinary science toward more integration among the biophysical and social sciences. To consider assessment science as a single new discipline is to miss the point. The context in which much of science is now conducted is increasingly collaborative, in part because the questions asked of science are increasingly complex. New technologies, which have allowed us to see connections among many systems at many scales, have prompted new questions about the consequences of change in any one of those systems.

These complex questions demand a lot of scientists. Their training, and the expert knowledge, may not have prepared them to make connections easily across so many traditional boundaries. Equally challenging is the application of elegant ecosystem theory to the practical, specific problems faced by managers and policy makers.

These complex questions demand even more of our research institutions. By and large, though, institutions have not been established that would be devoted to bioregional assessments. Most of the assessments we reviewed were ad hoc efforts by groups of experts pulled together for a temporary period. Once the assessment was finished, these assessment teams were dissolved. Like a donkey, they had no sense of the past and no hope for the future. So the next crisis will call up a new set of conscripts into the ragtag assessment army, again with little time or data to use in addressing some of society's most pressing natural resource problems and little opportunity to learn from experience.

Despite this lack of institutional connection, assessments seem to build on each other, within a region and beyond. This book has provided its contributors with an opportunity to cross-reference experience and knowledge that would not be available to them in any other way. Greater institutional capacity to use and build on assessment results would certainly improve the transfer of knowledge among all practitioners of assessments, in turn improving the way assessments are conducted and how they are applied.

Building Institutional Capacity for Bioregional Implementation and Governance

Institutions often are created in response to crises or problems. Thus, it should not be surprising that none of the bioregions we studied had institutions in place to carry forward recommendations at the bioregional scale when the assessments were completed. Generally, the geographic area covered in the assessment was governed by a patchwork of overlapping institutions that did not recognize the boundaries of the bioregion as useful for governance and did not perceive problems as interconnected among jurisdictions. In this context, assessments seem to dump new understanding that does not fit into the old framework of existing institutions.

Bioregional assessments create knowledge that is intended to guide policy making and management action. But when this knowledge is publicly developed through an assessment process and then ignored, management agencies are vulnerable to legal and political challenge. No amount of scientifically credible results can protect against decisions that apply little or none of the knowledge brought to light by the assessment.

Yet in many cases, the management institutions that receive assessment results do not immediately have the capacity to accommodate changes in management that may be suggested. As Frank Mazzotti warns, "No amount of good science can substitute for the lack of political will. The inability of institutions to escape the limitations of their own agendas and philosophies may be a formidable barrier to ecological improvement."

The assessments we studied vary tremendously in the degree to which institutions have developed to effectively use the results. Part of the variability is a function of how much time has elapsed since the first assessment results surfaced. It is not surprising that the Great Lakes have a much more sophisticated set of institutions at various spatial scales to apply assessment results.

In the Great Lakes, Michael Donahue (chapter 7) points out that the New Federalism during the early 1980s weakened the ability of federal agencies to exert a dominant influence in the region. At the same time, a complex set of regional commissions, local initiatives, and strongly held convictions about state's rights served to empower states and lower levels of government. It also created social capital, which allowed state and local governments to use the new sense of community to create public interest in solving regional problems. Bioregional assessments both contribute to and are a product of that growing sense of community and its cultural identification with the Great Lakes ecosystem.

As a result, there seems to be broad-based political support for ecosystem management in the Great Lakes, years ahead of other regions in our review. How has this evolved? In part, it is thanks to people such as Henry Regier, who have participated in (perhaps infiltrated?) many committees and commissions, forming intellectual bridges between what otherwise might have been separate, unrelated efforts. Their long-term contribution to "science in the service of society" has helped to build connections among the region's various decision-making authorities, so that eventually, with the Ecosystem Charter signed in 1996, bioregional governance in the Great Lakes seems to be working toward something, rather than against itself.

In other regions, existing political entities are beginning to reform themselves in order to implement decisions based on bioregional assessments. After the results of the assessments and subsequent plans have been obtained, the task of implementation has fallen to the states, counties, national forests, and other entities. In the case of FEMAT, interagency coordinating committees have begun to smooth relations among federal agencies in the Northwest while turning day-to-day implementation of the plan over to the national forests and BLM districts.

The NCCP demonstrates remarkable devolution of authority to allow a regional solution to conservation of a federally listed species. Rempel et al. (chapter 10) describe several new regional authorities to help ensure the long-term viability of the plan as well as the species while leaving it up to the counties to do much of the day-to-day implementation. For the Northern Forest Lands Council, on the other hand, a deliberate sunset clause ensured that implementation went back to the states without continuing regional oversight.

While the resistance to regional governance is understandable, bioregions that do not develop some oversight authority for coordinating local plans face the possibility of slipping back into the problems that created the need for an assessment in the first place.

Building Capacity for Bioregional Understanding and Change

For as difficult as it seems to have been to integrate an interconnected, ecosystem approach into science, management, and policy, it almost seems to have come as second nature to many members of the public. Developing public understanding of bioregional conditions and options has been relatively easy for those assessments that attempted it. This means not that everyone became an instant expert, but that the general principles for understanding bioregional connections made sense to many of the people who participated in assessments. Perry Hagenstein (chapter 9) talks about the north woods as a place that was intuitively understood by the people in New England. That sense of place allowed people to understand connections between land, water, and the communities in which they live. Concerned citizens did not need empirical analyses to teach them that changes in one part of the system will have consequences in other parts of the system. They already understood that, and so they were prepared to engage in discussions about change and consequences. County commissioners in the interior Columbia Basin engaged in discussions of cause and effect across the landscape and, on some points, argued with scientists over different interpretations of the data. It is difficult to imagine the same level of public engagement in discussions of single disciplines, such as civil engineering or forest genetics.

This capacity for public understanding of complex interconnections does not mean that consensus was reached or that policy making was made any easier. In public discussions, there can be as many opinions as there are self-interests, and the job of decision making can be overwhelming. The way public opinion is used in policy making is beyond the scope of this book. Our interest is in the way science is used in policy making, through assessments. It would appear, from our few examples, that the interested public not only is capable of understanding many of the concepts presented by science in assessments but is also capable of debating the application of those concepts. Don Erman (chapter 12) suggests that bioregional assessments can do more than allow the public to comment formally on proposed projects, as in NEPA. "By revealing the basic data as well as the tools used in analysis, we empower citizens to check the results or conduct their own analyses. Thus, expanding the availability of information and analysis methods should be as important as the conclusions individual scientists reach as part of a study."

An ecosystem approach in assessments may provide a useful framework for engaging public discussion. The choice to engage public discussion in decision making may be a policy choice, as similarly, Cortner argues that the decision to engage science in decision making is a policy choice. Bioregional assessments at their best demystify science. According to Regier, they are part of a narrative mindscape as old as traditional storytelling. According to Maggy Hurchalla (chapter 8), they are part and parcel to the democratic process. In SNEP and ICBEMP, they tried to bring science to the people. All these assessments attempted to develop new understanding about a place, built on knowledge from all corners of science and experience.

Development of a Sense of Place

When we define our place in the world, we often turn to the natural grandeur around us, to the New England woods in autumn color or to the vast wild River of Grass. The intuitively understood character of these places makes the idea of a bioregion more integrated into the lives of people who live and work there.

Some images, such as that of the Great Lakes, are easily identifiable, with boundaries visible on a map and defining of the region. This is the front yard for nearly forty million people, and when pollution threatened the lakes, the public response was unanimous concern. The image of the Sierra Nevada is more remote but just as grand; its special places have become icons in celebration of wilderness and natural beauty. However, as thoroughly as we thought we knew the place, the assessment of the Sierra Nevada has developed a deeper understanding of the frailties, as well as the grandeur, of this much loved mountain range.

Until recently, the coastal sage scrub in Southern California was merely the place you planted your feet (or built your house) to view the Pacific Ocean. In a remarkable reversal of fortune, this low-profile natural community now identifies a region that is struggling to find a way to conserve a remnant of wild land among some of the nation's most expensive urban real estate.

These assessments provided the opportunity for people to learn more about their bioregions and their place within them. As a result, a sense of place has developed, or deepened, in many of those who participated in the sometimes grueling process. Such a sense of place is an unexpected but significant product of bioregional assessments, as they bring together different experiences from around the region to negotiate a shared vision for the future.

Measures of Success in Bioregional Assessments

There are many possible ways to judge bioregional assessments. We have chosen three criteria, one each from policy making, science, and management: (1) Did the assessment lead to solution of the problem that caused the policy maker to request it? (Was it pragmatic?) (2) Did it significantly improve understanding in the bioregion about the resources being studied? (Was it contextual?) (3) Was the understanding integrated into managers' thinking to guide future action? (Was it integrated?) We have evaluated the assessments in this book by these criteria. Knowing that our evaluation is subjective and based on our incomplete understanding of each of the assessments, we nonetheless found it useful to consider each assessment against these three simple criteria.

By these criteria, some of the assessments stand out. In the Great Lakes, where bioregional assessments have been conducted for over a quarter of a century, institutions demonstrate a commitment to strategies of adaptive management and local involvement in their approach to multiple environmental problems. There seems to be a regional acceptance of the open-ended nature of knowledge generated about this heavily impacted ecosystem. Uncertainty is an acknowledged part of much of the region's management. With its pulses of assessments, actions following assessments, adjustment based on these results, increasingly sophisticated view of the problem, and bioregional governance, the Great Lakes is, in some ways, a model for other bioregions.

NCCP also meets all the criteria. It provides a conservation strategy for coastal sage scrub, has fundamentally changed and broadened people's understanding of the bioregion, and provides a template for management and development of the region. Many questions remain, but in general, the knowledge that is emerging from the NCCP about volunteer initiatives and local control of decision making may provide a model for new planning processes in California and throughout the country.

The other assessments do not fully meet our criteria. FEMAT and its predecessors provided a model of integrating science into policy. That work provided a new framework for federal forest management in the region. It also provided a model for species and ecosystem conservation that is now a reference point for broader conservation efforts such as the recovery of depleted salmon stocks. FEMAT, though, only partially has been integrated into the thinking of managers.

Knowledge generated by scientific assessment of the Everglades in South Florida has provided a blueprint for the slogan "Save the Everglades" and suggests that the restoration will not be easy, cheap, or certain. But policy makers and managers have been slow to take it to their bosoms.

The outcomes of ICBEMP, SNEP, and the Northern Forest Lands Council are still unfolding. All energetically engaged the public in open discussions of the condition and possible futures of their bioregion. ICBEMP and SNEP both developed new ways of looking at the resources of their bioregions and the risks and opportunities there. Although the policy makers who requested SNEP were out of office by the time the final report was released, its use is building in bits and pieces, as different agencies and groups grapple with the problems that continue to bedevil the Sierra Nevada. Four years into the making, ICBEMP is not yet finished, as the alternatives based on the scientific analysis are still being revised. Its classification of the different parts of the basin in terms of ecological integrity and socioeconomic resiliency is breathtaking in its scope, but it is not clear that this classification will provide a way to evaluate the resources and places of the interior Columbia Basin. As with SNEP, the policy makers who requested it have gone from the scene, replaced by a new set of policy makers, who, undoubtedly, are somewhat puzzled as to what to make of the huge beast they have inherited. The assessment of the Northern Forest Lands Council has challenged the people to plan for the future of the north woods, but it is not clear that the policy solutions that were developed will be sufficient to preclude additional large sales of the northern forest to developers.

Considering our criteria another way around suggests something about what bioregional assessments can and cannot do. By our measure, most of the assessments were able to improve regional understanding. By this we mean that they were able to synthesize a broad range of information about the bioregion's social and natural systems into an understanding that made sense to people. Many of these assessments brought a new level of scientific understanding to the people of the bioregion. It turns out that the science was the easy part.

However, we found that few of the assessments were able to easily integrate this bioregional understanding into a framework for guiding management action. The gap between knowing what is wrong and knowing what to do about it is still broad. However, it is interesting for us to realize that more of this integration is occurring as time goes by. Natural resource management increasingly is conceptualized with an ecosystem approach, and these assessments provide the necessary framework for understanding cumulative effects and consequences of actions. So we learned that integrating assessment results into management occurs slowly.

Ironically, by our measure, the most difficult thing for assessments was to assist in solving the problems that spawned them. In the cases where expectations were very specific, and the time frame very brief, it could be said that the assessment helped the policy maker address the problem at hand. But in most cases, the mandate given to the assessment did not consider larger, overarching problems that continued to nettle society. If the mandate was sweeping, the assessment took too long to satisfy the immediate needs of policy makers. If the mandate was specific, the assessment may have overlooked larger problems. It seems (and not so surprisingly, after all) that solving policy problems is the hardest part.

Overall Observations and Conclusions

Context

1. *Most bioregional assessments are crisis driven.* Assessments represent a big commitment of social capital and research brainpower; they are not the first tool to pull out when there is a resource problem, especially if a solution seems relatively straightforward. It takes a crisis for political organizations to risk the major upheaval that comes with taking a fresh look at large bioregions. When policy makers call for an assessment, they are bequeathing power to another entity; they will do that primarily when there is an immediate problem that they must solve but for which they do not have a ready solution. Only when it appears that some resource of direct interest to people may be threatened in the future, such as a traditional way of life, do policy makers appear willing to get out in front of the problem.

2. *Bioregional assessments are often surrogates for larger questions that society may not be ready to tackle.* The questions asked of bioregional assessments may be surrogates for larger social problems. Concern for dwindling open space, declining habitats and species, and fouled wetlands may be the local symptom of more pervasive problems such as unsustainable population growth and questions about the quality of life. It is important to recognize that these overarching problems are part of the context of bioregional assessments, although they may not be explicit in the mandate.

3. *Clear questions are needed from the policy makers; otherwise you have the makings of an Oklahoma land rush.* In some assessments, a clear question was asked of the assessment team. For example, FEMAT was asked to find strategies with a medium-high to high level of protection of species that provide the greatest possible contribution to economic and social well-being. In other assessments, the team was given broad discretion in choosing the questions to pursue. In the Interior Columbia Basin, the team was asked to develop a framework for ecosystem management and was left to decide what that meant. As the ICBEMP team stated, "First and foremost is the need for clear questions from decision makers. What decisions do they face and what information will improve those decisions? The issue is not so much about defining . . . the types of questions, but the science need is for clarity about types and of information needed. Given the

cost of an assessment, we can ill afford to embark on a data hunt" (Quigley et al. 1997).

Conducting Assessments

4. *Developing and applying a model of how bioregions work is at the heart of successful assessments.* We all—scientists, managers, policy makers, and the public—tend to see only part of the problem. Yet as Thomas and others have pointed out in this book, the integration of scientific specialties into a "model" of how bioregions work is at the heart of successful assessments. Using this model to project the implications of alternative futures for the bioregion so policy makers can understand, comprehensively, the likely effects of alternative policies enables a new way of thinking about problems and solutions.

5. *Frameworks, more than data, are the key to successful bioregional assessments.* Identifying concepts and frameworks that address bioregional problems, and that can be built on, are the key to bioregional assessments. If we can develop and apply acceptable frameworks for analysis, we can at least make informed estimates. Without the conceptual framework, however, no amount of data will provide the necessary understanding to guide management decisions. As Lance Gunderson says, robust responses are more important than correct responses. The classification of watersheds relative to strongholds for salmon stocks in the ICBEMP and the application of conservation biology tenets in NCCP are two examples of robust responses that allow informed estimates within accepted measures.

6. *Answers to the policy makers' questions will almost always require substantial extrapolation from existing research and the application of informed opinion; there is no way that research can outrun the questions of policy makers.* Controversies in resource management tend to occur in the gaps between scientific understanding and at scales off the radar screen of most research. In addition, advances in science lead to new questions. We should not expect that one day we will have a storehouse of knowledge that can be drawn on to answer the questions that need answering in bioregional assessments. Rather, we will always be asking scientists to reach beyond what is known to provide informed speculation about potential outcomes and solutions to problems.

7. *The shorter the time frame, the easier bioregional assessments are to do and the more likely they will be successful in addressing the immediate problem.* If an assessment cannot be completed within a year, perhaps it should not be started. Short bursts of focused problem solving are more likely to result in usable results than a protracted effort with shifting focus or, more deadly, a long, intermittent effort. As months stretch to years, some predictable conduct occurs among participants. Academic scientists, needing publications to justify their time away from research, turn the assessment into small disciplinary bits that can be easily published. Resource managers cannot wait forever for enlightenment; eventually, decisions must be made and actions taken. The policy makers who commissioned the assessment may turn to other problems or other ways of solving this problem, or they may be out of office; those who took their place will almost always be less

than interested in the results. The public loses interest and moves on. All of these things have happened during the course of protracted assessments. FEMAT, for all its warts, was finished quickly and was accepted by the policy makers in time to solve the problem facing them.

8. *A large budget, especially when combined with a long time frame, can doom the hopes for an integrated projection of alternative futures, but at the same time it will undoubtedly result in some diamonds among the many works that are produced.* The more money available and the longer the time frame allowed to complete the assessment, the more likely that scientists will pursue it as regular research, with associated modeling building and data gathering. Nothing brings out the blood lust in scientists like the chance to bring home a pile of money to their research institutions. A project budget can become a common pool from which researchers maximize their portion with little thought of the overall project needs, which works directly against the need for integration.

Two recent projects with the biggest budgets and longest time frames were SNEP and ICBEMP. Both had difficulty making an integrated projection of alternative futures from the assessment pieces developed by their scientists. Yet both are making major contributions through individual works that made up the assessment: ICBEMP through its classification of aquatic resources in the Columbia Basin and analysis of the threats from roads, and SNEP from its analysis of potential biodiversity areas, key watersheds, late-successional forests, and the inclusion of wildfire in strategic planning.

9. *The more that assessments push participants out of their conventional thinking, the more likely they are to succeed.* Assessments by their very nature call for scientists to look beyond their discipline and their highly focused analyses and to find connections among other points of view and broad landscapes. As pointed out by Gunderson, Donahue, Thomas, and others, assessments occur at a time of great change, when it becomes obvious that the traditional management and policy bromides do not work anymore. Part of the challenge is to force us all (policy makers, managers, scientists, and the public at large) to find our roles within that change.

Results and Outcomes

10. *If you can't answer the original question in less than five pages, in a way that your great aunt could understand, the assessment will have limited value in policy making.* Politicians like straight talk from the elders of the village; scientists prefer indirect, heavily qualified tomes. Successful assessments must provide direct answers to questions. The original one-sentence query from policy makers requesting an evaluation of the state of old growth and the ecosystems of the Sierra Nevada resulted in two thousand pages of reports and documents summarized in two booklets. Compare that to the Gang-of-Four Report (on which the mandate for SNEP was modeled), which gave a clear, short, disturbing message that policy makers could understand. SNEP can have great value in policy making once its results are distilled, but that distillation will be needed.

11. *If scientists want to be loved, they should not undertake assessments.* In

bioregional assessments, scientists ask questions that affect other people's lives. Scientists are used to questioning other people's theories in forums where the rules of evidence are well developed. Assessments throw scientists into the public arena, where personal attacks in the press, in Congress, and in the propaganda of interest groups may occur. Managers and policy makers are used to having their parentage questioned, but this treatment may be new to scientists.

12. *Thinking globally and acting locally are hard to do.* All the assessments that we reviewed mastered the task of assessing the status and future of resources over a very large area. Many uncovered problems that policy makers, managers, and the public would like to see fixed. But how to specify repairs remains a major issue. Some approaches, such as FEMAT, specify in detail management practices to be applied throughout the region. This ensures that management action applied at the local level will achieve the intended result but has the tendency to straitjacket managers and can result in inappropriate prescriptions on atypical parts of the landscape. Other assessments, such as that of the Northern Forest Lands Council and NCCP, specified goals at the regional level and then allowed localities to adopt plans to achieve the goals. This, in theory, allows managers and the public to be involved in tailoring practices to their local situation. But who will make sure that all these disparate efforts add up to the agreed upon regional goals? Regional authorities and clearinghouses can accomplish this task, in theory, but our track record is not good in applying the concentration and discipline needed to make this work. We are still learning how to translate global thinking into local action in implementing the plans that emerge from bioregional assessments.

13. *Finding ways to fix the problems that are discovered is everybody's business.* Excluding scientists, managers, policy makers, or the public from development of alternatives can greatly limit the value of the assessment. Suggesting solutions to the stated problem is an important part of most assessments and benefits from the full participation of scientists, managers, policy makers, and the public. At various times, assessments have excluded one or more of these sources of ideas. No one group has a monopoly on the best ideas. When groups are systematically excluded from developing solutions, resentment builds that may take a generation to overcome. As an example, FEMAT's inability to provide objective guidance on how to adjust its prescriptions to specific sites will haunt it for a long time.

14. *Leadership is everything in a bioregional assessment.* Assessment teams generally are ad hoc groups whose members may not know each other well. Team members, often with different perspectives on the cause of the problems and their possible solutions, find themselves working in an environment with few established rules and protocols. Such a situation can lead to suspicion, intrigue, and factions much like you might find in a junior high school. Without a powerful and respected leader, assessment teams can easily become overwhelmed by their differences and gradually disintegrate. For a team leader to be truly effective, he or she must be able not only to lead the assessment toward its goals but also to communicate with, and have the trust of, managers and policy makers.

15. *Sometimes there is a trade-off between public participation and the ability to make necessary changes to management and policy.* Many authors in this book have discussed the value of public participation. In the West, though, we are faced with the paradox that the major changes in federal resource management (to solve the problem of declining species and ecosystems) made little progress when the public held significant sway over the process (through the federal forest plans) and made great progress (at least in terms of reducing litigation and the threat of litigation) when the public's influence was greatly reduced. When parts of the public sense they are about to lose their established advantage, they (understandably) try to prevent the change from occurring. We have not yet, in the Northwest at least, figured out how to make major shifts in resource conservation through a public participation process in which the public has significant influence.

16. *Broad-based public participation and support can help enormously in crafting and implementing bioregional plans, but this support cannot always be attained.* As many studies point out, broad-based public support is important for the implementation of plans that result from bioregional assessments. Without public support, there is little hope that the regional land-use plan that emerged from NCCP can be successfully implemented. Its strength, according to Jim Whalen, is that a strong central coalition successfully marginalized the extreme voices on either side. FEMAT, on the other hand, demonstrates a case in which intense polarization did not allow the definition of a middle ground around which a collation could be built. Both sides disliked the preferred option because they said it was biased. The lack of public support has made it difficult to obtain needed budgets from Congress to implement the preferred option as originally envisioned.

17. *Scientists generally will not stick around to help substantially with implementation, yet managers will often have great difficulty in applying the new concepts and frameworks that emerge from bioregional assessments without help from the scientists who developed them.* Most of the plans that result from the assessments call for general prescriptions to be applied until more site-specific analysis is done. Yet Judy Nelson and Martha Hahn both point out that such analysis is very difficult without the assistance, at least initially, of the scientists whose thinking is represented in the assessments. In reference to implementing the ICBEMP, Hahn states, "We need a partnership between managers and scientists in the nuts and bolts of making it all work on the ground." The ICBEMP scientists have spent months and, in many cases, years away from their regular duties; it would take an act of God to bring them back now.

Managers need advisors that they can call on a continuing basis as they try to develop the way of thinking that the scientists employed in the assessment, yet scientists rapidly disappear after their work on an assessment is done, leaving managers puzzled about this foundling they have inherited. It could easily become a full-time job for a group of scientists for a number of years as a bioregion tries to get its new approach to management off the ground. Volunteering to help implement a bioregional assessment would be akin to a death sentence for any scientist who wished to attain national prominence in his or her field: such work is rarely publishable and then only in the applied journals. The theoretical work, or solid experimental analysis, that takes a scientist to the top of his or her discipline would be largely missing from such an assignment.

18. *Assessment results must be translatable to the finer scales at which managers work for managers to use fully the results.* Bioregional assessments often deal with millions or tens of millions of acres. Thus, they often need to abstract from the detailed descriptions of the landscape that managers employ. Still if they develop concepts and data that are not translatable into these detailed descriptions, it will be difficult for the results to be used. As an example, the Columbia Basin effort used one-kilometer pixels to describe the vegetation of the sixty million acres of its assessment. Each kilometer pixel was assigned one vegetation class based on the plurality of vegetation within it. Managers, of course, use much finer-scale representations of vegetation for their planning and operations. Thus, it may be difficult for them to apply the models and results of the Columbia Basin work.

19. *Managers and policy makers will be tempted to use scientists as human shields to protect themselves from critics, an approach that can be treacherous for all concerned.* Once the assessment is completed, the decision made, and implementation begun, challenges will begin to plans that emerge from the assessment. Scientists can expect that managers and policy makers will meet these challenges by holding the scientists between themselves and their critics much as people held a talisman between themselves and evil in the Middle Ages. Such an approach is understandable but fraught with peril for all concerned. In so doing, the managers and policy makers become increasingly dependent on the scientist's views; the scientists can become increasingly dissatisfied with the whole process and retreat to their laboratories.

What Does the Future Hold?

20. *Bioregional assessments will continue, taking many different forms.* In the last few years, we have learned something that has been known in the Great Lakes region for much longer, that problems in natural resource management must be solved with an ecosystem approach within an ecologically defined region. And we have seen growing interest from society in scientifically credible plans for solving resource problems. Bioregional assessments bring scientific credibility to regional resource planning. The process can be used to build understanding about river basins, natural communities, or an area of cultural significance. Assessments will take different forms, according to their context, but they will all share a need for credible information to solve a regional problem.

21. *Most bioregional assessments will continue to be driven by crisis.* While there is much rhetoric in many quarters about getting out of a crisis approach to managing bioregions, many forces work against this. First, as pointed out earlier, it takes a crisis for policy makers to call for most assessments. Second, as John Gordon points out, it takes crises to shake loose funds for bioregional assessments. Third, assessments could be seen as causing the problem rather than pro-

viding an early warning of its future arrival. Anticipatory assessments will remain the exception rather than the rule.

22. *States and regional authorities, rather than federal agencies, will need to lead bioregional assessment across ownerships.* Despite examples of successful collaborations, many private landowners distrust federal agents who may be examining their lands as part of a larger assessment, and who may suggest changes to land use and management. Among our case studies, the two that were led by federal agencies (FEMAT and ICBEMP) focused on federal lands, especially in developing alternative policies, and were reluctant to directly address the role that non-federal lands will need to play in any of the solutions to resource problems. The other assessments, led by states, localities, universities, or ad hoc groups, considered nonfederal lands as a matter of course and therefore developed a more realistic and complete assessment of the bioregion.

23. *Building institutional capacity for bioregional assessments will take a major change in our strategic thinking about science and research.* Most research institutions divide natural resource issues into small units for detailed examination. The model for advancement in science is one of individual work on a particular specialty that then can be published in a specialized journal. The integrative work required for assessments is neither supported nor encouraged. Further, it can be costly in terms of future budgets for research institutions to be identified with assessments that challenge the establishment. It is much safer for research institutions to allow members to participate (the glory of having faculty on all those commissions) without the danger of being held responsible for the results.

A major change in thinking of our scientific intelligentsia must occur for the United States to develop the capacity for bioregional assessments. Otherwise, the assessments will remain a largely ad hoc endeavor.

24. *Bioregional government will take many forms.* Most of the assessments in this book call for some kind of bioregional coordinating body to implement the results of the bioregional assessments and monitor their effects. The Great Lakes provides an outstanding example of what can be accomplished in this regard. Examination of the crazy quilt of governing bodies in each bioregion suggests a major challenge. States, counties, and municipalities do not give up their prerogatives easily. It may be easier if these existing governing bodies take on new responsibilities, rather than superimposing new institutions over existing ones. Who or what will add the parts into an integrated whole remains a big question.

The argument for a bioregional government would be easier to make if we could use one bioregion for all our ecological issues. But bioregional boundaries will change with changing questions, or more likely, we will have a number of overlapping bioregions to address different bioregional issues, as in the Northwest, where the salmon crisis has overlapped concern for the spotted owl. Perhaps gearing up our existing institutions with some regional oversight is the best we can do.

25. *The day of scientists as philosopher kings is over; the day of scientists as lead-*

ers is just beginning. At one time, it may have been common for scientists to diagnose a problem and propose a single solution, such as the ISC's conservation plan for the northern spotted owl. Scientists took sole responsibility for establishing the facts and applying social values, and, as a result, they took on the role of policy makers. We are evolving now to a role for scientists that focuses on their ability to develop frameworks for understanding conservation of bioregional resources and using these frameworks to help find solutions to regional problems, leaving selection of the best solution to others.

While scientists are being asked to function as philosopher kings less and less, they are being asked to provide conservation leadership more and more. Leadership can be defined as the activity of mobilizing people to work on problems, to face and incorporate problematic realities, to grapple with tough issues, and to make the changes in attitude and action that constitute solutions (Jeff Sirmon, former deputy chief of USDA Forest Service, personal communication). Bioregional assessments, at their best, do this kind of work. Scientists, being less committed to established procedures than policy makers and managers, are well situated to do what Franklin calls "wild science," to examine the assumptions underlying established procedures and suggest fundamental changes to solve problems that have resulted from those procedures. Almost every bioregional assessment in this book has provided new frameworks for resource management—scientists have made a major contribution in each case.

LITERATURE CITED

California Spotted Owl Federal Advisory Committee. 1997. *Final Report—Evaluation of the "revised draft environmental impact statement (RDEIS) for managing California spotted owl habitat in the Sierra Nevada national forests of California."* C. Philpot, chair. Washington, DC: USDA.

Estill, Elizabeth. 1995. Paper presented at the workshop at the Crossroads of Science, Management, and Policy, November 6–8, 1995, Portland, OR.

Everest, F., D. Swanston, C. Shaw, W. Smith, K. Julin, and S. Allen. 1997. *Evaluation of the use of scientific information in developing the 1997 forest plan for the Tongass National Forest.* General Technical Report PNW-GTR-415, Portland: USDA Forest Service.

Jasanoff, Sheila. 1990. *The fifth branch: Science advisors as policy-makers.* Cambridge, MA: Harvard University Press.

Johnson, K. N., J. Sessions, J. Franklin, and J. Gabriel. 1998. Integrating wildfire into strategic planning for Sierra Nevada forests. *Journal of Forestry* 96:42–48.

Lackey, Robert T., and Roger L. Blair. 1995. *Science, policy, and acid rain: Lessons learned.* Paper presented at the workshop At the Crossroads of Science, Management, and Policy, November 6–8, 1995, Portland, OR.

Lee, K. N. 1993. *Compass and gyroscope: Integrating science and politics for the environment.* Washington, DC: Island Press.

Trombulak, Stephen C. 1994. The northern forest: Conservation biology, public policy, and a failure of regional planning. *Endangered Species Update* (October):7–16.

USDA Forest Service, Pacific Southwest Region. 1996. *Revised draft environmental impact statement for managing California spotted owl habitat in the Sierra Nevada national forests of California: An ecosystem approach.* San Francisco: USDA Forest Service Regional Office.

About the Contributors

James Addis was born in Sandusky, Ohio, and has observed the changes in western Lake Erie throughout his life. He is trained as an aquatic ecologist and limnologist. He has served as chair of most of the major committees and advisory panels for the Great Lake Fisheries Commission and is presently director of the Bureau of Integrated Science Services. Throughout his career, he has been interested in bringing laypeople together to develop natural resource policy through a dialogue about the scientific foundations for management options.

James K. Agee is professor of forest ecology at the University of Washington. He graduated from the University of California in 1973 and worked with the National Park Service as a forest ecologist and research scientist. In 1988, he became chair of the Forest Resource Management Division and later the Division of Ecosystem Sciences at the University of Washington until 1993. He is the author of *Fire Ecology of Pacific Northwest Forests,* published by Island Press.

Paula Burgess is Governor John Kitzhaber's natural resources assistant in Oregon. She is leading Kitzhaber's collaborative approach to restoring Oregon's salmon runs and improving water quality. The approach is broadly supported in Oregon by the forest industry, agriculture, cities and counties, and environmentalists.

David E. Capen is an associate professor of natural resources at the University of Vermont, where he has been a member of the faculty since 1976. He has degrees from the University of Tennessee (B.S.F.), University of Maine (M.S.), and Utah State University (Ph.D.). His research interests focus on avian habitat studies, spatial analysis, and biodiversity modeling.

Kristin Aldred Cheek is currently a research and training associate with the Bolle Center for People and Forests and the Institute for Tourism and Recreation Research at the University of Montana's School of Forestry. While coauthoring her contribution to this book, she was a master's student in the College of Forestry at Oregon State University, where she focused on the social and economic aspects of forest communities and forest policy.

Hanna J. Cortner is a professor at the School of Renewable Natural Resources, University of Arizona, where she teaches and does research in the area of natural resources policy and administration. She has held visiting appointments with the USDA Forest Service and the U.S. Army Corps of Engineers and has served as an executive assistant for an elected county supervisor. Her current research is focusing on the institutional challenges of ecosystem management.

Robert R. Curry is an emeritus professor of environmental geology at the University of California at Santa Cruz and serves as director of research for the Watershed Institute of California State University at Monterey Bay. He has focused his research on the Sierra Nevada and is coauthor of the Sierra Nevada Natural Regions study for the Department of Interior.

Michael Donahue is the executive director of the Great Lakes Commission, an interstate compact agency that promotes informed public policy decisions on an array of Great Lakes issues through information sharing, policy research and development, and advocacy. He is an adjunct professor at the University of Michigan, has lectured extensively in the United States and Canada, and has authored more than 150 professional papers, journal articles, and book chapters on resource policy and institutional design and analysis issues.

Don C. Erman is a UC Davis professor of aquatic ecology in the Wildlife, Fish & Conservation Biology Department and director of the UC-wide Centers for Water and Wildland Resources. In addition to his duties as professor and director, he served as the UC representative on the California Council on Biodiversity. From 1993 to 1996, he held the position of science team leader of the congressionally mandated Sierra Nevada Ecosystem Project.

Jerry F. Franklin earned his B.S. and M.S. in forest management and statistics at Oregon State University and his Ph.D. in botany and soils at Washington State University. He spent thirty-five years as a research scientist with the U.S. Forest Service, primarily in the Pacific Northwest. Currently, he is professor of ecosystem analysis at the University of Washington.

John C. Gordon is dean and Pinchot Professor of Forestry and Environmental Studies at the Yale School of Forestry and Environmental Studies. Along with research interests in carbon, nitrogen, and forest productivity, he has strong interests in the application of ecosystem principles to resource management and is involved in studies of federal land management methods and policies. He has been involved in several of the bioregional assessments discussed in this book.

Russell T. Graham is a research forester for the Rocky Mountain Research Station, USDA Forest Service, in Moscow, Idaho. His primary research is in silviculture and landscape analysis and assessment. He has worked in the Rocky Mountains for twenty-three years.

Sarah Greene is a forest ecologist with the Pacific Northwest Research Station, USDA Forest Service in Corvallis, Oregon. She manages an interagency natural areas program in Oregon, Washington, and Alaska and two Forest Service experimental forests.

Lance H. Gunderson has twenty years' experience as a wetlands ecologist in the ecosystems of southern Florida, including the Big Cypress and the Everglades. He spent over a decade as a researcher with the U.S. National Park Service and is now a research professor in the Department of Zoology at the University of Florida, researching the application of ecological theory to resource management policy and practice.

Perry R. Hagenstein has been an independent natural resources economics and policy consultant since 1976 in Wayland, Massachusetts. He is chairman of the Board of Trustees of the New England Natural Resources Center and has served on numerous committees of the National Research Council and the National Academy of Sciences that concern natural resources and conservation issues.

Martha G. Hahn is Idaho state director of the Bureau of Land Management. Previously, she worked as the associate state director for BLM in Colorado. Prior to that, she was the vice president for conservation at the Grand Canyon Trust.

Richard W. Haynes is a forest economist and program manager with the U.S. Forest Service, Pacific Northwest Research Station, Portland, Oregon. He oversees Forest Service social and economics research in Washington, Oregon, and Alaska. His own research involves developing the analytical methods used to make long-term projections of timber market activity, using those models in Forest Service planning and policy analysis efforts, and developing economic characterizations as part of ecoregion assessments.

Margaret Herring is a science editor specializing in issues of natural resources throughout the West. She is trained as an ecologist and worked as a research biologist before turning to science communication. She holds degrees from the University of Virginia and the University of California, Santa Cruz.

Richard Holthausen is the national wildlife ecologist for the USDA Forest Service. He is located at the Rocky Mountain Research Station in Flagstaff, Arizona. In addition to his work on the terrestrial staff of FEMAT, he was cochair of the Northern Spotted Owl Recovery Team and has worked extensively on the Interior Columbia Basin Ecosystem Management Project.

John Howard is a county commissioner in Union County, Oregon, and was president of the Association of Oregon Counties (AOC) in 1994. His interest in public land management issues and policy is demonstrated by his role in the creation of the Blue Mountain Institute and the Grand Ronde Model Watershed program, and as a member of AOC and National Association of County Officials (NACO) Public Lands committees.

Maggy Hurchalla was born in Miami, Florida, in 1940 and graduated from Swarthmore College in 1962. She served for twenty years as a county commissioner in Martin County, Florida. She is presently a member of the Governor's Commission for a Sustainable South Florida.

Susan D. Jewell is the senior wildlife biologist for the U.S. Fish and Wildlife Service's Arthur R. Marshall. Loxahatchee National Wildlife Refuge in the northern Everglades. She holds a B.S. in wildlife biology from the University of Vermont and an M.S. in zoology from the University of Connecticut. She has been a biologist in the Everglades since 1986, including in Everglades National Park.

K. Norman Johnson is currently a professor in the College of Forestry, Oregon State University, where he teaches forest policy and forest management. He has participated in a number of bioregional assessments, including the Gang of Four, FEMAT, and SNEP. He has also served as chairman of the committee of scientists advising the secretary of agriculture on improvement in land and resource planning for the National Forests.

Michael L. Jones is an associate professor of fisheries and wildlife at Michigan State University in East Lansing. He is a specialist in Great Lakes fishery management issues, with a background in modeling and adaptive management. He has worked extensively with agencies, academicians, and stakeholder groups in the Great Lakes bioregion for the past sixteen years. He is a Canadian and earned his Ph.D. at the University of British Columbia.

Jon D. Kennedy is a career Forest Service employee who retired in 1997 after forty-one years of service, thirty-three of those in California. Most recently he represented the Forest Service in the California state capital of Sacramento working with the state legislature, the governor's office, agencies, and special interest groups. He served as a consultant for the SNEP effort.

Marc Luesebrink serves as a special assistant to the secretary at the California Resources Agency. Prior to joining the Resources Agency, he worked at the law firm of Siemon, Larsen, and Marsh in Irvine, California, and at the National Audubon Society, Starr Ranch Sanctuary. He has a B.S. in conservation and resource studies from the University of California at Berkeley and a master's degree in environmental studies from Yale University.

Dennis T. Machida is the executive officer of the California Tahoe Conservancy, a state agency. He was assistant secretary for legal affairs at the Resources Agency for the state of California and in that capacity served as a special consultant to the SNEP project. He is an attorney and was a public affairs fellow for the Coro Foundation.

Frank J. Mazzotti is an assistant professor of wildlife ecology at the University of Florida's Everglades Research and Education Center. He has been teaching and researching ecological issues in South Florida for over twenty years. He continues his long-term research program on the ecology of the endangered American crocodile in Florida and has developed a research and education program on conservation planning and landscape ecology.

Laura Falk McCarthy staffed and coauthored the Northern Forest Lands Study. She directed New Hampshire's Forest Resources Steering Committee in its work to determine how to implement the Northern Forest Lands Council's recommendations. She now lives in Santa Fe, New Mexico, where she is development director at the Forest Trust, a nonprofit dedicated to preserving forest ecosystems and improving the lives of rural people.

Andrew H. McLeod is director of Rhode Island's Department of Environmental Management. Previously, he was deputy secretary of operations at the California Resources Agency and assistant secretary at the California Resources Agency. He was a boyhood resident of Rhode Island.

Margaret A. Moote is a senior research specialist at the Udall Center for Studies in Public Policy at the University of Arizona, where she researches issues in natural resource policy and designs and mediates public participation forums.

Dennis D. Murphy is currently a research professor in the Department of Biology at the University of Nevada at Reno and is team leader for the Lake Tahoe Watershed Assessment in California and Nevada. Previous to serving as chair of the Scientific Review Panel to the nation's first Natural Community Conservation Planning effort in Southern California, he served on the Integrated Spotted Owl Scientific Committee. He received the California Governor's Award for Leadership in Economics and the Environment for the work described in his chapter section.

Judy E. Nelson is the chief of the Branch of Biological Science, Oregon State Office, Bureau of Land Management. She was district manager of the Lakeview and Eugene districts in Oregon from 1988 to 1997, during the time frame leading up to FEMAT and during the implementation of the NFP. She has also worked in BLM's national office and the Nevada State Office.

Logan A. Norris is department head and professor in the Department of Forest Science at Oregon State University. He worked in watershed management research for the USDA Forest Service until he took his present position in 1983. He is a fellow of the Society of American Foresters and currently is chair of the Independent Multidisciplinary Science Team for the Oregon Salmon Plan.

John C. Ogden is senior ecologist in the Executive Office of the South Florida Water Management District. He is a member of the Science Coordination Team of the South Florida Ecosystem Restoration Working Group and is chair of the Plan Evaluation Team for the congressionally authorized Central and South Florida Project Restudy. He is an author on over sixty refereed and technical publications pertaining to ornithology and wetland ecosystems and is a fellow of the American Ornithologists' Union.

Thomas M. Quigley is program manager of the Managing Disturbance Regimes Research Program and science team leader for the Columbia Basin Ecosystem Management Project. He received his Ph.D. in range science from Colorado State University in 1985. He is stationed at the Pacific Northwest Research Station in La Grande, Oregon.

Henry A. Regier was born in 1930 in a pioneer's log home in Northern Alberta, where survival depended on a sense of place. At Queen's, Cornell, and Toronto he explored ecostudies (a nested conceptual system of ecology in economics in ekistics in ecumenics in ecosophy), in which context or sense of place is key. He has worked in ecostudies at Cornell and Toronto and iteratively with some five nested levels of governance from local to global.

Ron Rempel has conducted research on big game and threatened and endangered species during his twenty-five-year career with the California Department of Fish and Game. For the past ten years he has been overseeing the development of habitat conservation plans throughout California and managed the department's Natural Community Conservation Program. He is currently the chief of the department's Environmental Services Division and regional manager for the Southern California Coastal Region.

James Sedell is a research ecologist with the Pacific Northwest Research Station, USDA Forest Service, in Corvallis, Oregon. He chaired the aquatics group that designed the aquatic conservation strategy for FEMAT, was cochair of the field team assigned to develop a national Forest Service strategy for management of anadromous fish, and was coleader of the aquatic and watershed group for the ICBEMP assessment.

Margaret A. Shannon is a natural resource social scientist specializing in the study of natural resource policy and administration. She was a member of the

FEMAT Social Science Team and has conducted research on over a dozen bioregional science–policy assessments. She continues to study these science–policy
processes in terms of public participation, sociology of science, and implementation capacity. She teaches at the University of Buffalo School of Law.

Peter A. Stine earned his B.S. and M.S. at the University of California at
Berkeley, and his Ph.D. in geography at the University of California at Santa
Barbara. He is currently employed at the California Science Center (National
Biological Service), where he assists the Resources Agency in the development of
the California Environmental Resources Evaluation System (CERES).

Frederick J. Swanson is ecosystem team leader and research geologist for
the USDA Forest Service, Pacific Northwest Research Station (Corvallis, Oregon)
and professor (Courtesy) at Oregon State University. His principal interests concern interactions of ecological and geological components of landscape and use
of that understanding in land management.

Jack Ward Thomas worked for twenty-seven years with the U.S. Forest
Service and served for three years as its chief. He earned his Ph.D. in forestry at
the University of Massachusetts and served as president of the Wildlife Society,
where he was awarded the Aldo Leopold Medal. He is currently Boone and
Crockett Professor at the University of Montana.

Mary G. Wallace is a Ph.D. candidate in the Political Science Department at
the University of Arizona. At the time her contribution was written, she was
senior research specialist at the Water Resources Research Center at the University
of Arizona.

James E. Whalen is a wildlife biologist (by education) who has been in the
real estate development business for thirteen years. He is now the principal of J.
Whalen Associates, a development and consulting firm specializing in public policy and environmental and public finance issues and represents landowners, public utilities, jurisdictions, and others facing regulatory and development challenges. He has been involved in Southern California's groundbreaking conservation planning efforts since their inception seven years ago.

Henry L. Whittemore is responsible for the oversight of forestry operations on the Hancock Timber Resource Group's properties throughout the
Northeast. He has extensive land conservation experience and has negotiated
many transactions to protect important public values on privately owned land
throughout northern New England. He holds a B.A. in history and environmental studies from Williams College and a master's degree in forestry from Yale
University School of Forestry and Environmental Studies.

Index

Acceptance and sequence of events, 52
Accessible, making a report, 318–319
Accountability, 2
Acid rain, 50
Activist science, 79
Adaptive management:
  areas, adaptive management, 95, 108,
    125
  California natural community
    conservation planning, Southern,
    244–245
  Columbia River Basin ecosystem
    management project, 292
  ecosystem management, 54
  flexibility, 124–125
  four-phase cycle of adaptive renewal,
    30–32, 37–38
  information needs, 113
  integrated/multidisciplinary method
    for natural resources management,
    35
  iterative communication process, 63
  local information, the use of, 109
  public lands, 44
  uncertainty in natural and social
    systems, 35–36, 73
Adirondack Park, 205–206
Advocacy, 52, 64–65, 79
Agency unification, 123
Agricultural systems, 36
AIDS (acquired immune deficiency
    syndrome), 33
Alpha phase, 29
Analytical approaches/techniques,
    evaluating, 6, 76–77
  see also Challenges for bioregional
    assessments; Theory and practice of
    bioregional assessments

Aquatic Conservation Strategy, 96
Army Corps of Engineers, U.S., 34, 177,
    193
Arthur R. Marshall Loxahatchee National
    Wildlife Refuge, 172, 193
Assessments, 1–2, 49, 59–62
  see also Bioregional assessments;
    Uncertainty in natural and social
    systems; *individual subject*
    headings
Assurances and Southern California
    natural community conservation
    planning, 265–267
Asymmetric interactions between levels,
    30
Autecological studies of sensitive animals
    and plants, 252

Babbitt, Bruce, 107–108, 233
Bargaining-conflict containment mode,
    76
Baseline and Columbia River Basin
    ecosystem management project,
    289–290
Below cost arguments, 22
Biological perspective, evaluation of
    options from a, 95–96
Biological results of FEMAT, 107
Bioregional assessments:
  alternative to court-ordered/
    crisis-driven decision making, 1
  cultural context, broader, 139–140
  emergence of, 55–56
  reviewing, 2–7
  science, bioregional, 4, 55–57
  science incorporated into resource
    policy development, xiii
  what are, 135–136

Bioregional assessments (*continued*)
    see also Challenges for bioregional
        assessments; Uncertainty in natural
        and social systems; *individual subject
        headings*
Bottom-up approaches, *see* Collaborative
        groups working together
Breedlove, Dennis, 308
Broad-perspective synthesis based on
        qualitative interpretations, 60,
        73–75
Brussard, Peter, 248
Bureau of Land Management (BLM):
    challenges for bioregional assessments,
        13–15
    Columbia River Basin ecosystem
        management project, 271–276,
        286, 289, 292
    Northwest Forest Plan (NFP), 125, 127
    Option 9 forest plan, 101
Bush, George, 15, 89

California Biodiversity Council, 334
California Environmental Resources
        Evaluation System (CERES), 330
California natural community
        conservation planning,
        Southern:
    collaboration between scientists and
        resource managers, 2
    conclusions, 245
    data gap, bridging the, 240–245
    history of NCCP, 231–233
    management review, 255–261
    new approach, 234–237
    pilot program, 233
    policy review, 262–268
    science and NCCP, 237–239
    science review, 248–254
    see also Sierra Nevada ecosystem
        project
Canada-U.S. University Seminars
        (CUSIS), 148
Case studies, synthesis of the collective
        experience within, 6–7
    see also individual case studies
Challenges for bioregional assessments, 3,
        7–8
    assessment, a new kind of, 15–17
    conclusions, 23–24

failures of past planning systems, 11–13
Forest Ecosystem Management
        Assessment Team (FEMAT), 118
lessons, an opportunity to consider, 17,
        19–21
limits to science and management,
        21–22
new way of doing business, a, 13
owls to ecosystems, from spotted,
        14–15
plans and promises, 17–19
scientists and science in bioregional
        assessments, 64–66
temptations and consequences,
        22–23
Chandler, David, 147–148
Charter for the assessment, crafting an
        explicit, 58
Checkerboard management, 125–126
Churchman, C. W., 150
Classical science, 99
Clinton, Bill:
    California natural community
        conservation planning, Southern,
        263
    Forest Conference in 1993, 90–91, 130,
        273
    Forest Ecosystem Management
        Assessment Team (FEMAT), xii, 15,
        16, 93–95, 98, 107–111
    Option 9 forest plan, 18
Coastal sage scrub community, 240–244,
        251–253
Collaborative groups working together:
    alternative strategy for involving the
        public, 12–13
    California natural community
        conservation planning, Southern, 2
    Columbia River Basin ecosystem
        management project, 6–7
    cross-disciplinary exchange of
        information, 19
    Forest Ecosystem Management
        Assessment Team (FEMAT), 129
    lessons, learning, 20
    policies, natural resource, xi–xiii
    Sierra Nevada ecosystem project, 334
    universities, 53–54
Columbia River Basin ecosystem
        management project:

collaboration between scientists and resource managers, 6–7

conducting and reviewing the assessment, 280–282

Endangered Species Act (ESA), 35

identifying and recognizing the problem, 58

implementing the assessment, 283–284

management review, 293–298

organization, 276–278

policy review, 299–302

political and geographical context, 271–276

public participation, 278–280, 295–296, 300–301

results, the, 282–283

retrospective, 284–286

science review, 288–292

Combining agencies, 123

Command and control approach, 33–34

Commissions, interstate and binational, 141–142

Communicating with the public, 79–80, 128

Community capacity to absorb and respond to changes, 97

Comparative outcomes of policy decisions, non-comparability of the, 43

Complex systems, dynamics of, 28–32

Comprehensive plan, need for a, 87–89

Computer simulation, 55

Conducting an assessment, 59–62

Conflict industry, 12–13

Congress, U.S.:

Agriculture Committee and the Merchant Marine and Fisheries Committee, 93

Columbia River Basin ecosystem management project, 272, 277, 279, 286

Congressional Research Service, 51

deficit, federal, 110

development and conservation, relationship between, 236

Everglades-South Florida assessments, 177

funding-planning link, 123

Interagency Scientific Committee, 3

Northern Forest Lands, 204–205, 213

Northwest Forest Plan, 122

Powell, John W., 45

Scientific Panel on Late-Successional Forest Ecosystems, 89

Sierra Nevada ecosystem project, 306–308, 310, 315, 317, 328, 330–331, 333

social relevance, focus on problems of, 55

timber harvesting activities, 88

West, the, 2

see also Legislation; Regulations and laws

Conservation:

California natural community conservation planning, Southern, 236–237, 249–253

development and, interactions between, 55, 236

easements, 221–222

phase, 28–29

regional, 131

Conservatism, political, 256

Consistency check, science, 63

Context, importance of, 115

Cooperation, 20, 111

see also Collaborative groups working together

Creative destructive phase, 29

Credibility checks, 17, 18–19, 56–57, 60–62

Crisis-driven decision making, 1

Cross-disciplinary exchange of information, 19, 111

Cross-scale linkages, 32

Culture:

bioregional assessments in a broader cultural context, 139–140

educational and organizational, 50–51

science and policy, 78–79

Cycles, adaptive, 30–32, 37–38

Dasmann, R. F., 135

Data-quality issues:

California natural community conservation planning, Southern, 240–245

Sierra Nevada ecosystem project, 323, 335–336

technology hindered by, 60

Debriefings, periodic, 73
Decentralized decision making, 73
Decision making:
  crisis-driven, 1
  decentralized, 73
  distinction between scientists and
    decision makers, 63
Default prescriptions, 18
Deficit, federal, 110
Defining the charge for the assessment, 58
Delays, 19, 20
Democratic governance model, 76, 80,
  114
Development and conservation,
  interactions between, 55, 236
Devolution of governance in Great
  Lakes–St. Lawrence River Basin,
  140–143
Diamond International, 204
Discipline differences, 65
Discontinuities in processes and
  structures, 30
Discursive designs, 77, 80
Dispersal characteristics and landscape
  corridor use, 251–252
District Resource Plans, 109
Diversity of plant/animal species, 256
Divisiveness, reducing, 129–130
Dobbs, David, 211
Dombeck, Michael, 16
"Don't Define the Problem" (Lloyd), 74
Douglas fir forests, 92
Durability of problems, 51
Duration of the assessment, 60
Dwyer, William, 14, 16, 88–89, 92,
  105–106

Eastside Ecosystem Coalition of Counties
  (EECC), 299–300
  see also Columbia River Basin
  ecosystem management project
Ecologists, 65
Economic issues:
  California natural community
    conservation planning, Southern,
    256
  ecosystem management, 110
  Forest Ecosystem Management
    Assessment Team (FEMAT), 97,
    110–111, 127

Ecosystem management:
  costs to the taxpayer, 110
  Forest Ecosystem Management
    Assessment Team (FEMAT), 16
  Forest Service (FS), 13
  Great Lakes–St. Lawrence River Basin
    assessments, 143–144, 147–148,
    162–164
  hierarchy, ecosystem, 30
  information needs, 113
  interactive science, 54
  lessons, an opportunity to consider, 17
  owl, northern spotted, 14–15
  public lands, 44
  rudimentary paradigms for, 113–114
  scarcity of land, 44
  Scientific Assessment Team (SAT), 15
  succession, ecosystem, 28–29
Educational culture, 50–51
Elitism, science, 118
Elliott-Fisk, Deborah, 309, 310
El Niño/Southern Oscillation, 32
Emerging Era, The, 148–150, 157–158,
  160–161, 163
Endangered Species Act (ESA), 5
  California natural community
    conservation planning, Southern,
    231, 233, 234
  Columbia River Basin ecosystem
    management project, 35
  Forest Ecosystem Management
    Assessment Team (FEMAT), 92
  Forest Service and Bureau of Land
    Management, 13
  Oregon Endangered Species Task Force,
    87–88
  political context of the program, 14
  salmon, 131
  Seattle Audubon Society, 88
Environmental Protection Agency (EPA),
  172
Environment impact statement (EIS), 12,
  18, 98, 103–106, 281–282, 333
Ethical issues, 65
*Everglades: The Ecosystem and Its
  Restoration* (Davis & Ogden), 175,
  187
Everglades-South Florida assessments:
  adaptive management, 36
  flexibility, 35

history of change, 171–174
initial responses, 177–178
institutional gridlock, 5
management review, 192–195
policy review, 196–200
pre-drainage Everglades, 169–171
present and future, the, 183–185
problems and opportunities,
    178–183
resilience in resource systems, 34, 36
science review, 187–191
shortcomings of water policies, 29
two assessments, 175–176
Exotic species, invasions by, 33
Experimentation, 35, 108, 113, 125
Exploitation phase, 28–30, 60
Extractive management, 113

Failures of past planning systems, 11–13
Faulty models, 49–50
FEMAT, *see* Forest Ecosystem
    Management Assessment Team
Field examples helping to test ideas and
    ease implementation, 115
*Finding Common Ground: Conserving the
    Northern Forest*, 219
Fire management, 322
Fish and Wildlife Service, U.S. (FWS):
    California natural community
        conservation planning, Southern,
        235, 263
    Everglades-South Florida assessments,
        172
    habitat conservation planning efforts,
        237
    judicial system, xii
    Northwest Forest Plan (NFP), 124
    owl, northern spotted, 89
    successful science assessment, 110
Fish species, anadromous, 14, 16, 92, 124,
    272
Flexibility, 18, 35, 124–125
Florida, *see* Everglades-South Florida
    assessments
Forecasting future problems, 51
Forest Conference in 1993, president's,
    90–91, 130, 273
Forest Ecosystem Management Assess-
    ment Team (FEMAT), xii
    comprehensive plan, need for a, 87–89

Forest Conference in 1993, president's,
    90–91
formation of, 15–16
geographic integration, 62
identifying and recognizing the
    problem, 58
implementing the assessment, 106–107,
    112–113
lessons learned, 2, 111–115
management review, 6, 121–126
manager frustration, 6
Option 9 forest plan, 18, 100–103,
    107–108
policy review, 127–132
processes involved, 91–92
public involvement, 80, 97–98, 114, 128
results, the, 100–103
role of the public/managers/policy
    makers, 97–99
science in, the practice of, 99–100
science review, 117–120
strategies and developing options,
    choosing, 94–97
studies before, influence of, 92–94
success, measuring, 107–111
supplemental environmental impact
    statement, 103–106
transition from assessment to policy, 62
turning point in history of Pacific
    Northwest forests, 4
Forests, *see* Columbia River Basin
    ecosystem management project;
    Northern Forest Lands; Old-growth
    forests; Timber harvesting activities
Forest Service (FS):
    Bureau of Land Management, common
        management strategy with the, 127
    Columbia River Basin ecosystem
        management project, 271–276,
        286, 289, 292
    discrediting projects and agendas of,
        xi–xii
    ecosystem management, 13
    National Forest Management Act
        (NFMA), 104
    Northern Forest Lands assessments,
        203, 205
    owl, northern spotted, xii, 88–89
    planning system, three-tier, 110
    Scientific Assessment Team, 14–15

Forest Service (*continued*)
   Sierra Nevada ecosystem project, 333
   timber harvesting activities, 14–15
   utilitarian land ethic, 65
Francis, George, 139
Franklin, Jerry, 14, 95
Funding-planning link, 123, 128, 337

Gang of Four, *see* Scientific Panel on
   Late-Successional Forest
   Ecosystems
Genesis mindscapes, 149–150
Genetic studies, 252–253
Geographic information systems (GIS),
19
Geographic integration, 62
Gilpin, Michael, 248
Gnatcatcher, California, 5, 235
God Squad, 15, 89
Golding, Susan, 258
Gordon, John, 4, 14, 196, 308, 309
Government:
   coordination, 279
   devolution of governance, 140–143
   nested levels of assessment and
     governance, 138–139
   see also Congress, U.S.; Legislation;
     Policies, natural resource; Political
     context of the program; Regulations
     and laws
Graber, David, 310
Great Lakes–St. Lawrence River Basin
   assessments:
   commissions, interstate and binational,
     141–142
   cultural context, 139–140
   devolution of governance, 140–143
   ecosystem management, 143–144,
     147–148, 162–164
   evolving process of more than 25
     years, 4
   governance, nested levels of assessment
     and, 138–139
   international agreements, 2
   management review, 157–161
   mindscapes, competing and
     collaborating, 148–150
   policy review, 162–165
   reciprocally responsible action,
     145–147
   research institutions, 66

   science review, 153–156
   scientific information services,
     136–138, 144–145
Great Lakes United (GLU), 148
Green constituencies, 18
Greene, Sarah, 4
Gundrerson, Lance, 4

Habitat conservation planning efforts, 92,
   232, 235–237, 262–263
Hagenstein, Perry, 5
Hierarchical approach, 118
Hierarchical mindscapes, 149
Hierarchy, ecosystem, 30
Hierarchy of models/data/judgments with
   different levels, 77
Hierarchy theory, 60
High-resolution data, 60
History and assessments:
   broad perspective, taking a, 73–74
   cultures, educational and
     organizational, 50–51
   faulty models, 49–50
   institutional model for research access
     and assessment organizations, 51–52
   Kuhn, Thomas, 43
   old growth forests/SAF task force/ and
     Scientific Panel, 45–46
   Powell, John W., 44–45
   science policy issues in natural
     resources, 47–49
   Scientific Panel on Late-Successional
     Forest Ecosystems, 46–47
   universities, a collaborative role for,
     53–54
Holling's conceptual model of adaptive
   cycles, 28, 37–38
Human elements, keeping in mind the,
   75–76, 127
Hurchalla, Maggy, 6

Idaho, *see* Columbia River Basin
   ecosystem management
   project
Identifying and recognizing the problem,
   58
Ignorance response, 33
Implementing the assessment:
   Columbia River Basin ecosystem
     management project, 283–284, 297
   field examples, 115

Forest Ecosystem Management
  Assessment Team (FEMAT),
  106–107, 112–113
Northwest Forest Plan (NFP), 122–123
political context of the program, 80
scientists and science in bioregional
  assessments, 63–64
Incentives and Southern California
  natural community conservation
  planning, 262–265
Independent mindscapes, 149
Industrial ecology, 44
Information:
  ecosystem /adaptive management, 113
  limited, 118
  local, using, 109
  scientific information services,
    136–138, 144–145
Informed citizenry, 38
Initiation of a bioregional assessment,
  57–58
Institutions:
  capacities of, developing, 336–337
  gridlock, 5
  inertia, 66
  reframing our, 38
  research access and assessment
    organizations, institutional model
    for, 51–52
Integrated assessments, 35, 62, 77,
  111–112, 119
Integrated Scientific Assessment for
  Ecosystem Management in the
  Interior Columbia Basin (ISA), 288
  see also Columbia River Basin ecosys-
    tem management project
Interactive science, 54
Interagency Scientific Committee (ISC),
  3–4, 13–14, 46, 88–89, 92–93
Interior, U.S. Department of the, 89, 233
Interior Columbia Basin Ecosystem
  Management Project (ICBEMP),
  60, 62, 273
  see also Columbia River Basin
    ecosystem management project
International agreements, 2
International Association of Great Lakes
  Research (IAGLR), 147
Iterative communication process, 63

Jamison, Cy, 14, 15

Johnson, Norm, 6
Jones, Michael, 6
Jorman, K., 14
Journal of Forestry, 117
Judicial system:
  Audubon v. Evans, 88
  *Babbitt v. Sweet Home Chapter of
    Communities for a Greater Oregon*
    (1995), 236
  Bennett v. Spear, 236
  Everglades-South Florida assessments,
    194
  Forest Ecosystem Management
    Assessment Team (FEMAT), 99
  Forest Service (FS), xi–xii
  judges, federal, 20
  *Palila v. Hawaii Department of Land
    and Natural Resources* (1979), 236
  property rights, 236, 237
  Seattle Audubon Society v. Mosley, 104
  timber harvesting activities, xi–xii, 14
Justice, U.S. Department of, 104–105

Kilgore, Bruce M., 308
Kissimmee River, 192
  see also Everglades-South Florida
    assessments
Kitzhaber, John, 130–131
Knudson, Tom, 305
K-strategists, 28
Kuhn, Thomas, 43

Lake Okeechobee, 34, 197
  see also Everglades-South Florida
    assessments
Land:
  ecosystem/adaptive management on
    public, 44
  multiple-use land management
    agencies, 12
  scarcity of, 44
  uses and land use patterns, dealing with
    different, 125–126
  utilitarian land ethic, 65
  values, 256
Late-successional habitats/reserves, 107,
  323–324
  see also Forest Ecosystem Management
    Assessment Team (FEMAT); Old-
    growth forests; Scientific Panel on
    Late-Successional Forest Ecosystems

Laws, *see* Judicial system; Legislation;
    Regulations and laws
Leadership, 6, 115
Leahy, Patrick, 204, 205, 213
Learning, assessments as tools for, 79–80
Legislation:
    Everglades Protection Act of 1994, 194,
        199
    Federal Advisory Committee Act
        (FACA), 21, 98–99, 108, 290–291,
        307, 317, 333
    Freedom of Information Act, xii
    National Environmental Policy Act
        (NEPA), 12, 62, 88, 89, 92, 109, 331,
        333
    National Forest Management Act
        (NFMA), 12, 13, 87, 88, 92, 104, 109
    Natural Communities Conservation
        Planning Act of 1991, 232
    Water Resources Development Act of
        1992, 195
    Water Resources Development Act of
        1996, 199
    see also Endangered Species Act (ESA)
Lessons, learning:
    challenges for bioregional assessments,
        17, 19–21
    Columbia River Basin ecosystem
        management project, 297–298
    Forest Ecosystem Management
        Assessment Team (FEMAT), 2,
        111–115, 119–120
    Great Lakes–St. Lawrence River Basin
        assessments, 164
Limits to science and management,
    21–22
Lloyd, Iris, 74
Local information, the use of, 109
Local officials, 21, 327

Magic options, 52
Maine, 206, 215–216
    see also Northern Forest Lands
        assessments
Management review:
    California natural community
        conservation planning, Southern,
        255–261
    Columbia River Basin ecosystem man-
        agement project, 293–298

    Everglades-South Florida assessments,
        192–195
    Forest Ecosystem Management
        Assessment Team (FEMAT), 6,
        121–126
    Great Lakes–St. Lawrence River Basin
        assessments, 157–161
    Northern Forest Lands, 218–223
    Sierra Nevada ecosystem project,
        326–329
Managers, resource:
    collaboration between scientists and,
        xi–xiii
    Forest Ecosystem Management
        Assessment Team (FEMAT), 98
    frustration felt by, 6
    limits to science and management,
        21–22
    new natural resource management
        paradigm, 43–44
    1950s, xi
    1980s, xii
    resources, conflict over managing, 2
Market-based property-rights systems, 37
Martin, John, 48
Memorandum of Understanding (MOU),
    221
Methodologies, scientific, 78
Millar, Constance, 310
Miller, George, 306
Mindscapes competing and collaborating
    in bioregional assessment, 148–150
Montana, *see* Columbia River Basin
    ecosystem management project
Moral issues, 65
Muir, John, 305
Multiple assessments, analysis of, 60
Multiple Species Conservation Program
    (MSCP), 255, 258–259, 265–266
Multiple stable states, 30
Multiple-use land management agencies,
    12
Murphy, Dennis, 5, 248
Murrelets, marbled, 94

National Academy of Sciences, 51, 55, 292
National Acid Precipitation Assessment
    Program (NAPAP), 72
National Marine Fisheries Service, 131
Native Americans, 136, 139, 172, 279

Natural·Communities Conservation
    Planning (NCCP), Southern
    California's, 5, 62, 231–233,
    237–239
  see also California natural community
    conservation planning, Southern
NCCP, *see* Natural Communities
    Conservation Planning, Southern
    California's
Negotiation as fundamental to the
    elements/outcomes/procedures
    of fact production, 76
Nelson, Judy, 6
NEPA, *see* National Environmental Policy
    Act under Legislation
Nevada, *see* Columbia River Basin
    ecosystem management project
New Brunswick, 36
New England, *see* Northern Forest Lands
    assessments
New Hampshire, *see* Northern Forest
    Lands assessments
New York, 206
  see also Northern Forest Lands
    assessments
NFMA, *see* National Forest Management
    Act under Legislation
Nonlinear coevolution between people
    and ecosystems, 37
Northern Forest, The (Dobbs & Ober),
    205
Northern Forest Lands assessments:
    advocacy, 64
    bioregion, the, 205–207
    conclusions, 211–212
    epilogue, 212–213
    Forest Service (FS), 203, 205
    getting underway, 204–205
    human elements, keeping in mind the,
      75–76
    management review, 218–223
    policies, natural resource, 207–208,
      224–228
    public involvement, 80
    science, the, 208–211
    science review, 214–217
    transition from assessment to policy,
      62
    two-part assessment, 5
Northwest Forest Plan (NFP), 121–126

  see also Forest Ecosystem Management
    Assessment Team (FEMAT)
Northwest Forest Resource Council
    (NFRC), 105
Noss, Reno, 248

Ober, Richard, 211
Office of Technology Assessment, 51
Ogden, John, 5
Oil crisis of 1979, 32
Old-growth forests:
    owl, northern spotted, 13, 14
    Scientific Assessment Team (SAT), 16
    Scientific Panel on Late-Successional
      Forest Ecosystems, xii, 4, 16, 30,
      45–47, 89, 93
    Sierra Nevada ecosystem project,
      323–324
    Society of American Foresters' (SAF),
      45–46
  see also Forest Ecosystem Management
    Assessment Team (FEMAT)
O'Leary, John, 248
Open process, 20–21, 128
Optimization approach, 117–118
Option 9 forest plan, 18, 100–103,
    107–108
Orange County, 235–236
  see also California natural community
    conservation planning, Southern
Oregon:
    Endangered Species Task Force,
      87–88
    Plan for Salmon and Watersheds, 131
  see also Columbia River Basin
    ecosystem management project;
    Forest Ecosystem Management
    Assessment Team (FEMAT)
*Organic Machine: The Remaking of the
    Columbia River* (White), 87
Organizational culture, 50–51
Outputs, forest, 107–108
Outputs and FEMAT, 107–108
Owl, northern spotted:
    Clinton, Bill, 94
    ecosystem management, 14–15
    Forest Service (FS), xii, 88–89
    Interagency Scientific Committee
      (ISC), 3, 13, 92–93
    judicial system, 104

Owl, northern spotted (*continued*)
  Northern Spotted Owl Recovery Plan,
    93
  Oregon Endangered Species Task Force,
    87–88

Pacific Northwest forests, *see* Forest
    Ecosystem Management
    Assessment Team (FEMAT)
Panarchy, 31
Panels, the use of, 119
Panetta, Leon, 306, 307, 309
Paradigm shift in science, 43–44
Peer review, 21, 56, 61–62, 119, 316–317
Place-based management, 163
Polarization, 12–13
Policies, natural resource:
  advocacy, 64–65
  California natural community
    conservation planning, Southern,
    262–268
  collaboration between scientists and
    resource managers, xi–xiii
  Columbia River Basin ecosystem
    management project, 299–302
  comparative outcomes of policy
    decisions, non-comparability
    of the, 43
  decision makers, distinction between
    scientists and, 63
  Everglades-South Florida assessments,
    196–200
  Forest Ecosystem Management
    Assessment Team (FEMAT), 127–132
  Great Lakes–St. Lawrence River Basin
    assessments, 162–165
  implementing the assessment, 63–64
  new natural resource management
    paradigm, 44
  Northern Forest Lands assessments,
    207–208, 224–228
  phases and transitions, 29
  piecemeal approach to policy
    formulation, 117
  science having haphazard effect on,
    47–49
  science incorporated into resource
    policy development, xiii
  scientists catapulted into central roles
    in, xii–xiii

Sierra Nevada ecosystem project,
    331–338
  technofix options, 52
  transition from assessment to policy,
    62–63
  see also Forest Ecosystem Management
    Assessment Team (FEMAT); Political
    context of the program; *individual
    case studies*
Political context of the program:
  analytical approaches/techniques,
    · evaluating, 76–77
  begin and end with policy *vs.* scientific
    problems, 72–73
  broad perspective, taking a, 73–75
  California natural community
    conservation planning, Southern,
    256–257
  Columbia River Basin ecosystem
    management project, 271–276
  conclusions, 80–81
  criteria, political context, 71–72
  define the assessment team's role,
    78–79
  Endangered Species Act (ESA), 14
  Forest Ecosystem Management
    Assessment Team (FEMAT), 111
  human elements, keeping in mind the,
    75–76
  learning, assessments as tools for, 79–80
  promises, politicians', 20
  restricting assessments, 52
  risks for scientists, 65
  scientists understanding the political
    world in which assessments are
    made, 6
Population viability assessments, 61, 252
Powell, John W., 44–45, 47
Private property rights, 236, 237
Probable sale quantity (PSQ), 129
Procedural correctness, 17, 18–19
Property rights, 236, 237
Public involvement:
  adaptive management, 125
  California natural community
    conservation planning, Southern,
    258
  collaborative groups working together,
    12–13
  Columbia River Basin ecosystem

management project, 278–280, 295–296, 300–301
debate, the need for public, 121–123
Forest Ecosystem Management Assessment Team (FEMAT), 80, 97–98, 114, 128
Great Lakes–St. Lawrence River Basin assessments, 159
informed citizenry, 38
NEPA and NFMA, 12
Northern Forest Lands and FEMAT, 80
Sierra Nevada ecosystem project, 310–311, 330
Public lands, ecosystem management/adaptive management on, 44

Quantitative flavor of the assessment, 119
Questions to be asked at onset for carrying out assessments, 49, 51
Quick fixes, 22–23

Railroads, 125
Range management, 322
Rational-analytical model, 76–77
Reciprocally responsible action (RRA), 145–147
Record of Decision, 129
Reframing our institutions, 38
Regional conservation strategies, 131
Regulations and laws:
    costs of compliance with, 22
    failures of past planning systems, 11–12
    Forest Ecosystem Management Assessment Team (FEMAT), 94
    new natural resource management paradigm, 44
    overlaps in agency jurisdiction and conflicting case law, 23
Reiger, Henry, 4
Release phase, 29
Remedial action plans (RAPs), 139–140
Remembrance, the property of, 31, 32
Renewal in resource systems, 30–32, 37, 38
Reorganization phase, 29–30
Report on the Lands of the Arid Region (Powell), 44
Research access organization, 51–52, 66
Research agenda and Southern California

natural community conservation planning, 250–253
Reserve areas, 94–95, 100–103, 124, 130, 306
Resilience in resource systems, 30–38
Resources:
    adaptive management, 35
    Columbia River Basin ecosystem management project, 295
    conflict over managing, 2
    predictable and sustainable level of, 128–129
    reducing/preventing degradation of natural, 188–189
    resilience, assessing and managing for, 30–38
    tension between development and conservation of natural, 1
Revolt, 31, 32
Riparian management, 92, 94–95, 100–101, 124, 126
Risk, assessing, 19, 65–66, 94, 124
River of Grass Evaluation Methodology (ROGEM), 178
Robertson, Dale, 88, 309
Robertson, W. B., Jr., 170
Romm, Jeffrey, 308
R-strategists, 28
Rudman, Warren, 204, 205

Sage scrub community, coastal, 240–244, 251–253
Salmon, 131, 272
San Diego County, 235–236
    see also California natural community conservation planning, Southern
Scale, issues of, 60, 288–289, 294–295
Scarcity, responses to, 44
"Science Advocacy Is Inevitable: Deal With It" (Shannon, Meidinger & Clark), 64
Science-based assessments:
    activist science, 79
    bioregional science, 4, 55–57
    California natural community conservation planning, Southern, 237–239
    Columbia River Basin ecosystem management project, 300
    consistency check, 63

Science-based assessments (*continued*)
curiosity, scientific, 72
Forest Ecosystem Management
Assessment Team (FEMAT), 99–100
Great Lakes–St. Lawrence River Basin
assessments, 136–138, 144–145
information services, scientific,
136–138, 144–145
interactive science, 54
limits to science and management,
21–22
methodologies, scientific, 78
Northern Forest Lands, 208–211
paradigm shift in science, 43–44
successful, 52, 107–111
see also Bioregional assessments;
History and assessments; Political
context of the program; Scientists
and science in bioregional
assessments; *individual subject
headings*
Science Integration Team (SIT), 276–278,
281–283
see also Columbia River Basin
ecosystem management project
Science review:
California natural community
conservation planning, Southern,
248–254
Columbia River Basin ecosystem
management project, 288–292
Everglades-South Florida assessments,
187–191
Forest Ecosystem Management
Assessment Team (FEMAT),
117–120
Great Lakes–St. Lawrence River Basin
assessments, 153–156
Northern Forest Lands, 214–217
Sierra Nevada ecosystem project,
321–325
Scientific Assessment Team (SAT), 14–16,
93–94
Scientific Panel on Late-Successional
Forest Ecosystems, xii, 4, 16, 30,
45–47, 89, 93
Scientific Review Panel (SRP), 240–241,
248–249
see also California natural community
conservation planning, Southern

Scientists and science in bioregional
assessments:
central roles in natural resource policy,
catapulted into, xii–xiii
challenges for scientists and science,
64–66
collaboration between resource
managers and scientists, xi–xiii
conclusions, 67
emergence of bioregional assessments,
55–56
implementing the assessment, 63–64
implications for, 66–67
political world in which assessments
are made, understanding the, 6
role of scientists, 57–64
Sierra Nevada ecosystem project, 327
traditional work of scientists, 56–57
Scrub community, coastal sage, 240–244,
251–253
Seattle Audubon Society (SAS), 88, 105
Secrecy, 118
Self-policing process in science, 56–57
Sensitivity analyses, 61
Sequence of events and acceptance, 52
Sesco, Jerry A., 308
"Sierra in Peril, The" (Knudson), 305
Sierra Nevada ecosystem project, 6, 305
accessible, making a report, 318–319
assembling the pieces, 314–315
conclusions, 319–320
Congress, U.S., 306–308, 310, 315, 317,
330–331, 333
identifying and recognizing the problem, 58
management review, 326–329
personal interactions of the team
members, 312–313
policy review, 331–338
public in the process, 310–311
review, the role of, 316–318
science review, 321–325
staffing, 313–314
structure, creating the, 308–310
Silent Spring (Carson), xi
Silviculturists, 65
Simulation, computer, 55
Snow, C. P., 50
Social perspective, evaluation of options
from an, 97

Social relevance, focus on problems of, 55
Society of American Foresters (SAF),
    45–46
Space, Jim, 308
Species-specific analysis techniques, 61
Species viability, 104
St. Lawrence River, *see* Great Lakes–St.
    Lawrence River Basin assessments
Stable states, multiple, 30
Staff, professional, 51, 313–314
Standardized/national protocols for
    regional assessments, 73
Stasis mindscapes, 149
State officials, 21, 130–131, 327–328
State of the Lake Ecosystems Conferences
    (SOLECs), 144
Structure of Scientific Revolutions (Kuhn),
    43
Subsidy arguments, 22
Successful science assessment, 52,
    107–111
Succession, ecosystem, 28–29
Summary of Scientific Findings, 292
Supplemental environmental impact
    statement (SEIS), 103–106
Supreme Court, U.S., *see* Judicial system
Surprises, 32–33
Surrogate species, 238
Surveys of sensitive animals/plants, 252
Sustainable development, 236
Swanson, Frederick, 4

Team participation, requisites for, 58–59,
    78–79
Technical aspects/constraints of
    conducting an assessment,
    59–60, 75, 77, 114
Technofix options, 52
Technologies, new, 33
Tension between developing and
    conserving natural resources, 1
Theory and practice of bioregional
    assessments:
    analytical approaches/techniques,
        evaluating, 76
    complex systems, dynamics of, 28–32
    multiple assessments, analysis of, 60
    reframing our institutions, 38
    resilience, assessing and managing for,
        36–38

River of Grass Evaluation Methodology
    (ROGEM), 178
uncertainty, 27, 33–36, 38
unknown, confronting the, 32–33
see also Challenges for bioregional
    assessments
Thermodynamics, 32
Thomas, Jack W., 3–4, 88, 91, 115, 273
Timber harvesting activities:
    community capacity to absorb and
        respond to changes, 97
    Forest Conference in 1993, president's,
        90–91, 130, 273
    Forest Service (FS), 14–15
    God Squad, 15
    judicial system, xi–xii, 14
    long-term health of forests, 128
    Option 9 forest plan, 18, 100–103
    outputs, forest, 107–108
    owl, northern spotted, 13
    predictable and sustainable level of
        resources, 129
    risk, assessing, 124
    Scientific Panel on Late-Successional
        Forest Ecosystems, 89, 93
    Seattle Audubon Society, 88
    Time, limited, 118
Tongass National Forest, 62, 63
Top-down approach, 122
Transition from assessment to policy,
    62–63
Transparent process, 20–21
Trees, invasion of alien, 33
    see also Forest *listings*
Tribal coordination, 279
Trout, 124
Trust, 37, 112–113, 158–159
Tukey, John, 43

Uncertainty in natural and social systems:
    adaptive management, 35–36, 73
    bioregions, 4
    Columbia River Basin ecosystem
        management project, 291–292
    credibility of assessments, 60–62
    Forest Ecosystem Management
        Assessment Team (FEMAT), 112
    Great Lakes–St. Lawrence River Basin
        assessments, 154–155
    narrowing, 99

Uncertainty in natural and social systems
      (*continued*)
   pervasiveness of, 27
   public and scientific uncertainty,
      relationship between, 48–49
   theory and practice of bioregional
      assessments, 32–33, 38
   treating, 57
Understanding and trust, 37
Unexpected events, 32–33
Universities, a collaborative role for, 53–54
Unmanageable slivers, 129
Urbanization of natural lands, 235
Utah, *see* Columbia River Basin ecosystem
      management project
Utilitarian land ethic, 65

Validity of evidence, 111
Value issues associated with forest
      management, 122
Values, science and debate over, 48
Vermont, *see* Northern Forest Lands
      assessments
Viability analyses/regulation, 13, 61, 96,
      252
Vulnerability of system to disturbances,
      31

Washington, *see* Columbia River Basin
      ecosystem management project;
      Forest Ecosystem Management
      Assessment Team (FEMAT)
Washington, Thomas, 148
Water resource systems, 36, 109
   see also Everglades-South Florida
      assessments
Weber, Barbara C., 308
West, the, 2
   see also California natural community
      conservation planning, Southern;
      Columbia River Basin ecosystem
      management project; Forest
      Ecosystem Management Assessment
      Team (FEMAT); Sierra Nevada
      ecosystem project
Wilson, Pete, 232, 305
Wisdom of the Spotted Owl, The (Yaffee),
      xii
Worldviews, changes in and differing, 44,
      111, 112
Worster, Donald, 45
Wyoming, *see* Columbia River Basin
      ecosystem management project

Zoological Society of San Diego, 258